VOLUME NINETY FIVE

SEMICONDUCTORS AND SEMIMETALS

2D Materials

SERIES EDITORS

VOLUME NINETY FIVE

SEMICONDUCTORS AND SEMIMETALS

2D Materials

Edited by

FRANCESCA IACOPI
Environmental Futures Research Institute and School of Engineering, Griffith University, Nathan, QLD, Australia

JOHN J. BOECKL
Air Force Research Laboratory, Materials and Manufacturing Directorate, Wright-Patterson AFB, Dayton, OH, United States

CHENNUPATI JAGADISH
Research School of Physics and Engineering, Australian National University, Canberra, ACT, Australia

ELSEVIER

AMSTERDAM • BOSTON • HEIDELBERG • LONDON
NEW YORK • OXFORD • PARIS • SAN DIEGO
SAN FRANCISCO • SINGAPORE • SYDNEY • TOKYO
Academic Press is an imprint of Elsevier

Academic Press is an imprint of Elsevier
50 Hampshire Street, 5th Floor, Cambridge, MA 02139, USA
525 B Street, Suite 1800, San Diego, CA 92101-4495, USA
The Boulevard, Langford Lane, Kidlington, Oxford OX5 1GB, UK
125 London Wall, London, EC2Y 5AS, UK

First edition 2016

ISBN: 978-0-12-804272-4
ISSN: 0080-8784

Publisher: Zoe Kruze
Acquisition Editor: Poppy Garraway
Editorial Project Manager: Shellie Bryant
Production Project Manager: Vignesh Tamil
Cover Designer: Maria Ines Cruz

Typeset by SPi Global, India

CONTENTS

CONTRIBUTORS

G.R. Bhimanapati
The Center for 2-Dimensional and Layered Materials; NSF I/UCRC Center for Atomically Thin Multifunctional Coatings, The Pennsylvania State University, University Park, PA, United States. (ch3)

T. Björkman
Åbo Akademi University, Turku, Finland. (ch1)

R.A. Burke
US Army Research Laboratory, Sensors and Electron Devices Directorate, Adelphi, MD, United States. (ch6)

A. Castellanos-Gomez
Instituto Madrileño de Estudios Avanzados en Nanociencia (IMDEA-Nanociencia), Madrid, Spain. (ch7)

M.L. Chin
US Army Research Laboratory, Sensors and Electron Devices Directorate, Adelphi, MD, United States. (ch6)

M.E. Dávila
Instituto de Ciencia de Materiales de Madrid-ICMM-CSIC, Madrid, Spain. (ch4)

J.F. Dobson
Queensland Micro- and Nanotechnology Centre, Griffith University, Nathan, QLD, Australia. (ch1)

M. Dubey
US Army Research Laboratory, Sensors and Electron Devices Directorate, Adelphi, MD, United States. (ch6)

A. Ferreira
University of York, York, United Kingdom. (ch2)

N.R. Glavin
Air Force Research Laboratory, Wright Patterson AFB, Dayton, OH, United States. (ch3)

T. Gould
Queensland Micro- and Nanotechnology Centre, Griffith University, Nathan, QLD, Australia. (ch1)

J.O. Island
Kavli Institute of Nanoscience, Delft University of Technology, Delft, The Netherlands. (ch7)

J. Jung
University of Seoul, Seoul, South Korea. (ch2)

G. Le Lay
Aix Marseille Université, CNRS, PIIM UMR 7345, Marseille, France. (ch4)

S. Lebègue
Laboratoire de Cristallographie, Résonance Magnétique et Modélisations (CRM2, UMR CNRS 7036), Institut Jean Barriol, Université de Lorraine BP 239, Vandoeuvre-lès-Nancy, France. (ch1)

N. Leconte
University of Seoul, Seoul, South Korea. (ch2)

L.C. Lew Yan Voon
School of Science and Mathematics, The Citadel, Charleston, SC, United States. (ch4)

Y.-C. Lin
The Center for 2-Dimensional and Layered Materials; The Center for Atomically Thin Multifunctional Coatings, The Pennsylvania State University, University Park, PA, United States. (ch5)

A.L. Mazzoni
US Army Research Laboratory, Sensors and Electron Devices Directorate, Adelphi, MD, United States. (ch6)

S. Najmaei
US Army Research Laboratory, Sensors and Electron Devices Directorate, Adelphi, MD, United States. (ch6)

B.M. Nichols
US Army Research Laboratory, Sensors and Electron Devices Directorate, Adelphi, MD, United States. (ch6)

J.A. Robinson
The Center for 2-Dimensional and Layered Materials; NSF I/UCRC Center for Atomically Thin Multifunctional Coatings, The Pennsylvania State University, University Park, PA, United States. (ch3,5)

P.B. Shah
US Army Research Laboratory, Sensors and Electron Devices Directorate, Adelphi, MD, United States. (ch6)

K. Zhang
The Center for 2-Dimensional and Layered Materials; The Center for Atomically Thin Multifunctional Coatings, The Pennsylvania State University, University Park, PA, United States. (ch5)

J. Zhao
Laboratory of Materials Modification by Laser, Ion and Electron Beams (Dalian University of Technology), Ministry of Education, Dalian, China. (ch4)

PREFACE

1. HISTORICAL: GRAPHENE AS THE "FATHER" OF 2D MATERIALS

In 2004, a dozen years ago, a new chapter of *Science* opened with the experimental *rediscovery* of graphene when Andre Geim and Konstantin Novoselov, at the University of Manchester, isolated graphene flakes via mechanical exfoliation from a block of graphite. The two scientists have also pioneered the formalization of graphene as an atomically thin and flat carbon material with electronic and mechanical properties unprecedented in the world of bulk, or even thin-film materials (Novoselov et al., 2004). Graphene may in fact have already been observed long before, for instance in 1965, when Badami reported on the graphitization of hexagonal silicon carbide (Badami, 1965). However, at that time Nanotechnology was still in its infancy, and a single layer of atomically thin material, carbon included, was not expected to be stable. Times were mature only much later, roughly at the same time as Geim's and Novoselov's report in *Science*, when Walt De Heer's group at Georgia Tech also reported the synthetic achievement of an atomically thin, purely two-dimensional carbon layer via thermal decomposition of silicon carbide at high temperature (Berger et al., 2004). In fact, the early 21st century scientific community was so mature for the acceptance of 2D materials that Geim and Novoselov were awarded a Nobel Prize just only a few years after their first publication on graphene. Meanwhile, the number of published papers on graphene started increasing exponentially to reach nearly 200,000 entries in a little over 10 years (source: Web of Knowledge).

Graphene can be rightly considered the "Father" of all 2D materials. The scientific community has started considering the possibility of experimentally obtaining alternate 2D materials only since the rediscovery of graphene. Nevertheless, an enormous progress has been equally made within a few years in the area of alternative 2D materials, leading to the discovery of a plethora of nearly atomically thin, atomically flat materials.

2. HAS GRAPHENE DISAPPOINTED?

Great expectations were associated with graphene almost since its rediscovery, to the point that it has been dubbed a wonder material, as well as, in some instances, accused of being a hype material, associated too often with exaggerated claims. Not surprisingly, as the material properties derived from a perfect covalent sp^2 carbon bond matrix such as graphene are truly extraordinary under many different aspects, from an electrical and electronic perspective with ultrahigh carrier mobility ($\sim$200,000 $cm^2/V\ s$), micrometer-scale mean free path, electron-hole symmetry and quantum Hall effect, down to a mechanical viewpoint, with a reported elastic modulus around 1 TPa and an outstanding fracture strength (Balandin, 2011; Chen et al., 2008; Nair et al., 2008). Not to mention other aspects such as high thermal conductivity, extremely low surface energy and permeability, the possibility for low-loss surface plasmons, and, last but not least, its optical "transparency." It is hard to imagine any other single material characterized by so many outstanding and practically useful properties. So why has graphene not been considered yet for integration in any high-end application (ie, excluding composites or simple protective coatings)?

There is a latent feeling of disappointment behind this question, and the large scale of scientific and technological interest from those early days of this "wonder" material is now starting to fade. However, this disappointment is only generated by a wrongly defined problem. The ground-breaking nature of this material, the scale and extreme pace of international attention and research effort, and the exaggerated perceptions are the reason why it is being so harshly judged:

- First of all, the extraordinary properties quoted for graphene can only be achieved for an ideal, almost defect-less graphene.
- Ideal graphene may be obtained over rather limited areas in the lab, and scientific studies focus more on the best results rather than averaged results.
- Any high-end application will require consistent high quality and control over large areas over repeated runs, which at the moment is still far from established in most cases.
- There is a terminology and standardization issue when talking about graphene. Graphene can be obtained by a large number of different methodologies, each of which can have vastly different properties (number of layers, surrounding environment and intercalation, amount of defects, etc.). Not all "graphene" is really graphene.
- The current theoretical models do not take into account all of the different flavours of graphene (disordered graphene), and as such, cannot be predictive. The latest developments in this area are presented in Chapter 2.

- Additionally, graphene is a 2D material, and as such, its material properties have a different meaning than those of a thick 3D material. Graphene may have an outstanding electrical conductivity, but it is still only one-atom thick! Graphene is mechanically very strong or stiff, but as a consequence it is also very flexible. Also, graphene is only transparent because it is so thin. And so the list of misunderstandings goes on....
- Last but not least, graphene is a semimetal, and as such, it cannot be fully considered as either a metal or a semiconductor. Graphene has a long way to go to replace silicon, and more to the point, its properties may be ultimately best used as a complement rather than a replacement of silicon.

In summary, graphene will certainly be a key material in future technologies, but the road is still long, significant work is still needed in terms of controlled synthesis and full understanding of the graphene system and its interaction with the environing materials and ambient. Also, graphene will not replace the device materials we use now, though it will augment a plethora of future devices which will be entirely engineered ex novo for 2D materials. As graphene science and technology is much further ahead than the rest of the 2D materials, and counts already on a large number of books and reviews, in this book we decided only to feature a chapter on the advances of the graphene theory. The bridging of the theory developed around an ideal graphene lattice with the experimental often disordered graphene is nowadays a strong necessity for practical progress.

3. INTEREST AND SPECIFICITY OF 2D MATERIALS

Two-dimensional materials are a class of recently discovered crystalline substances which can exist as atomic-thin sheets over large in-plane areas. Some of the 2D materials can exist in nature as a macroscopically stacked form such as graphene in graphite, from which single sheets can be exfoliated. Some others do not exist in nature as layered materials, can only be produced by synthetic routes, and are generally not stable in the environment. All of them though are characterized by the fact that they possess (almost) only in-plane bonds, showing exclusively very weak interactions in the z-direction, typically Van der Waals type. Hence the name 2D materials.

The interest in 2D materials is from both a scientific and technological aspect. From a scientific point of view, the study of graphene and other 2D crystals has opened a completely novel chapter of the study of condensed matter. Their discover has allowed science to refine the basic theory and models of Van der Waals interactions, and the detailed understanding of the profound consequences of 2D confinement on all of the physical and chemical properties of materials. 2D confinement strongly modifies the band structure of the materials, making, for example, conductive materials out of substances that otherwise would have been semiconducting or insulating, and strongly affecting charge transport phenomena in general. Novel extraordinary properties in such materials are discovered almost on a daily basis, attracting growing attention from the global scientific community.

From a technological point of view, the interest clearly lies in harnessing such novel properties in devices with added functionalities for scopes as broad as electronics, photonics, energy, sensing, and more. 2D materials offer a combination of properties not obtainable from conventional thin-film materials. Yet, though nanomaterials, they have strong similarities to thin films (the thinnest!), simplifying significantly device design. Although in reality this is only just an apparent advantage. A 2D monolayer is essentially a thin film made almost exclusively of surfaces, and we all know too well how the control (passivation) of surfaces has been one of the nightmares for the fabrication, for example, of electronic devices. How about when, additionally, controlling film doping means controlling the doping of a surface that anything in the surroundings can influence? Moreover, if the film/surface itself is not stable? These are just some of the big challenges ahead from a technological standpoint.

4. SEMICONDUCTORS, SEMIMETALS … AND INSULATORS

Essentially, all material types are represented in the 2D family, which is a fast-growing group of materials all characterized by their monolayer nature. Quite a few different classes of 2D materials have been identified experimentally or are at least theoretically predicted, as discussed in detail in Chapter 1. At the moment, some materials are more mature than others, so we have decided to have extended chapters only for the materials or classes of materials which have been extensively studied by at least a number of groups worldwide. Here we will go briefly through the different materials, with no pretension of being exhaustive.

The first obvious class is that of the so-called Dirac 2D materials. These are materials presenting a Dirac cone band structure typology, semimetals with their valence and conduction bands around their Brillouin points shaped as inversed cones touching at their vertex, with "massless" fermions around that point. This class includes graphene, which is also the most stable Dirac material, thanks to the favorable energetic state of its sp^2 hybridization. Following down from carbon the column of Group IV elements, we find the two next monoelement Dirac 2D materials: silicene and germanene (Chapter 4). These allotropes of Si and Ge are unstable and exclusively synthetic. They are reported to physically buckle because of their instability, introducing as a consequence a bandgap in their band structure, and hence are of large potential interest for electronics.

A well-studied and quite long-known 2D insulator is hexagonal boron nitride (h-BN), sometimes referred to as the insulating counterpart of graphene, given their matching hexagonal structures (Chapter 3). h-BN is another stable material in its monolayer form and can be obtained by exfoliation, though its bulk form is usually a powder substance.

A large class of 2D semiconductors are the transition metal dichalcogenides (Chapter 5). Though of recent discovery, already an extensive body of literature has been produced on these layered materials which are often obtained by mechanical or chemical exfoliation.

There are a few more monoelement 2D materials, of which only phosphorene, a semiconducting material obtained by exfoliation of black phosphorous, is somewhat more mature (Chapter 7). Some other predicted but not yet fully demonstrated monoelement materials are stanene or tinene (from Sn), borophene (from B), and antimonene (from Sb).

Another emerging class of 2D materials is the MXenes, which are transition metal carbides and carbonitrides. MXenes are obtained by etching away the A elements (usually metals) from a MAX compound (Naguib et al., 2012). As such, they are exclusively synthetic and a very special type of monolayer. Their characteristics vary from metallic to semiconducting with a small bandgap.

Many more 2D materials and maybe even classes may be discovered over the next few years. In the meantime, one area to consider will certainly be the organic route toward 2D materials. Conjugated polymers confined in two dimensions could become an alternate and cheaper route toward an "organic graphene" (Perepichka and Rosei, 2009). At the moment those materials still appear too disordered as compared to other 2D materials. However, if successful, this route could certainly open up a plethora of complementary technological applications, thanks to its versatility and bottom-up approach.

5. CHALLENGES AND OPPORTUNITIES, FROM HYPE TO HOPE

There is no doubt that 2D materials over the next years will lead further to the discovery of more novel and exciting fundamental phenomena (Chapter 1). As such, this area will remain a great playground for many years to come for the scientific community. However, the larger community always rightly expects a societal impact accompanying such great discoveries. Opportunities are plenty, but there will certainly be many challenges ahead.

One group of technological challenges regards quite obviously the synthesis of 2D material in large volumes or over large areas with a consistent quality. Much of current 2D materials science is based on exfoliation of layered materials and flake transfer to the substrate of interest, which is useful to some of the bulk applications such as energy storage, but is not compatible with large-scale nanodevice production. Also, some of the nonlayered materials can be obtained in situ by van der Waals epitaxy like silicene and germanene, and others are not stable if exposed to ambient. Also, appropriate international standards for measurements, assessment, and benchmarking of 2D materials will need to be developed before any step toward industrial applications can be made. To date, definite standards related to graphene are still lacking, including appropriate and widely accepted language defining all of the different types of graphene available.

Another critical aspect and intrinsic challenge of all 2D materials is the fact that they are made up solely of surfaces. This needs to drive a substantial shift in mentality and approach, considering the fact that technologists have become familiar in dealing with thin films, while surfaces and interfaces were often considered like a headache, or an uncontrollable nuisance at the very least.

First of all, it is extremely difficult to keep a surface absolutely clean and well controlled. Also, it is not trivial to do so when any dangling bonds or reactive molecules present in the environment around the 2D materials can potentially affect and change the materials response. Intercalation and heterostacking are the answers. The science related to molecular intercalation as a way to control electrical and optical properties of layered materials has been extensively developed around graphene (Riedl et al., 2009), and much more work in fundamental understanding and fine engineering will be needed to extend this knowledge to other 2D materials.

Also, the intimate understanding of how junctions between two or more vertically stacked 2D materials can be achieved in a controlled fashion will be key to many of the applications (Chapter 6 by Dubey et al.). Surfaces, interfaces, and 2D heterojunctions will have to shift from being parasitic aspects to becoming central characteristics of the future components for nanodevices. This also inevitably means that rather than reengineering older device concepts such as MOSFETs to adapt them to 2D materials, the future devices will have to be designed specifically for 2D materials, in order to extract maximum advantage from this new class of materials.

In summary, the road to a societal impact of 2D materials is still long; however, there is reasonable hope that this will happen in due course, provided enough effort will be dedicated to the challenges mentioned above. Significant impact can be expected in a broad range of aspects, from efficient electronics allowing for ultimate miniaturization, to applications in sensing, energy, medical care, photovoltaics, and many others. In the meantime, 2D materials will continue to be for Condensed Matter Physics and Chemistry what the large particle accelerators are for High Energy Physics: a means to probe the frontiers of science and to develop a fundamental model of materials and their properties.

We would like to sincerely acknowledge all of the authors of this book for their valuable contributions, our research group members as well as our instrumental colleagues and collaborators in the 2D materials area, particularly those who have lent their expertise to help finalizing this volume: Prof. Uli Zuelicke from the McDiarmid Institute in Wellington, New Zealand; Dr. Ruth Pachter from the Air Force Research Laboratories, Wright-Patterson Air Force Base, Ohio, USA; Assoc. Prof. Igor Aharonovic from the University of Technology Sydney, Australia; Prof. Nunzio Motta from the Queensland University of Technology, Brisbane, Australia; Assoc. Prof. Madhu Bhaskaran from the Royal Melbourne Institute of Technology, Australia; Dr. Alexandra Carvalho from the National University of Singapore; and Dr. Yuerui Lu from the Australian National University in Canberra, Australia.

F. Iacopi*, J.J. Boeckl†, C. Jagadish‡,
*Environmental Futures Research Institute and School of Engineering, Griffith University, Nathan, QLD, Australia
†Air Force Research Laboratory, Materials and Manufacturing Directorate, Wright-Patterson AFB, Dayton, OH, United States
‡Research School of Physics and Engineering, Australian National University, Canberra, ACT, Australia

REFERENCES

Badami, D., 1965. X-Ray studies of graphite formed by decomposing silicon carbide. Carbon 3, 53–57.

Balandin, A.A., 2011. Thermal properties of graphene and nanostructured carbon materials. Nat. Mater. 10, 569–581.

Berger, C., et al., 2004. Ultrathin epitaxial graphite: 2D electron gas properties and a route toward graphene-based nanoelectronics. J. Phys. Chem. B 108 (52), 19912–19916.

Chen, J.-H., Jang, C., Adam, S., Fuhrer, M., Williams, E., Ishigami, M., 2008. Charged-impurity scattering in graphene. Nat. Phys. 4, 377–381.

Naguib, M., Mashtalir, O., Carle, J., Presser, J.V., Lu, J., Hultman, L., Gogotsi, Y., Barsoum, M.W., 2012. Two-dimensional transition metal carbides. ACS Nano 6 (2), 1322–1331.

Nair, R., Blake, P., Grigorenko, A., Novoselov, K., Booth, T., Stauber, T., Peres, N., Geim, A., 2008. Fine structure constant defines visual transparency of graphene. Science 320, 1308.

Novoselov, K.S., Geim, A.K., Morozov, S.V., Jiang, D., Zhang, Y., Dubonos, S.V., Grigorieva, I.V., Firsov, A.A., 2004. Electric field effect in atomically thin carbon films. Science 306, 666–669.

Perepichka, D.F., Rosei, F., 2009. Chemistry. Extending polymer conjugation into the second dimension. Science 323, 216–217.

Riedl, C., Coletti, C., Iwasaki, T., Zakharov, A.A., Starke, U., 2009. Quasi-free-standing epitaxial graphene on SiC obtained by hydrogen intercalation. Phys. Rev. Lett. 103 (24), 246804.

CHAPTER ONE

2D Structures Beyond Graphene: The Brave New World of Layered Materials and How Computers Can Help Discover Them

T. Gould*^,1, S. Lebègue†, T. Björkman‡, J.F. Dobson*

*Queensland Micro- and Nanotechnology Centre, Griffith University, Nathan, QLD, Australia
†Laboratoire de Cristallographie, Résonance Magnétique et Modélisations (CRM2, UMR CNRS 7036), Institut Jean Barriol, Université de Lorraine BP 239, Vandoeuvre-lès-Nancy, France
‡Åbo Akademi University, Turku, Finland
[1]Corresponding author: e-mail address: t.gould@griffith.edu.au

Contents

1. INTRODUCTION

Since its isolation in 2004, the exemplar two-dimensional (2D) material graphene has been the subject of incredibly intense study. As the body of work on graphene has been covered in numerous review articles, beginning with Gein and Novoselov's (2007) seminal work, we will not discuss it here. However, we will note an important property of graphene that is rarely mentioned in reviews: *it can't do everything!*

Semiconductors and Semimetals, Volume 95
ISSN 0080-8784
http://dx.doi.org/10.1016/bs.semsem.2016.04.001

Fortunately, graphene is only one of many "2D materials"—materials characterized by two macroscopic dimensions and one nanoscopic dimension (typical thicknesses range from around 0.3 to $o(1)$ nm). These various "layered" materials offer vast promise for future technological developments as their low effective dimension often grants them properties not found (or rarely found) in bulk 3D materials.

Furthermore, the unique way in which 2D materials bind allows them to be assembled into heterostructures—stacks of different layers combined like LEGO bricks (Geim and Grigorieva, 2013). This is made possible because the different layers bind via van der Waals (vdW, dispersion) forces which can allow neighboring lattices to be mismatched with minimal weakening of the binding energy. These heterostructures can combine the properties of their constituent layers, offering the promise of materials with *multiple* useful properties contained in a small volume. Thus, devices based on functional heterostructures can be effectively miniaturized down to the nanoscale.

In this chapter we will first survey "third-generation" layered materials, pointing to the wide range of possibilities they offer to applied material science. We will follow this with a discussion on van der Waals and related forces that hold 2D materials together. We will focus on the unusual van der Waals properties that 2D materials can exhibit, why this makes them challenging for conventional theories, and how theory can be used to model them. We will then discuss the role computers and theory play in the comprehension and prediction of 2D materials. Finally, we will conclude with a discussion of the role computers will play in 2D material science, and where it will likely take us in the future.

2. THIRD-GENERATION MATERIALS

After the initial discovery of graphene, a second generation of 2D materials quickly appeared, comprising hexagonal BN, BCN, transition metal dichalcogenides (most famously MoS_2), and functionalized graphene. Their properties have been reviewed numerous times (Butler et al., 2013; Castro Neto et al., 2009; Flynn, 2011; Ganatra and Zhang, 2014; Geim and Novoselov, 2007; Xu et al., 2013a). Here we will focus on even more recent compounds, giving rise to a "third generation" of materials.

From the perspective of this chapter, it is interesting to note that many of the "third-generation" materials listed below were first predicted by computer simulations before being isolated in the laboratory. Indeed in some cases theoretically predicted materials (eg, arsenene) have not been fully isolated. This highlights the important role that ab initio theory and computation plays, and will increasingly play, in the world of van der Waals materials. Good electronic structure techniques can rapidly screen for or predict useful materials (see later sections). Experimental effort can then be focused on the most likely candidates.

A brief summary of known third-generation materials appears below. This is by no means a comprehensive list but should serve as a taste of the present state of 2D material discovery.

- *Silicene, germanene, and stanene*: Since silicon, germanium, and tin are in the same column of the periodic table as carbon, it seemed natural that the corresponding two-dimensional layers (respectively, silicene, germanene, and stanene/tinene) could exist. The stability of these systems was shown by theoretical predictions (Cahangirov et al., 2009; Xu et al., 2013b), although the ground-state structure, unlike graphene, displayed a significant buckling caused by a tendency for sp^3 over sp^2 bonds. Silicene and germanene are believed to be tunable topological insulators (Ezawa, 2012; Liu et al., 2011), making them potentially useful for certain nanotechnologies. In fact, silicene (Vogt et al., 2012) and germanene (Derivaz et al., 2015; Dvila et al., 2014) were obtained experimentally on metal substrates, together with stanene (Saxena et al., 2015).

  Furthermore, in the same way that graphene can be chemically modified to give graphane, silicene, germanene, and stanene were predicted to become silicane, germanane, and stanane (Xu et al., 2013b) when interacting with hydrogen atoms. Indeed silicane is known in a bulk form since 1996 (Yamanaka et al., 1996) and was isolated in the form of few layers recently (Nakano et al., 2012; Okamoto et al., 2010), followed by germanane (Bianco et al., 2013).
- *Phosphorene, arsenene, and antimonene*: Black phosphorus has been known for more than a century. However, it was only in 2014 that a single or few layers of it were isolated, giving rise to phosphorene (Xia et al., 2014). It was demonstrated that phosphorene has a carrier mobility and an on–off ratio of the corresponding transistors (Li et al., 2014) which is intermediate between that of graphene and that of transition metal dichalcogenides, making it promising for certain applications.

  Also it was recently shown (Zhang et al., 2015) using theoretical calculations that two-dimensional arsenic and two-dimensional antimony could exist in the form of isolated layers (arsenene and antimonene) and can be tuned to direct bandgap semiconductors under a small biaxial strain. In a different work, the crystal structure of phosphorene has been used as the backbone to predict the existence of some new 2D IV-VI semiconductors (Singh and Hennig, 2014).

- *Indium and gallium chalcogenides*: Indium and gallium chalcogenides are well-known compounds existing with various stoichiometries such as InX, In_2X_3, and In_3X_4 (X=S, Se, or Te) in the case of indium. Recently, these compounds triggered the interest of researchers working in the field of two-dimensional materials, and it was reported that single or few layers of InSe and In_2Se_3 can exist (Jacobs-Gedrim et al., 2014; Lei et al., 2014) and are suitable for photodetection. Also, indium chalcogenides were studied using ab initio calculations: while In_3Se_4 and In_3Te_4 are metallic (Debbichi et al., 2014), InX and In_2Se_3 are semiconductors (Debbichi et al., 2015; Zlyomi et al., 2014).

  As shown by theory, monolayers of GaX are indirect bandgap semiconductors (Zlyomi et al., 2013) presenting a Lifshitz transition when doped with holes.
- *Silicon dioxide, clays, and minerals*: Although known already in various forms, SiO_2 can also exist as a two-dimensional solid: several 2D polymorphs have been obtained (Shaikhutdinov and Freund, 2013), all corresponding to a honeycomb lattice made of SiO_4 tetrahedra. Also, a particular attention was given to the fact that 2D SiO_2 could form a 2D glass (Ben Romdhane et al., 2013; Björkman et al., 2013; Huang et al., 2012; Lichtenstein et al., 2012). Closely related to silicon dioxide are certain clay and mineral compounds, for instance, those which are a stack of silica and metal hydroxide layers, bonded by weak interactions. Several authors have already reported (Chalasani et al., 2013; Włodarczyk et al., 2013) the possibility to have these materials in a 2D form, which is particularly interesting since they are notoriously abundant in nature.
- *MXenes*: MXenes (Naguib et al., 2011, 2014) are a family of two-dimensional materials that are obtained from MAX phase ceramics (Barsoum, 2000; Barsoum et al., 1997) by etching out the A-group element. They have the general chemical formula $M_{N+1}X_N$ where M is an early transition metal and X is C and/or N, and each layer is formed by bonded alternating hexagonal planes of M and X, with M being the terminating plane on each side. Among MXenes, Ti_3C_2 was synthesized first (Naguib et al., 2011), followed by other compounds such as Ti_2C, $(Ti_{0.5}Nb_{0.5})_2C$, V_2C, Nb_2C, Ti_3CN, $(V_{0.5}Cr_{0.5})_3C_2$, Nb_4C_3, and Ta_4C_3 (Ghidiu et al., 2014; Naguib et al., 2012, 2013).

Using ab initio predictions, several authors (Kurtoglu et al., 2012; Shein and Ivanovskii, 2012) have predicted new stable MXenes, while other calculations suggested that Ti_2C and Zr_2C have a magnetic order (Zhao et al., 2014). MXenes possess a range of interesting properties, such as stiffness (Guo et al., 2015; Naguib et al., 2014; Wang et al., 2014), hydrophilicity (Ling et al., 2014; Naguib et al., 2012), and metallicity (Lashgari et al., 2014).

- *Cu_2X*: Although Cu_2S has been well studied (Zheng et al., 2011) in its bulk form, it was only very recently that it could be obtained in the form of few layers (Ding et al., 2014; Romdhane et al., 2015), showing some interesting properties such as an electronic bandgap of 0.9 eV (Romdhane et al., 2015) as obtained with the theoretical *GW* approximation.

  Also, it is known that bulk Cu_2Se and bulk Cu_2Te exist in the form of layered materials (Chi et al., 2014; Choi et al., 2015; Liu et al., 2013; Nguyen et al., 2013) and therefore could possibly exist as well in isolated two-dimensional form.

This list of new (third-generation) 2D compounds is by no means complete and is extended almost on an everyday basis. It is also important to note that all these materials can be in principle stacked to form van der Waals heterostructures (Geim and Grigorieva, 2013), extending even further the range of properties available. The range of possible heterostructured materials is essentially infinite.

Describing accurately the weak interactions involved in these heterostructures is one of the current challenges in condensed matter physics, as it is necessary to include these interactions to understand and predict the properties of 2D systems. With such a large range of possibilities, it will increasingly become vital at the developmental state to use computers alongside conventional experiments. But this requires good approximations. We thus dedicate the rest of this chapter to first describing what these weak van der Waals dispersion forces are and why they can be unusual in layered materials, and then to surveying state of the art in the approximation of weak interactions. We finally survey other computational developments of interest to 2D material science.

3. VAN DER WAALS FORCES

It has been known for more than a century that any two well-separated electronic systems will feel an attractive force to one another in the form of an inverse power law $F(D) = pC_pD^{-(p+1)}$ (or $U(D) = -C_pD^{-p}$) in the distance D between the two systems. If neither system has a permanent charge or dipole (or, potentially, quadrapole), this force is the van der Waals or London dispersion force—arising from fluctuating second-order perturbations in the Coulomb interaction between two or more systems. The most familiar van der Waals power law relates the attractive potential energy between two well-separated atoms or molecules and their frequency-dependent electric dipolar polarizabilies $\alpha(\omega)$ through

$$U_{\mathrm{vdW}}(D) = \frac{-C_6^{AB}}{D^6}, \qquad C_6^{AB} = \frac{3}{\pi}\int_0^\infty d\omega\alpha^A(\omega)\alpha^B(\omega). \tag{1}$$

This vdW (London dispersion) potential forms the attractive part of the popular Lennard-Jones potential.

Typically van der Waals forces make a small change to properties (eg, leading to nonideal gas laws). However, in 2D layered systems the van der Waals force is essentially responsible for interlayer binding and thus determines the stability of larger structures. This is because individual 2D layers share only a small overlap of their electron clouds, which does not lead to covalent or ionic bonds because of the distances involved (eg, the intraplanar C—C bond length in graphene is 0.14 nm, while the interplanar C—C bond length is two and half times bigger at 0.33 nm). At contact distances, the small electron overlap leads to a net repulsive force from electrostatic interactions. The van der Waals force, caused by fluctuations within the layers, is the dominant attractive force contribution at conventional lattice spacings. It is thus *responsible for binding the layers together.* The contribution of different forces is illustrated in Fig. 1.

3.1 (Non)additivity of Atom-Wise vdW Forces

When evaluating the energetics of a system, the Lennard-Jones potential has long served as the "textbook" option. However, the Lennard-Jones and similar potentials using (1) form only one part of the long-range interaction between subsystems. In molecular systems this is often enough—especially as ionic and covalent bonding tends to dominate in the regimes where it is inappropriate. But for an accurate energetic description of layered materials we need to consider other "nonadditive" contributions.

Consider the prediction of the vdW interaction energy between two multiatom species (eg, a pair of 2D layers, or a layer and an adsorbate molecule). Perhaps the simplest approach conceptually is to add the long-ranged

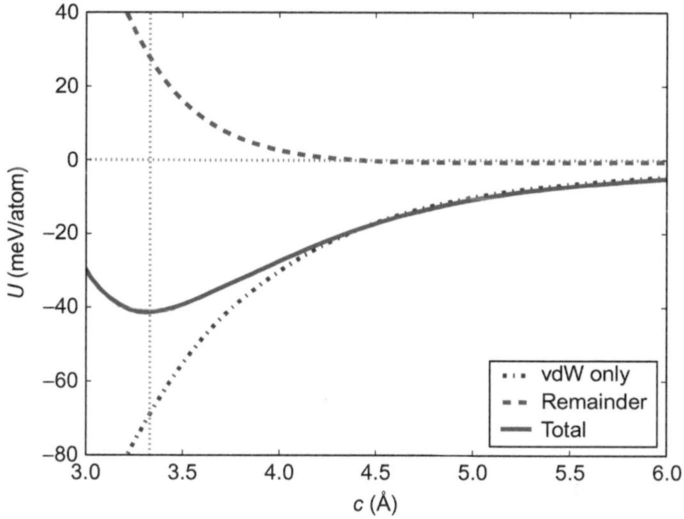

Fig. 1 Potential energy of BN broken into van der Waals and remaining electrostatic contributions.

attractive D_{ij}^{-6} contributions from every pair of atoms (i,j) where i lies in one species and j lies in the other. For atoms very close to one another, ie, those in the contact region, one has to modify the atom–atom interaction to make it repulsive. Furthermore an atom in a molecule may be less polarizable (and hence more weakly vdW interacting) than the free atom. Such effects nowadays can be dealt with in several quite sophisticated ways (Grimme, 2004, 2006; Grimme et al., 2010; Tkatchenko and Scheffler, 2009; Tkatchenko et al., 2012).

These methods already give reasonable results for the cohesion of layered systems, but have quantitative errors that can lead to poor predictive accuracy in layered systems. One reason is what has been termed (Dobson, 2014) the "type B" nonadditivity of vdW interactions, whereby, for example, a third polarizable atom interferes with (screens) the Coulomb interaction that leads to the R^{-6} interaction between a given two atoms. This leads to three-atom (Axilrod–Teller) interactions, four-atom interactions, etc., that cannot be reduced to a sum over atom pairs. For a series of isolated regions $\mathcal{R}_i$ (eg, nonoverlapping atomic volumes) we must thus write the energy of their union as

$$E[\cup\mathcal{R}_i] = \sum_i E[\mathcal{R}_i] + \sum_{(i,j)} U_2[\mathcal{R}_i, \mathcal{R}_j] + \sum_{(i,j,k)} U_3[\mathcal{R}_i, \mathcal{R}_j, \mathcal{R}_k] + \ldots \quad (2)$$

where the sum over pairwise U_2 interactions is the additive component, while remaining terms (eg, U_3 which contains the Axilrod–Teller terms) are collectively the nonadditive component.

There is also a second type of anomalous "type C" (Dobson, 2014) nonadditive interactions that appear in systems with zero electronic gap. These interactions yield, eg, $E \propto D^{-3}$ for graphene and $E \propto D^{-2.5}$ for metallic monolayers, compared with D^{-4} predicted by atom-pair summation or any finite sum over pairwise, triplet-wise, etc. interactions. These are described in more detail in Section 3.2.2.

3.2 Lifshitz Theory and Beyond

An alternative to pairwise summation lies in the popular Lifshitz approach. The Lifshitz approach to vdW interactions dates from the 1950s (Lifshitz, 1954). It was derived (Dzyaloshinskii et al., 1961) from a quantum electrodynamic approach, using a surface integral over the Maxwell stress tensor S to deduce directly the force on each interacting subsystem. Since S is bilinear in the fields, the force depends on nonzero fluctuation quantities like $\langle E^2 \rangle$ which can be obtained from the interacting linear response properties (electromagnetic Green function) via a fluctuation–dissipation theorem.

Alternatively (in the electromagnetically nonretarded, low temperature limit) the Lifshitz approach can be derived (Dobson and Gould, 2012) from the random-phase approximation (RPA; see Section 4.2 for further discussion) correlation energy approach. Here the RPA is used for the Coulomb interaction *between* the systems. In the Lifshitz approach the matter is described by a local approximation $\epsilon(\omega)$ for its nonlocal dielectric function, assumed constant inside each piece of matter: $\epsilon(\boldsymbol{r},\boldsymbol{r}';\omega) \approx \delta(\boldsymbol{r}-\boldsymbol{r}')\varepsilon(\omega)\theta$ where $\theta = 1$ inside the matter and is zero outside.

The original Lifshitz treatment was limited to the interaction between thick slabs of uniform matter with spatially abrupt parallel surfaces separated by dielectric of thickness D. The force between the slabs was found to vary as D^{-3} (ie, resulted from a potential varying as D^{-2}), which of course diverges near contact, which is a result of the spatially abrupt boundaries and the local form of response. This approach and various semi-empirical extensions to nonplanar geometries were the mainstay of the Casimir physics community for decades. The dielectric function in standard Lifshitz theory was taken from experimental or theoretical studies on an infinite solid.

One could certainly consider extending the Lifshitz approach to a nanothin layer such as BN, by treating it as a slab of continuous dielectric with a small but finite thickness. However, the response properties of a 2D layer are not just the same as those of the bulk solid, so the standard assumptions would need to be modified.

3.2.1 Layer Response Theory

Recently an alternative related "layer response theory" approach has been proposed (Dobson et al., 2016). It obtains realistic long-wavelength response properties of an isolated layer via efficient calculations using standard solid-state packages to provide the macroscopic dielectric function of a "stretched" solid. The solid consists of the target layer (eg, BN), repeated at a spacing D that is modestly greater than the equilibrium spacing D_0, sufficient that wavefunction overlap between layers is negligible. The remaining electromagnetic interaction between layers is removed analytically to reveal the response properties of an isolated layer. The nonlocal response to second order in the wavenumber is then approximated using full RPA (see later) calculation. These data are then used to approximate the RPA correlation energy of more complex structures involving the layer(s) in question. This looks promising and the method has so far reproduced the full RPA energy curve $E(D)$ within 1–2 meV per atom for several structures involving graphene and BN layers (Dobson et al., 2016). Much work remains to be done to apply this approach to more layer species.

One point that emerged from this study (Dobson et al., 2016) is that a single power law $E = -CD^{-p}$ is not valid for the vdW energy E of layered systems in the range of layer separations of most interest here, namely 0.3 nm $< D <$ 10 nm. This occurs because the internal Coulomb screening inside a layer is relatively ineffective: most field lines from a charge disturbance in the layer point into space and are not intercepted by other charges in the layer. This contrasts with 3D materials where all field lines are intercepted by matter, resulting in complete screening. This means that the interacting response of a 2D system is spatially nonlocal, whereas that of a 3D system is approximately local. This is the origin of the inadequacy of a single power to describe the energy vs separation curve.

3.2.2 Weird Asymptotic Results

As mentioned above, "type C" nonadditivity can introduce qualitative, as well as quantitative differences to the van der Waals force, at least in the asymptotic regime. This occurs in zero-gap systems, eg, those with Dirac cones (notably graphene and silicene) and in metallic systems. These systems

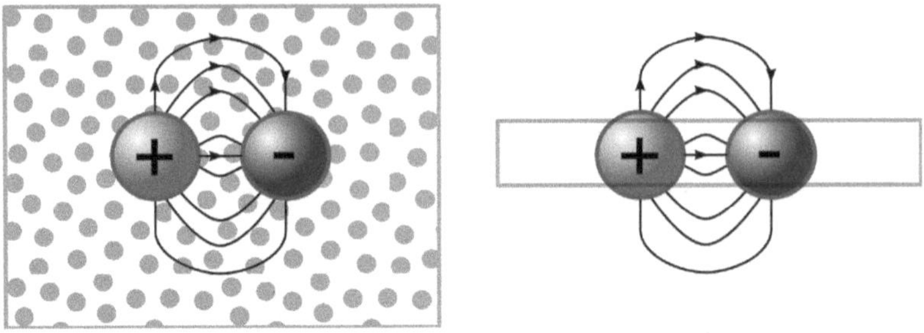

Fig. 2 Screening example in bulk (*left*) and 2D layer (*right*).

have such severe nonadditive contributions that the exponent of the power law is changed. These laws are essentially of *geometric* origin and come from (a) small or zero gaps enabling long-range collective behaviors with layers and (b) partial screening of fields caused by the nanometer scale of the layers, as illustrated in Fig. 2.

A brief mathematical picture can be obtained as follows. Let us consider the small wavenumber q and imaginary frequency ω noninteracting response of a Dirac cone and a 2D metal. These, respectively (Dobson et al., 2006), take the form

$$\chi_0^D(q,\omega) \approx \frac{-Kq^2}{\sqrt{v_f^2 q^2 + \omega^2}}, \qquad \chi_0^M(q,\omega) \approx \frac{-K' q^2}{\omega^2}. \tag{3}$$

Applying RPA screening $\chi = \chi_0 + \chi_0 w \chi$ and using the bulk 3D Coulomb potential $w^{3D} = 4\pi/q^2$ gives

$$\chi_{3D}^D(q,\omega) \approx \frac{-Kq^2}{\sqrt{v_f^2 q^2 + \omega^2} + 4\pi K}, \qquad \chi_{3D}^M(q,\omega) \approx \frac{-K' q^2}{\omega^2 + 4\pi K'} \tag{4}$$

so that the denominator approaches a constant as ω goes to zero: a characteristic which means that the screened material will behave similarly to an insulator. However, when we have layered systems we must not use w^{3D} and instead must use the 2D intralayer Coulomb potential $w = w^{2D} = 2\pi/q$. Now

$$\chi_{2D}^D(q,\omega) \approx \frac{-Kq^2}{\sqrt{v_f^2 q^2 + \omega^2} + 2\pi q K}, \qquad \chi_{2D}^M(q,\omega) \approx \frac{-K' q^2}{\omega^2 + 2\pi q K'}, \tag{5}$$

means that the denominator is small for $\omega \to 0$ and $q \to 0$. The layers are no longer similar to insulators for $q \to 0$.

The asymptotic van der Waals force between two layers is given, to leading order, by

$$U_{\rm vdW}^{AB}(D) \approx -\frac{1}{2}\int \frac{d\omega}{\pi}\int \frac{d\mathbf{q}}{(2\pi)^2}\chi^A(q,\omega)\chi^B(q,\omega)w^2 e^{-2qD}. \quad (6)$$

After some work (Dobson et al., 2006), including appropriate changes of various variables (eg, $\omega \rightarrow \sqrt{q}x$ and $q \rightarrow y/D$ in metallic systems), it is possible to show that (ignoring other possible effects; Dobson et al., 2014)

$$U_{\rm vdW}^{GG}(D) \rightarrow \frac{-C_3}{D^3}F(D), \qquad U_{\rm vdW}^{MM}(D) \rightarrow \frac{-C_{5/2}}{D^{5/2}}, \quad (7)$$

where $F(D)$ is a slow-varying function obeying $F(D \rightarrow \infty) = 1$ that can be approximated by an arctan (Gould et al., 2013a). Here G indicates graphene or another Dirac cone layer and M indicates a metal. By contrast, two insulators (I) interact as

$$U_{\rm vdW}^{II}(D) \rightarrow \frac{-C_4}{D^4} \propto \sum_{ij \in \rm layers} \frac{-C_{6ij}}{R_{ij}^6} \quad (8)$$

as predicted (qualitatively but not necessarily quantitatively) by an additive theory like Lennard-Jones. The van der Waals attraction is fundamentally altered by the partial screening giving rise to qualitative differences in the attractive force.

Furthermore, similar analyses can be carried out (Gould et al., 2008, 2009) on more general geometries, including exfoliation and cleavage (respectively, where a single layer is removed from the top of a bulk stack and where a bulk stack is split into two at its middle). Results are presented in Table 1. As the system becomes more bulk-like (cleavage is more bulk-like than exfoliation is more bulk-like than bilayer interactions) the power laws for the different layer types converge, as expected by the screening arguments, and all three types of systems become asymptotically equivalent for cleavage (albeit with different prefactors).

These unusual power laws caused by nonadditivity could affect self-assembly and other processes relying on long-range attractions. But their effect on bound heterostructures is likely to be much less significant. Quantitative tests using layer response theory suggest that they dominate only at around 10 nm, and probably contribute significantly less than 10% of the binding energy in bulk or bilayered graphene.

Table 1 Asymptotic Power Laws of Different Layer Types in Different Geometries

Layer	Bilayer	Exfoliation	Cleavage
	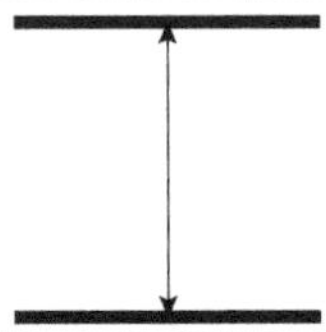	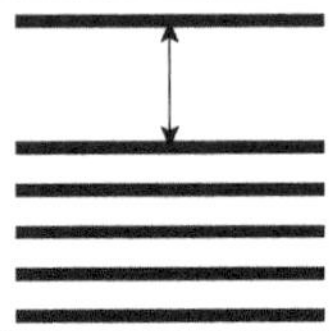	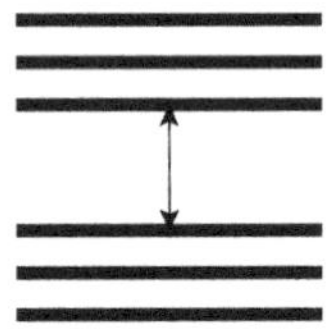
Insulators	D^{-4}	D^{-3}	D^{-2}
Graphene	D^{-3}	$D^{-3}\log(D/D_0)$	D^{-2}
Metals	$D^{-5/2}$	$D^{-5/2}$	D^{-2}

Stretching (in which all layers are separated equally) obeys the same power law as a bilayer.

4. CALCULATION OF 2D MATERIALS

While van der Waals forces alone are vitally important to the binding of layered materials, they are not sufficient to make quantitative predictions of materials properties in the vital near-contact regime, ie, for distances close to the equilibrium spacing. To obtain energies in this realm we need to use full details of the electronic structure using more sophisticated methods. In this section we will first outline some of the specific challenges facing ab initio calculations of 2D material, and discuss the present state of the art. We will then study the role computers have and will play in material understanding and discovery.

4.1 Energetic Properties

2D materials are typically quite strong within each layer (via strong ionic and covalent intralayer forces). Thus, arguably the most important energetic characteristics of a 2D material are those that act between layers (the weak interlayer van der Waals forces) as they are responsible for the ultimate stability of the material (for example, graphene has an intralayer elastic modulus of around 1000 Gpa, while the interlayer elastic modulus C_{33} is around 40 Gpa).

The interlayer energetics may be broadly characterized by three numbers: the interlayer lattice constance c, the interlayer elastic constant C_{33}, and the interlayer binding energy U_b (and the closely related cohesive and exfoliation energies). The first two of these can be found from experiments (albeit with only moderate accuracy for the elastic modulus), while the third is very difficult (see discussion later). We focus our attention in this chapter on these energetic properties because of their importance in layered materials.

A rough guide to the predictive quality of different approaches is provided in Table 2. This summarizes the approximate accuracy of different theories for different measurable quantities. It may be quite wrong for certain 2D materials, but should provide some guidance in interpreting numerical results.

4.1.1 Conventional Semi-local Approximations

The commonest ab initio electronic materials modeling methods (LDA, GGA (Becke, 1988; Lee et al., 1988; Perdew et al., 1996), hybrids (Becke, 1993)) are not suitable for our target layered systems because they do not correctly describe the van der Waals (dispersion) interaction, and indeed they miss it completely at large separations. Conventional semi-local

Table 2 Very Rough Guide to Predictive Qualities of Selected Functionals

Method	c_0	C_{33}[a]	U_b
RPA	Best	OK	Benchmark[b]
LDA	OK; sometimes very small	OK	Quite small
PBE	Unusable	Unusable	Unusable
PBE-D	Mostly OK	Poor	Poor
VV10	Mostly OK	OK	Too big
vdW-DF	Too big	Much too small	Small
vdW-DF2	Too big	Too small	Small
C09	OK	—	Too big
optB86b-DF2	Good	—	OK

[a]Elastic constants generally show wide variation in accuracy against benchmarks (which themselves vary). As such we make a note only when errors are systematic.
[b]RPA has known problems in energy predictions, but is presently likely to be the most accurate method for many materials.
We advise checking the original manuscripts for further cases and quantitative details.
Source: Data used for this table are taken from Björkman, T., 2014. Testing several recent van der Waals density functionals for layered structures. J. Chem. Phys. 141 (7), 074708. http://scitation.aip.org/content/aip/journal/jcp/141/7/10.1063/1.4893329. Björkman, T., Gulans, A., Krasheninnikov, A.V., Nieminen, R.M., 2012a. Are we van der Waals ready? J. Phys. Condens. Matter 24 (42), 424218. http://stacks.iop.org/0953-8984/24/i=42/a=424218. Björkman, T., Gulans, A., Krasheninnikov, A.V., Nieminen, R.M., 2012b. van der Waals bonding in layered compounds from advanced density-functional first-principles calculations. Phys. Rev. Lett. 108, 235502. http://link.aps.org/doi/10.1103/PhysRevLett.108.235502.

and hybrid approximations thus tend to be quite poor at predicting interlayer properties of layered systems. A remarkable exception is the LDA, which has proved to be quite good at predicting interlayer lattice parameters and OK for bulk moduli. However, its success is believed to rely on a cancellation of errors in the exchange and correlation energies which may not work so well for heterostructures. Also, it substantially underestimates binding energies.

4.1.2 van der Waals Approximations

Given the importance of van der Waals forces in layered materials, it is unsurprising that approximations (Andersson et al., 1996; Bučko et al., 2014; Dion et al., 2004; Dobson and Dinte, 1996; Gobre and Tkatchenko, 2013; Grimme, 2004, 2006; Grimme et al., 2010; Langreth et al., 2005; Tkatchenko and Scheffler, 2009; Tkatchenko et al., 2012; Vydrov and Van Voorhis, 2009, 2010a) that include them do better than their semi-local counterparts. Nonetheless, all low-level approaches fail to predict the long-range behavior of metallic and semi-metallic lattices. As we shall see in Section 4.2.1, they are also of inconsistent accuracy for other properties.

Of particular note is the "many-body" version of the Tkatchenko atom-based approach (Bučko et al., 2016; DiStasio et al., 2012; Tkatchenko et al., 2012) which can be considered an approximation to the RPA correlation energy based on a simplified description of each atom as a harmonic oscillator. The three-atom, four-atom, etc. ("type B" nonadditivity) effects are captured by this method. It does not give the type C effects in the asymptotic limit, but is quite useful for binding energetics at separations less than about 10 nm, although it is still not as reliable as the full RPA. Recent extensions have increased its applicability to systems with substantial ionic character (Bučko et al., 2014). It is computationally efficient and can be applied to a range of geometries well out of reach of the full RPA.

Another popular method is the vdW-DF class of density functionals (Andersson et al., 1996; Berland and Hyldgaard, 2014; Dion et al., 2004; Langreth et al., 2005; Vydrov and Van Voorhis, 2009, 2010a), which provide a correlation energy correction to semi-local density functionals. These seamless vdW functionals are efficient (Klimeš and Michaelides, 2012) and have been applied to a wide range of layered systems. Formally they lack both type B and type C nonadditivity and so are inapplicable to metals, or to small-gap systems at very large separations. Nevertheless, they give reasonable results for many-layered systems. Consistent and systematic description of the binding and elastic properties of our target systems has not been achieved, however (Björkman, 2014). Part of the problem appears to be the ambiguity in the choice of an exchange functional for use with vdW-DF. Recent work has shed further light on this issue (Berland and Hyldgaard, 2014).

4.2 RPA Calculations

Despite the plethora of new van der Waals approaches, layered materials continue to pose a particular challenge because of their peculiar screening properties and strong sensitivity to dispersion models for even basic properties. As demonstrated by Björkman and coworkers (Björkman, 2014; Björkman et al., 2012a,b), many van der Waals functionals fail to reproduce accurately some or all energetic properties in layered systems.

The most fundamental way to deal with these forces is high-level many-electron theory, the simplest version being the RPA. In the RPA the electron density fluctuations are acted on by a time-dependent mean field caused by the fluctuations at all other locations, leading *inter alia* to the many-atom interactions described in previous sections.

For energy calculations, the term "RPA" usually means random-phase approximation used in the adiabatic connection formula and fluctuation–dissipation theorem (ACFD-RPA) (Dobson and Gould, 2012; Eshuis et al., 2012; Ren et al., 2012). This allows electronic correlations to be treated seamlessly (Dobson and Wang, 1999) at the mean-field level. ACFD-RPA is available in various codes, including VASP and GPAW. However, it is very numerically demanding and is thus too difficult for highly complex systems, especially those that appear in heterostructures with lattice mismatches. Development of ACFD-RPA is ongoing, with new algorithms (Rocca et al., 2008) promising access to larger systems going into the future.

In an ACFD-RPA calculation, first the DFT response is calculated using a low-level approximation (typically PBE). This involves finding a set of orbitals $\phi_i \equiv \phi_{n\boldsymbol{k}}$ from the Kohn–Sham equation

$$\left[-\frac{1}{2}(\nabla - i\boldsymbol{k})^2 + v_s[n](\boldsymbol{r})\right]\phi_{n\boldsymbol{k}}(\boldsymbol{r}) = \epsilon_{n\boldsymbol{k}}\phi_{n\boldsymbol{k}}(\boldsymbol{r}). \tag{9}$$

Here discrete n labels the band while continuous $\boldsymbol{k}$ is the Bloch wavenumber and each orbital is assigned an occupation factor $f_i = f_{n\boldsymbol{k}} = \Theta(\epsilon_f - \epsilon_i)$. The orbitals are then used to obtain the noninteracting response function

$$\chi_0(\boldsymbol{r},\boldsymbol{r}';i\omega)=2\mathcal{R}\sum_{ij}(f_i-f_j)\frac{\phi_i^*(\boldsymbol{r})\phi_i(\boldsymbol{r}')\phi_j^*(\boldsymbol{r}')\phi_j(\boldsymbol{r})}{\epsilon_i-\epsilon_j+i\omega} \tag{10}$$

where the sums over i and j include sums over bands n and integrals over Bloch wavenumbers $\boldsymbol{k}$. We note that only transitions between occupied and unoccupied states appear in χ_0.

Second, the RPA interacting response is found (see previous discussion in Section 3.2 on screening in Lifshitz theory) by solving

$$\chi_\lambda = \chi_0 + \chi_0 \star v \star \chi_\lambda \tag{11}$$

where $\star$ indicates a convolution $f\star g\equiv\int d\boldsymbol{r}_2 f(\boldsymbol{r},\boldsymbol{r}_2)g(\boldsymbol{r}_2,\boldsymbol{r}')$. Finally, the ACFD is applied to find

$$E_c=\frac{1}{2}\int_0^1 d\lambda\int_0^\infty\frac{d\omega}{\pi}\mathrm{Tr}[(\chi_\lambda-\chi_0)\star v] \tag{12}$$

where $\mathrm{Tr}[f]=\int d\boldsymbol{r}f(\boldsymbol{r},\boldsymbol{r})$.

Typically, the second and final step can be combined into one step by analytically integrating over λ to yield

$$E_c=-\frac{1}{2}\int_0^\infty\frac{d\omega}{\pi}\mathrm{Tr}[\log(1-\chi_0\star v)+\chi_0\star v]. \tag{13}$$

In a planewave code, the symmetries of the system can be used to write equivalently

$$E_c=-\frac{1}{2}\int_0^\infty\frac{d\omega}{\pi}\int_{\mathrm{BZ}}\frac{d\boldsymbol{q}}{(2\pi)^3}\mathrm{Tr}\left[\log\left(\delta_{\mathbf{G}-\mathbf{G}'}-X^{\boldsymbol{q},\omega}_{\mathbf{G},\mathbf{G}'}\right)+X^{\boldsymbol{q},\omega}_{\mathbf{G},\mathbf{G}'}\right], \tag{14}$$

where

$$X^{\boldsymbol{q},\omega}_{\mathbf{G},\mathbf{G}'}=\frac{4\pi\chi_0(\mathbf{G},\mathbf{G}';\boldsymbol{q},i\omega)}{|\boldsymbol{q}+\mathbf{G}||\boldsymbol{q}+\mathbf{G}'|} \tag{15}$$

and $\chi_0(\mathbf{G},\mathbf{G}')=\int_{\mathrm{UC}}d\boldsymbol{r}d\boldsymbol{r}'e^{i(\mathbf{G}\cdot\boldsymbol{r}-\mathbf{G}'\cdot\boldsymbol{r}')}\chi_0(\boldsymbol{r},\boldsymbol{r}')$. Here the integral on $\boldsymbol{q}$ is taken over the Brillouin zone, integrals on $\boldsymbol{r}$ and $\boldsymbol{r}'$ are taken over the unit cell, the log acts on a matrix, and the trace is $\mathrm{Tr}[X]\equiv\sum_{\boldsymbol{G}}X_{\mathbf{G},\mathbf{G}}$.

The van der Waals force between layers separated at a distance D is defined via (13) as

$$U_{\mathrm{vdW}}(D)=E_c[\sum\text{layers}]-\sum E_c[\text{layers}] \tag{16}$$

in the absence of charged layers, ie, the difference in correlation energies between all layers together and the summed correlation energies of the isolated layers. If the ACFD-RPA calculations are performed very precisely, the asymptotic results will match those of Sections 3.2 and 3.2.2. For example, graphene has been shown to obey the correct asymptotic power law (Gould et al., 2013a; Lebègue et al., 2010).

RPA and extensions in practice are the only methods to date to give quantitatively reliable energetic predictions for all separations across a range of vdW-bound layered systems (Björkman et al., 2012a,b), and can thus be used for benchmarking (see below). However, the full RPA is numerically infeasible beyond the very simplest geometries. Even with methodological advances it is unlikely to be practical for complex heterostructures.

4.2.1 Benchmarking via RPA

It is often very difficult to experimentally determine important energetic properties in 2D materials. For example, even the interplanar elastic modulus C_{33} is known only poorly for many materials, including graphene. Binding energies, which come in different related flavors, pose even more of a challenge as they are both weak and indirectly measured (sometimes using inaccurate theory). Adding to the challenge is the fact that calculations are usually carried out at 0 K, while measurements are often carried out at room temperature, requiring additional interpretational issues to be dealt with by theory.

As an alternative, ab initio calculations using the RPA (discussed in Section 4.2) allow geometry to be controlled, while providing a theory that is expected to be accurate for energy differences. As such, RPA provides a natural benchmark for lower-level theories.

For graphite, RPA data for the interlayer bonding were provided by Lebègue et al. (2010). Using an efficient basis set extrapolation technique (Gulans, 2014), Björkman et al. (2012a) managed to extend the available RPA data for layered systems to a significantly larger set of 28 layered systems. This primarily contained transition metal dichalcogenides, but also hexagonal boron nitride (h-BN) and litharge, a layered form of lead oxide (PbO). Comparing to calculations with the vdW-DF type of functionals, it was found that the VV10, a van der Waals density functional by Vydrov and Van Voorhis (2010b), was able to very accurately reproduce the trends of the RPA data, albeit with a constant overestimate of approximately 50%.

This allowed RPA interlayer binding energies to be estimated for a set of approximately 100 layered systems by simply rescaling values calculated using the much less demanding VV10 method (Björkman et al., 2012b). These compounds were automatically identified in a database search through purely geometrical criteria further described in Section 4.3. These calculations lead to the observation that the van der Waals component of the binding in these layered systems appeared to mostly fall into a fairly narrow interval around 20 meV/Å^2. The study also found this to be independent of any other materials characteristics, such as layer thickness, detailed chemical composition, or whether the compound is insulating or metallic.

For atomic and molecular systems there has long been highly accurate quantum chemical calculations available for van der Waals bonded systems, such as for rare gas dimers (Slavček et al., 2003) or the celebrated S22 reference set (Jurecka et al., 2006). For solids, quantum chemical methods are much too expensive to apply and to date, these binding energies for 28 compounds calculated with RPA represent one of the largest sets of reference data for van der Waals bonded solids available.

Using these RPA data as a reference, further work was done on benchmarking a number of vdW-DF type of methods (Björkman, 2014; Björkman et al., 2012a), as well as the PBE-D method (Grimme, 2006). This study revealed that none of the methods investigated were capable of reproducing both RPA binding energies and equilibrium van der Waals bond lengths in agreement with experiments (Björkman et al., 2012a); functionals that were close to the RPA benchmark for binding energies had very large equilibrium separations between the layers. This weakness has since then been largely remedied by recent developments of the vdW-DF functionals (Berland and Hyldgaard, 2014; Hamada, 2014; Klimeš et al., 2010, 2011). These improved functionals give very good agreement with experiment for equilibrium geometries, but in order to achieve this, they still overestimate binding energies with respect to RPA.

These benchmarks also provided an interesting and instructive failure of the very popular PBE-D method (Grimme, 2006) to describe the interlayer interaction in the layered form of PbO, litharge. Unlike most other layered materials, in litharge the van der Waals bond is between the heavy Pb atoms, which are bonded to a middle layer of oxygen atoms. Since the atomic polarizability of Pb is very large, a method that attempts to sum such atomic polarizabilities will be very sensitive to the damping function being tuned just right for the situation at hand.

This is so, because the Pb atoms in litharge have a very much reduced polarizability from the charge transfer to oxygen atoms and sharing of the electrons over the whole layer. In most cases involving heavier atoms this sensitivity is hidden; since the heaviest atoms are seldom directly adjacent across a van der Waals bonded gap in a structure, there are usually lighter atoms in between that will mollify the effect of an overestimate of the atomic polarizability. See Fig. 3 for a comparison of the PbO structure to that of the layered transition metal dichalcogenide MoS_2. But in litharge the PBE-D functional gives an interlayer binding energy of 125.5 meV/Å^2, an overestimate of an order of magnitude compared to the RPA value of 20.3 meV/Å^2. When applied to layered solids the imported atomic parameters of PBE-D also give other, less drastic problems, such as a very clear hierarchy of interlayer binding energies among the interlayer binding energies of the transition metal dichalcogenides such that sulfides $<$ selenides $<$ tellurides. This clear ordering is a natural consequence of the atomic polarizabilities, but is absent in the RPA calculations or the vdW-DF functionals, which are based on the wavefunctions and densities of the solids.

By treating the RPA data as a benchmark functional, it is also possible to develop a functional to match the reference data as closely as possible. This route was taken by Björkman (2012) in an attempt to remedy the large and systematic overestimate of the VV10 functional observed in the previous study (Björkman et al., 2012a). For functional development purposes, the VV10 functional has the benefit that it follows the conventional route of first defining an exchange functional and then fitting the smaller correlation contribution to this (Vydrov and Van Voorhis, 2010b). For layered systems, two functionals, AM05-VV10sol and PW86R-VV10sol, were designed based on the AM05 (Armiento and Mattsson, 2005) and PW86R (Murray et al., 2009)

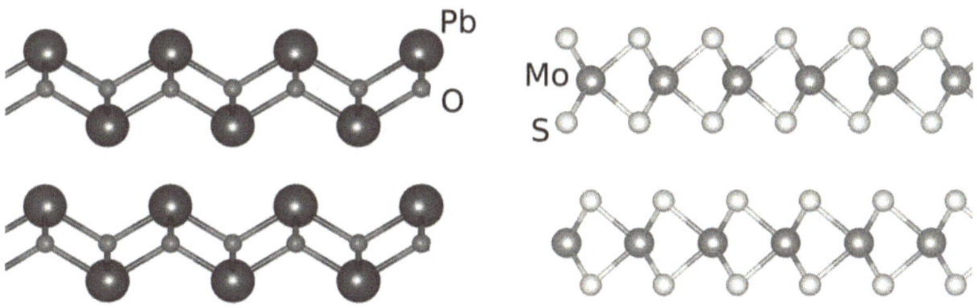

Fig. 3 The crystal structures of PbO (*left*) and MoS_2 (*right*) showing the difference in atomic arrangements, where PbO has the heavier atoms at either side of the van der Waals gap and MoS_2 has the more common arrangement with lighter atoms decorating a core of heavier atoms.

exchange functionals by fitting the VV10sol parameters to reproduce the RPA binding energies of the 28 layered solids. While the AM05-VV10sol and PW86R-VV10sol work well in layered systems, it has since their construction been found that other functionals in the vdW-DF class perform even better, without being constrained to a specific class of systems (Björkman, 2014), and the VV10sol methods may thus be regarded as mostly of theoretical interest.

An interesting feature of the RPA benchmark for layered systems that appears particularly difficult for other methods to reproduce is the significantly lower binding energy of BN compared to graphite. While both materials fall in the typical window identified, graphite has a binding energy of 18.3 meV/Å^2 and BN 14.5 meV/Å^2, which is 25% lower. The vdW-DF methods generally produce the correct ordering, with BN having lower binding energy than graphite, but the energies are very close. Both LDA and the PBE-D methods yield stronger binding for BN than for graphite, and this has also been found to be the case for the Tkatchenko–Scheffler method(Gao and Tkatchenko, 2015).

4.3 Computational Material Discovery of New Two-Dimensional Compounds

From an experimentalist point of view, the study of even a single two-dimensional system is a time-consuming process which includes its synthesis and its characterization using various spectroscopies, making almost impossible a systematic exploration of the properties of a large number of compounds. However, with the recent progress made in computational material science and with the increasing power of supercomputers, it has become possible to calculate the properties of a large number of compounds, with the hope that these properties will be superior to the properties of the compounds already known.

Nevertheless, isolating a relevant (even large) pool of candidate materials out of the millions of possible combinations that can be obtained when crossing a quasi-infinite number of chemical formulas with the various possible crystal structures is not a trivial task. Various methods have been developed to meet this goal, depending on the type and on the amount of information available. One can start, for example, from a set of known crystal structures and perform chemical replacements using known structures, which has led to various high-throughput projects (Curtarolo et al., 2012; Jain et al., 2013). Another way of proceeding is to explore the potential energy surface using only the chemical formula as input, as done in methods based on evolutionary/genetics algorithms (Oganov and Glass, 2006) to explore the potential energy surface. The idea is that starting from a first pool (called "generation" in the field of genetic algorithms) of structures generated randomly, the following generations of structures are modified using algorithms derived from evolution theory to obtain one or several structures corresponding to a minimum of enthalpy.

Yet another method consists of performing datamining in existing crystallographic databases (for example, in the Inorganic Crystal Structure Database—ICSD (Bergerhoff and Brown, 1987) or in the Crystallography Open Database—COD (Gražulis et al., 2009)) using a set of defined criteria. Datamining methods are becoming more and more popular in nanosciences and have been used, for instance, to identify new scintillator materials (Ortiz et al., 2009), novel superconductors (Klintenberg and Eriksson, 2013), and topological insulators (Klintenberg, 2010) with success.

In a recent article (Lebègue et al., 2013), Lebègue and coworkers presented an algorithm to search systematically in the ICSD for layered compounds. The philosophy of their approach can be summarized as follows: if layered compounds can be identified, then there is a chance that new two-dimensional compounds can be obtained by isolating one or few layers of each layered material. The algorithm was designed as follows:

1. Since layered materials are expected to have packing ratios (defined as the total covalent volume divided by the volume of the cell) between those close packed structures and molecular solids, a first screening was made to keep only compounds having a packing ratio between 0.15 and 0.50 (the total covalent volume is defined as the sum of the volumes of each sphere used to represent the covalent volume of each atom, calculated from the covalent radius (Barbalace, 2012)).
2. Then, noticing that layered materials which are suitable to obtain two-dimensional compounds correspond to a stack of layers bonded to each other by weak van der Waals interactions, the second step of the procedure selected compounds having a gap in distance of more than > 2.5 Åalong one of the crystallographic direction.
3. In the last step of the procedure, the algorithm would check that no covalent bond is present across the layers by comparing the sum of the covalent radii of the atoms in different layers with the distance between the layers. Notice that to simplify the procedure, non-stoichiometric compounds as well as compounds with a low symmetry were excluded from the search.

As a result of applying the procedure described earlier, the algorithm naturally identified graphite, hexagonal boron nitride, as well as a large list of dichalcogenides (MoS_2, WSe_2, etc.) as layered materials. More importantly, the same algorithm found around 40 compounds that satisfy the criteria defined previously for being considered as layered materials, and from which potential new two-dimensional materials could be obtained. As can be seen from Table 3, the chemical composition as well as the electronic structure of the monolayers found are quite diverse, ranging from relatively simple chemical formula like SrIF to complex one like P_2CuSe_6Bi and from metallic systems to large bandgap insulators. Some systems are found to present a magnetic order, like FeTe and $CrSiTe_3$ which are found to be ferromagnetic and $FeBr_3$ which is found to be antiferromagnetic. For each compound, the density of states and the bandstructures were computed using density functional theory at the generalized gradient level (GGA) (Lebègue et al., 2013). For example in Fig. 4 we show the bandstructure of monolayers of $ZnIn_2S_4$ and $Zn_2In_2S_5$, which are found to be metallic.

An important point is that the bulk form of each compound presented in Table 3 already exists and was characterized by X-ray diffraction. Therefore, the corresponding monolayers can possibly be obtained by exfoliation or chemical synthesis. As an example, it was shown (Ben Romdhane et al., 2015) that Cu_2S, which is listed in Table 3, can indeed be obtained in the form of bilayers over a substrate of carbon. This demonstrates once more that computational methods can nowadays be truly predictive in the design of new materials.

5. THE FUTURE ROLE OF COMPUTERS

Computers will play a fundamental role in heterostructure material science. There are an essentially infinite range of possible heterostructures, each of which could be interesting. But experiments can be expensive in time, money, and effort, making it impossible (or at last undesirable) to test all potentially interesting heterostructures. To fully harness the endless possibilities of heterostructure design, it will thus be necessary to employ computer-aided design processes: both to screen candidate heterostructures for desirable properties and to eliminate candidate heterostructures due to poor properties.

As discussed above (Section 4.3), computers can be used to scan through *known* materials for possible 2D candidates, expanding the available range of building blocks. Computers have also been used to discover and predict

Table 3 List of Possible Two-Dimensional Compounds Found by Datamining into the ICSD

2D Chemical Formula	3D ICSD Number	Gap (eV)	Magnetism	2D Chemical Formula	3D ICSD Number	Gap (eV)	Magnetism
HgI_2	150345	1.8		$KC_6FeO_3N_3$	280850	4.5	
PbIF	150193	2.3		$PbSb_2Te_4$	250250	0.8	
ZrClN	151468	1.9		MgI_2	281551	3.6	
BaIF	155006	4.3		BiIO	391354	1.5	
SrIF	155009	4.5		$FeBr_3$	410924	0.5	AFM
$AlCl_2$	155670	Metal		$MgPSe_3$	413165	2.1	
Ag_2ReCl_6	156662	Metal		IYGa	417149	Metal	
Ni_2Te_2Sb	158485	Metal		PTe_2Ti_2	418978	Metal	
$Bi_{14}Te_{13}S_8$	159356	0.9		ScP_2AgSe_6	420302	1.8	
$MgBr_2$	165972	4.8		$CrSiTe_3$	626809	0.6	FM
Cu_2S	166578	Metal		$FePSe_3$	633094	0.05	
P_2AgSe_6Bi	170640	1.4		FeS	633302	Metal	
P_2CuSe_6Bi	170642	0.8		FeTe	633877	Metal	FM
YI_3	170773	3.1		$Sb_2Ge_2Te_5$	637823	0.2	
GaS	173940	2.4		SbSiNi	646436	Metal	
VCl_2	246905	Metal		PbO	647260	2.5	

Continued

Table 3 List of Possible Two-Dimensional Compounds Found by Datamining into the ICSD—cont'd

2D Chemical Formula	3D ICSD Number	Gap (eV)	Magnetism	2D Chemical Formula	3D ICSD Number	Gap (eV)	Magnetism
VBr_2	246906	Metal		CdI_2	655780	2.5	
VI_2	246907	Metal		GaSe	660262	1.8	
PFeLi	247089	Metal		$ZnIn_2S_4$	660273	Metal	
$PbBi_2Te_4$	250249	1.0		$Zn_2In_2S_5$	660333	Metal	

The ICSD number of the corresponding bulk material is given at the right of the chemical formula of the monolayer, as well the calculated electronic bandgap (in eV) for the monolayer as obtained with density functional theory. FM refers to a ferromagnetic ordering, while AFM refers to an antiferromagnetic ordering within the layer.

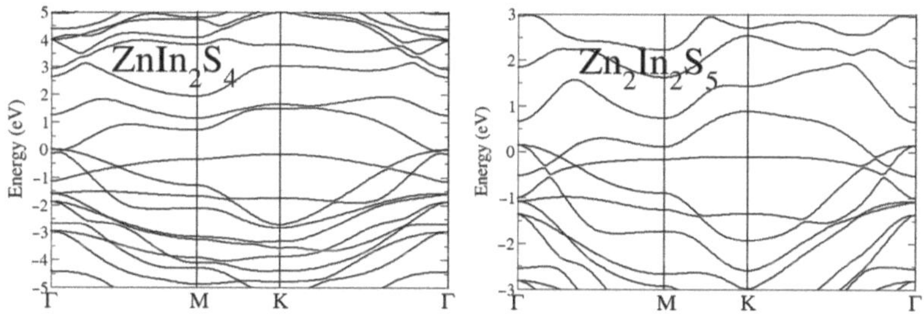

Fig. 4 Band structure along the high-symmetry directions Γ-M-K-Γ for $ZnIn_2S_4$ (*left*) and $Zn_2In_2S_5$ (*right*). The Fermi level is at 0 eV.

layered compounds. As electronic structure approximations, algorithms, and databases get better, automated scanning could potentially be combined with automated design—allowing heterostructured materials to be "proposed" by computers, despite some components never having been isolated.

But truly computationally driven material science must be based around theories which are accurate, especially for energetics. So far, only the computationally difficult RPA appears to be able to "guarantee" quantitative accuracy. Research is thus ongoing into making sure heterostructure calculations can be carried out accurately and reliably—this involves both benchmarking and the development new ab initio approaches, such as van der Waals approaches.

This chapter focused almost exclusively on zero temperature (0 K) calculations. Molecular modeling approaches and simplified DFT approaches will likely remain necessary for taking into account thermal effects such as phonons and band-structure smearing. It remains an open question whether or not current approaches (Gould et al., 2013b), based on simple, additive van der Waals theories, are up to this task. If not, new approaches must be investigated that properly account for these unusual, but important vdW forces.

6. CONCLUSIONS

Layered materials, especially heterostructures made from the three generations of 2D materials, are likely to revolutionize technology. In this chapter we have discussed the third generation of materials, discovered by computers and experiment (Section 2). We have summarized some key theoretical aspects of the vitally important van der Waals forces (Section 3).

We have discussed (Section 4) the advantages and disadvantages of different ab initio approaches, with the aim of providing practical guidance (Section 4.2.1) for studies of third-generation layered materials and heterostructures. Finally, we have discussed the role computers presently play (Section 4.3) and are likely to play (Section 5) in material discover and engineering.

REFERENCES

Andersson, Y., Langreth, D.C., Lundqvist, B.I., 1996. van der Waals interactions in density-functional theory. Phys. Rev. Lett. 76, 102–105. http://link.aps.org/doi/10.1103/PhysRevLett.76.102.

Armiento, R., Mattsson, A.E., 2005. Functional designed to include surface effects in self-consistent density functional theory. Phys. Rev. B 72, 085108. http://link.aps.org/doi/10.1103/PhysRevB.72.085108.

Barbalace, K., 2012. Environmentalchemistry.com. http://EnvironmentalChemistry.com/yogi/periodic/covalentradius.html.

Barsoum, M.W., 2000. The MN+1AXN phases: a new class of solids: thermodynamically stable nanolaminates. Prog. Solid State Chem. 28 (1–4), 201–281. http://www.sciencedirect.com/science/article/pii/S0079678600000066.

Barsoum, M.W., Brodkin, D., El-Raghy, T., 1997. Layered machinable ceramics for high temperature applications. Scr. Mater. 36 (5), 535–541. http://www.sciencedirect.com/science/article/pii/S1359646296004186.

Becke, A.D., 1988. Density-functional exchange-energy approximation with correct asymptotic behavior. Phys. Rev. A 38 (6), 3098–3100.

Becke, A.D., 1993. A new mixing of Hartree-Fock and local density-functional theories. J. Chem. Phys. 98 (2), 1372–1377. http://link.aip.org/link/?JCP/98/1372/1.

Ben Romdhane, F., Björkman, T., Rodríguez-Manzo, J.A., Cretu, O., Krasheninnikov, A.V., Banhart, F., 2013. In-situ growth of cellular two-dimensional silicon oxide on metal substrates. ACS Nano 7, 5175–5180.

Ben Romdhane, F., Cretu, O., Debbichi, L., Eriksson, O., Lebègue, S., Banhart, F., 2015. Quasi-two-dimensional Cu_2S crystals on graphene: in-situ growth and ab-initio calculations. Small 11, 1253.

Bergerhoff, G., Brown, I.D., 1987. In: Allen, F.H., Bergerhoff, G., Sievers, R. (Eds.), Crystallographic Databases. International Union of Crystallography.

Berland, K., Hyldgaard, P., 2014. Exchange functional that tests the robustness of the plasmon description of the van der Waals density functional. Phys. Rev. B 89, 035412. http://link.aps.org/doi/10.1103/PhysRevB.89.035412.

Bianco, E., Butler, S., Jiang, S., Restrepo, O.D., Windl, W., Goldberger, J.E., 2013. Stability and exfoliation of germanane: a germanium graphane analogue. ACS Nano 7 (5), 4414–4421. http://dx.doi.org/10.1021/nn4009406.

Björkman, T., 2012. van der Waals density functional for solids. Phys. Rev. B 86, 165109. http://link.aps.org/doi/10.1103/PhysRevB.86.165109.

Björkman, T., 2014. Testing several recent van der Waals density functionals for layered structures. J. Chem. Phys. 141 (7), 074708. http://scitation.aip.org/content/aip/journal/jcp/141/7/10.1063/1.4893329.

Björkman, T., Gulans, A., Krasheninnikov, A.V., Nieminen, R.M., 2012. Are we van der Waals ready? J. Phys. Condens. Matter 24 (42), 424218. http://stacks.iop.org/0953-8984/24/i=42/a=424218.

Björkman, T., Gulans, A., Krasheninnikov, A.V., Nieminen, R.M., 2012. van der Waals bonding in layered compounds from advanced density-functional first-principles calculations. Phys. Rev. Lett. 108, 235502. http://link.aps.org/doi/10.1103/PhysRevLett.108.235502.

Björkman, T., Kurasch, S., Lehtinen, O., Kotakoski, J., Yazyev, O.V., Srivastava, A., Skakalova, V., Smet, J.H., Kaiser, U., Krasheninnikov, A.V., 2013. Defects in bilayer silica and graphene: common trends in diverse hexagonal two-dimensional systems. Sci. Rep. 3, 3482.

Bučko, T., Lebègue, S., Ángyán, J.G., Hafner, J., 2014. Extending the applicability of the Tkatchenko-Scheffler dispersion correction via iterative Hirshfeld partitioning. J. Chem. Phys. 141 (3), 034114. http://scitation.aip.org/content/aip/journal/jcp/141/3/10.1063/1.4890003.

Bučko, T.A., Lebègue, S., Gould, T., Ángyán, J.G., 2016. Many-body dispersion corrections for periodic systems: an efficient reciprocal space implementation. J. Phys. Condens. Matter 28 (4), 045201. http://stacks.iop.org/0953-8984/28/i=4/a=045201.

Butler, S.Z., Hollen, S.M., Cao, L., Cui, Y., Gupta, J.A., Gutirrez, H.R., Heinz, T.F., Hong, S.S., Huang, J., Ismach, A.F., Johnston-Halperin, E., Kuno, M., Plashnitsa, V.V., Robinson, R.D., Ruoff, R.S., Salahuddin, S., Shan, J., Shi, L., Spencer, M.G., Terrones, M., Windl, W., Goldberger, J.E., 2013. Progress, challenges, and opportunities in two-dimensional materials beyond graphene. ACS Nano 7 (4), 2898–2926. http://dx.doi.org/10.1021/nn400280c.

Cahangirov, S., Topsakal, M., Aktürk, E., Şahin, H., Ciraci, S., 2009. Two- and one-dimensional honeycomb structures of silicon and germanium. Phys. Rev. Lett. 102 (23), 236804. http://link.aps.org/doi/10.1103/PhysRevLett.102.236804.

Castro Neto, A.H., Guinea, F., Peres, N.M.R., Novoselov, K.S., Geim, A.K., 2009. The electronic properties of graphene. Rev. Mod. Phys. 81 (1), 109–162. http://link.aps.org/doi/10.1103/RevModPhys.81.109.

Chalasani, R., Gupta, A., Vasudevan, S., 2013. Engineering new layered solids from exfoliated inorganics: a periodically alternating hydrotalcite-montmorillonite layered hybrid. Sci. Rep. 3, 3498. http://dx.doi.org/10.1038/srep03498.

Chi, H., Kim, H., Thomas, J.C., Shi, G., Sun, K., Abeykoon, M., Bozin, E.S., Shi, X., Li, Q., Shi, X., Kioupakis, E., Van der Ven, A., Kaviany, M., Uher, C., 2014. Low-temperature structural and transport anomalies in cu_2se. Phys. Rev. B 89 (19), 195209. http://link.aps.org/doi/10.1103/PhysRevB.89.195209.

Choi, J.H., Zhu, W., Ho, K.M., Wang, D., Zhang, Z., 2015. Energetics and atomic structures of Cu2Te overlayers on CdTe(111). J. Phys. Chem. C 119 (9), 4843–4847. http://dx.doi.org/10.1021/jp511776e.

Curtarolo, S., Setyawan, W., Wang, S., Xue, J., Yang, K., Taylor, R.H., Nelson, L.J., Hart, G.L., Sanvito, S., Buongiorno-Nardelli, M., Mingo, N., Levy, O., 2012. AFLOWLIB.ORG: a distributed materials properties repository from high-throughput ab initio calculations. Comput. Mater. Sci. 58, 227–235.

Debbichi, L., Eriksson, O., Lebègue, S., 2014. Electronic structure of In3Se4 and In3Te4 monolayers from ab-initio calculations. Ann. Phys. 526 (9-10), 402–407. http://onlinelibrary.wiley.com/doi/10.1002/andp.201400159/abstract.

Debbichi, L., Eriksson, O., Lebègue, S., 2015. Two-dimensional indium selenides compounds: an ab initio study. J. Phys. Chem. Lett. 6 (15), 3098–3103. http://dx.doi.org/10.1021/acs.jpclett.5b01356.

Derivaz, M., Dentel, D., Stephan, R.A., Hanf, M.C., Mehdaoui, A., Sonnet, P., Pirri, C., 2015. Continuous germanene layer on Al(111). Nano Lett. 15 (4), 2510–2516. http://dx.doi.org/10.1021/acs.nanolett.5b00085.

Ding, Y., Singh, V., Goodman, S.M., Nagpal, P., 2014. Low exciton phonon coupling, high charge carrier mobilities, and multiexciton properties in two-dimensional lead, silver, cadmium, and copper chalcogenide nanostructures. J. Phys. Chem. Lett. 5 (24), 4291–4297. http://dx.doi.org/10.1021/jz5023015.

Dion, M., Rydberg, H., Schröder, E., Langreth, D.C., Lundqvist, B.I., 2004. Van der Waals density functional for general geometries. Phys. Rev. Lett. 92 (24), 246401.

DiStasio, R.A., von Lilienfeld, O.A., Tkatchenko, A., 2012. Collective many-body van der Waals interactions in molecular systems. Proc. Natl. Acad. Sci. U.S.A. 109 (37), 14791–14795. http://www.pnas.org/content/109/37/14791.abstract.

Dobson, J.F., 2014. Beyond pairwise additivity in London dispersion interactions. Int. J. Quantum Chem. 114 (18), 1157–1161. http://dx.doi.org/10.1002/qua.24635.

Dobson, J.F., Dinte, B.P., 1996. Constraint satisfaction in local and gradient susceptibility approximations: application to a van der Waals density functional. Phys. Rev. Lett. 76, 1780–1783. http://link.aps.org/doi/10.1103/PhysRevLett.76.1780.

Dobson, J.F., Gould, T., 2012. Calculation of dispersion energies. J. Phys. Condens. Matter 24, 073201.

Dobson, J.F., Wang, J., 1999. Successful test of a seamless van der Waals density functional. Phys. Rev. Lett. 82 (10), 2123–2126.

Dobson, J.F., White, A., Rubio, A., 2006. Asymptotics of the dispersion interaction: analytic benchmarks for van der Waals energy functionals. Phys. Rev. Lett. 96, 073201.

Dobson, J.F., Gould, T., Vignale, G., 2014. How many-body effects modify the van der Waals interaction between graphene sheets. Phys. Rev. X 4, 021040. http://link.aps.org/doi/10.1103/PhysRevX.4.021040.

Dobson, J.F., Gould, T., Lebègue, S., 2016. Layer response theory: energetics of layered materials from semi-analytic high-level theory. Phys. Rev. B 93, 165436. http://journals.aps.org/prb/abstract/10.1103/PhysRevB.93.165436.

Dvila, M.E., Xian, L., Cahangirov, S., Rubio, A., Lay, G.L., 2014. Germanene: a novel two-dimensional germanium allotrope akin to graphene and silicene. New J. Phys. 16 (9), 095002. http://iopscience.iop.org/1367-2630/16/9/095002.

Dzyaloshinskii, I.E., Lifshitz, E.M., Pitaevskii, L.P., 1961. General theory of van der Waals' forces. Sov. Phys. Usp. 4 (2), 153–176.

Eshuis, H., Bates, J.E., Furche, F., 2012. Electron correlation methods based on the random phase approximation. Theor. Chem. Acc. 131, 1–18. http://dx.doi.org/10.1007/s00214-011-1084-8.

Ezawa, M., 2012. Valley-polarized metals and quantum anomalous Hall effect in silicene. Phys. Rev. Lett. 109 (5), 055502. http://link.aps.org/doi/10.1103/PhysRevLett.109.055502.

Flynn, G.W., 2011. Perspective: the dawning of the age of graphene. J. Chem. Phys. 135 (5), 050901. http://scitation.aip.org/content/aip/journal/jcp/135/5/10.1063/1.3615063.

Ganatra, R., Zhang, Q., 2014. Few-layer MoS2: a promising layered semiconductor. ACS Nano 8 (5), 4074–4099. http://dx.doi.org/10.1021/nn405938z.

Gao, W., Tkatchenko, A., 2015. Sliding mechanisms in multilayered hexagonal boron nitride and graphene: the effects of directionality, thickness, and sliding constraints. Phys. Rev. Lett. 114, 096101. http://link.aps.org/doi/10.1103/PhysRevLett.114.096101.

Geim, A., Grigorieva, I., 2013. Van der Waals heterostructures. Nature 499 (7459), 419–425.

Geim, A.K., Novoselov, K.S., 2007. The rise of graphene. Nat. Mater. 6 (3), 183–191. http://dx.doi.org/10.1038/nmat1849.

Ghidiu, M., Naguib, M., Shi, C., Mashtalir, O., Pan, L.M., Zhang, B., Yang, J., Gogotsi, Y., Billinge, S.J.L., Barsoum, M.W., 2014. Synthesis and characterization of two-dimensional Nb4C3 (MXene). Chem. Commun. 50 (67), 9517–9520. http://pubs.rsc.org/en/content/articlelanding/2014/cc/c4cc03366c.

Gobre, V.V., Tkatchenko, A., 2013. Scaling laws for van der Waals interactions in nanostructured materials. Nat. Commun. 4. http://dx.doi.org/10.1038/ncomms3341.

Gould, T., Simpkins, K., Dobson, J.F., 2008. Theoretical and semiempirical correction to the long-range dispersion power law of stretched graphite. Phys. Rev. B 77 (16), 165134.
Gould, T., Gray, E., Dobson, J.F., 2009. van der Waals dispersion power laws for cleavage, exfoliation, and stretching in multiscale, layered systems. Phys. Rev. B 79 (11), 113402.
Gould, T., Dobson, J.F., Lebègue, S., 2013. Effects of a finite Dirac cone on the dispersion properties of graphite. Phys. Rev. B 87, 165422. http://link.aps.org/doi/10.1103/PhysRevB.87.165422.
Gould, T., Liu, Z., Liu, J.Z., Dobson, J.F., Zheng, Q., Lebègue, S., 2013. Binding and interlayer force in the near-contact region of two graphite slabs: experiment and theory. J. Chem. Phys. 139 (22), 224704. http://scitation.aip.org/content/aip/journal/jcp/139/22/10.1063/1.4839615.
Gražulis, S., Chateigner, D., Downs, R.T., Yokochi, A.F.T., Quirós, M., Lutterotti, L., Manakova, E., Butkus, J., Moeck, P., Le Bail, A., 2009. Crystallography Open Database—an open-access collection of crystal structures. J. Appl. Crystallogr. 42 (4), 726–729. http://dx.doi.org/10.1107/S0021889809016690.
Grimme, S., 2004. Accurate description of van der Waals complexes by density functional theory including empirical corrections. J. Comput. Chem. 25 (12), 1463–1473. http://dx.doi.org/10.1002/jcc.20078.
Grimme, S., 2006. Semiempirical GGA-type density functional constructed with a long-range dispersion correction. J. Comput. Chem. 27 (15), 1787–1799. http://dx.doi.org/10.1002/jcc.20495.
Grimme, S., Antony, J., Ehrlich, S., Krieg, H., 2010. A consistent and accurate ab initio parametrization of density functional dispersion correction (DFT-D) for the 94 elements H-Pu. J. Chem. Phys. 132 (15), 154104. http://link.aip.org/link/?JCP/132/154104/1.
Gulans, A., 2014. Towards numerically accurate many-body perturbation theory: short-range correlation effects. J. Chem. Phys. 141 (16), 164127. http://scitation.aip.org/content/aip/journal/jcp/141/16/10.1063/1.4900447.
Guo, Z., Zhou, J., Si, C., Sun, Z., 2015. Flexible two-dimensional Tin+1Cn (n=1, 2 and 3) and their functionalized MXenes predicted by density functional theories. Phys. Chem. Chem. Phys. http://pubs.rsc.org/en/content/articlelanding/2015/cp/c5cp00775e.
Hamada, I., 2014. van der Waals density functional made accurate. Phys. Rev. B 89, 121103. http://link.aps.org/doi/10.1103/PhysRevB.89.121103.
Huang, P.Y., Kurasch, S., Srivastava, A., Skakalova, V., Kotakoski, J., Krasheninnikov, A.V., Hovden, R., Mao, Q., Meyer, J.C., Smet, J., Muller, D.A., Kaiser, U., 2012. Direct imaging of a two-dimensional silica glass on graphene. Nano Lett. 12, 1081–1086.
Jacobs-Gedrim, R.B., Shanmugam, M., Jain, N., Durcan, C.A., Murphy, M.T., Murray, T.M., Matyi, R.J., Moore, R.L., Yu, B., 2014. Extraordinary photoresponse in two-dimensional In2Se3 nanosheets. ACS Nano 8 (1), 514–521. http://dx.doi.org/10.1021/nn405037s.
Jain, A., Ong, S.P., Hautier, G., Chen, W., Richards, W.D., Dacek, S., Cholia, S., Gunter, D., Skinner, D., Ceder, G., Persson, K.A., 2013. Commentary: the materials project: a materials genome approach to accelerating materials innovation. APL Mater. 1 (1), 011002.
Jurecka, P., Sponer, J., Cerny, J., Hobza, P., 2006. Benchmark database of accurate (MP2 and CCSD(T) complete basis set limit) interaction energies of small model complexes, DNA base pairs, and amino acid pairs. Phys. Chem. Chem. Phys. 8, 1985–1993. http://dx.doi.org/10.1039/B600027D.
Klimeš, J., Michaelides, A., 2012. Perspective: advances and challenges in treating van der Waals dispersion forces in density functional theory. J. Chem. Phys. 137 (12), 120901. http://link.aip.org/link/?JCP/137/120901/1.
Klimeš, J., Bowler, D.R., Michaelides, A., 2010. Chemical accuracy for the van der Waals density functional. J. Phys. Condens. Matter 22 (2), 2201.

Klimeš, J., Bowler, D.R., Michaelides, A., 2011. Van der Waals density functionals applied to solids. Phys. Rev. B 83 (19), 195131.
Klintenberg, M., 2010. The search for strong topological insulators. arxiv:1007.4838.
Klintenberg, M., Eriksson, O., 2013. Possible high-temperature superconductors predicted from electronic structure and data-filtering algorithms. Comput. Mater. Sci. 67, 282–286.
Kurtoglu, M., Naguib, M., Gogotsi, Y., Barsoum, M.W., 2012. First principles study of two-dimensional early transition metal carbides. MRS Commun. 2 (4), 133–137. http://journals.cambridge.org/article_S2159685912000250.
Langreth, D.C., Dion, M., Rydberg, H., Schr Ader, E., Hyldgaard, P., Lundqvist, B.I., 2005. Van der Waals density functional theory with applications. Int. J. Quantum Chem. 101 (5), 599–610. http://dx.doi.org/10.1002/qua.20315.
Lashgari, H., Abolhassani, M.R., Boochani, A., Elahi, S.M., Khodadadi, J., 2014. Electronic and optical properties of 2D graphene-like compounds titanium carbides and nitrides: DFT calculations. Solid State Commun. 195, 61–69. http://www.sciencedirect.com/science/article/pii/S003810981400252X.
Lebègue, S., Harl, J., Gould, T., Ángyán, J.G., Kresse, G., Dobson, J.F., 2010. Cohesive properties and asymptotics of the dispersion interaction in graphite by the random phase approximation. Phys. Rev. Lett. 105 (19), 196401.
Lebègue, S., Bjorkman, T., Klintenberg, M., Nieminen, R.M., Eriksson, O., 2013. Two-dimensional materials from data filtering and ab-initio calculations. Phys. Rev. X 3, 031002.
Lee, C., Yang, W., Parr, R.G., 1988. Development of the Colle-Salvetti correlation-energy formula into a functional of the electron density. Phys. Rev. B 37, 785–789. http://link.aps.org/doi/10.1103/PhysRevB.37.785.
Lei, S., Ge, L., Najmaei, S., George, A., Kappera, R., Lou, J., Chhowalla, M., Yamaguchi, H., Gupta, G., Vajtai, R., Mohite, A.D., Ajayan, P.M., 2014. Evolution of the electronic band structure and efficient photo-detection in atomic layers of InSe. ACS Nano 8 (2), 1263–1272. http://dx.doi.org/10.1021/nn405036u.
Li, L., Yu, Y., Ye, G.J., Ge, Q., Ou, X., Wu, H., Feng, D., Chen, X.H., Zhang, Y., 2014. Black phosphorus field-effect transistors. Nat. Nanotechnol. 9 (5), 372–377. http://www.nature.com/nnano/journal/v9/n5/full/nnano.2014.35.html.
Lichtenstein, L., Büchner, C., Yang, B., Shaikhutdinov, S., Heyde, M., Sierka, M., Włodarczyk, R., Sauer, J., Freund, H.J., 2012. The atomic structure of a metal-supported vitreous thin silica film. Angew. Chem. Int. Ed. Engl. 51 (2), 404–407. http://dx.doi.org/10.1002/anie.201107097.
Lifshitz, E., 1954. The theory of molecular attractive forces between solids. J. Exp. Theor. Phys. 29, 94–110.
Ling, Z., Ren, C.E., Zhao, M.Q., Yang, J., Giammarco, J.M., Qiu, J., Barsoum, M.W., Gogotsi, Y., 2014. Flexible and conductive MXene films and nanocomposites with high capacitance. Proc. Natl. Acad. Sci. U.S.A. 111 (47), 16676–16681. http://www.pnas.org/content/111/47/16676.
Liu, C.C., Feng, W., Yao, Y., 2011. Quantum spin Hall effect in silicene and two-dimensional germanium. Phys. Rev. Lett. 107 (7), 076802. http://link.aps.org/doi/10.1103/PhysRevLett.107.076802.
Liu, H., Yuan, X., Lu, P., Shi, X., Xu, F., He, Y., Tang, Y., Bai, S., Zhang, W., Chen, L., Lin, Y., Shi, L., Lin, H., Gao, X., Zhang, X., Chi, H., Uher, C., 2013. Ultrahigh thermoelectric performance by electron and phonon critical scattering in Cu2Se1-xIx. Adv. Mater. 25 (45), 6607–6612. http://onlinelibrary.wiley.com/doi/10.1002/adma.201302660/abstract.
Murray, É.D., Lee, K., Langreth, D.C., 2009. Investigation of exchange energy density functional accuracy for interacting molecules. J. Chem. Theory Comput. 5 (10), 2754–2762.

Naguib, M., Kurtoglu, M., Presser, V., Lu, J., Niu, J., Heon, M., Hultman, L., Gogotsi, Y., Barsoum, M.W., 2011. Two-dimensional nanocrystals produced by exfoliation of Ti3AlC2. Adv. Mater. 23 (37), 4248–4253. http://onlinelibrary.wiley.com/doi/10.1002/adma.201102306/abstract.

Naguib, M., Mashtalir, O., Carle, J., Presser, V., Lu, J., Hultman, L., Gogotsi, Y., Barsoum, M.W., 2012. Two-dimensional transition metal carbides. ACS Nano 6 (2), 1322–1331. http://dx.doi.org/10.1021/nn204153h.

Naguib, M., Halim, J., Lu, J., Cook, K.M., Hultman, L., Gogotsi, Y., Barsoum, M.W., 2013. New two-dimensional niobium and vanadium carbides as promising materials for Li-ion batteries. J. Am. Chem. Soc. 135 (43), 15966–15969. http://dx.doi.org/10.1021/ja405735d.

Naguib, M., Mochalin, V.N., Barsoum, M.W., Gogotsi, Y., 2014. 25th Anniversary article: MXenes: a new family of two-dimensional materials. Adv. Mater. 26 (7), 992–1005. http://onlinelibrary.wiley.com./doi/10.1002/adma.201304138/abstract.

Nakano, H., Nakano, M., Nakanishi, K., Tanaka, D., Sugiyama, Y., Ikuno, T., Okamoto, H., Ohta, T., 2012. Preparation of alkyl-modified silicon nanosheets by hydrosilylation of layered polysilane (Si6H6). J. Am. Chem. Soc. 134 (12), 5452–5455. http://dx.doi.org/10.1021/ja212086n.

Nguyen, M.C., Choi, J.H., Zhao, X., Wang, C.Z., Zhang, Z., Ho, K.M., 2013. New layered structures of cuprous chalcogenides as thin film solar cell materials: Cu_2Te and Cu_2Se. Phys. Rev. Lett. 111 (16), 165502. http://link.aps.org/doi/10.1103/PhysRevLett.111.165502.

Oganov, A.R., Glass, C.W., 2006. Crystal structure prediction using ab initio evolutionary techniques: principles and applications. J. Chem. Phys. 124 (24), 244704.

Okamoto, H., Kumai, Y., Sugiyama, Y., Mitsuoka, T., Nakanishi, K., Ohta, T., Nozaki, H., Yamaguchi, S., Shirai, S., Nakano, H., 2010. Silicon nanosheets and their self-assembled regular stacking structure. J. Am. Chem. Soc. 132 (8), 2710–2718. http://dx.doi.org/10.1021/ja908827z.

Ortiz, C., Eriksson, O., Klintenberg, M., 2009. Data mining and accelerated electronic structure theory as a tool in the search for new functional materials. Comput. Mater. Sci. 44, 1042–1049.

Perdew, J.P., Burke, K., Ernzerhof, M., 1996. Generalized gradient approximation made simple. Phys. Rev. Lett. 77, 3865–3868.

Ren, X., Rinke, P., Joas, C., Scheffler, M., 2012. Random-phase approximation and its applications in computational chemistry and materials science. J. Mater. Sci. 47, 7447–7471.

Rocca, D., Gebauer, R., Saad, Y., Baroni, S., 2008. Turbo charging time-dependent density-functional theory with Lanczos chains. J. Chem. Phys. 128 (15). http://scitation.aip.org/content/aip/journal/jcp/128/15/10.1063/1.2899649.

Romdhane, F.B., Cretu, O., Debbichi, L., Eriksson, O., Lebègue, S., Banhart, F., 2015. Quasi-2D Cu2S crystals on graphene: in-situ growth and ab-initio calculations. Small 11 (11), 1253–1257. http://onlinelibrary.wiley.com/doi/10.1002/smll.201400444/abstract.

Saxena, S., Choudhary, R.P., Shukla, S., 2015. Stanene: atomically thick free-standing layer of 2D hexagonal tin. arXiv:1505.05062 [cond-mat]. http://arxiv.org/abs/1505.05062.

Shaikhutdinov, S., Freund, H.J., 2013. Ultrathin silica films on metals: the long and winding road to understanding the atomic structure. Adv. Mater. 25, 49–67.

Shein, I.R., Ivanovskii, A.L., 2012. Graphene-like titanium carbides and nitrides Tin+1Cn, Tin+1Nn (nÂ =Â 1, 2, and 3) from de-intercalated MAX phases: first-principles probing of their structural, electronic properties and relative stability. Comput. Mater. Sci. 65, 104–114. http://www.sciencedirect.com/science/article/pii/S0927025612004235.

Singh, A.K., Hennig, R.G., 2014. Computational prediction of two-dimensional group-IV mono-chalcogenides. Appl. Phys. Lett. 105 (4), 042103. http://scitation.aip.org/content/aip/journal/apl/105/4/10.1063/1.4891230.

Slavček, P., Kalus, R., Paška, P., Odvárková, I., Hobza, P., Malijevský, A., 2003. State-of-the-art correlated ab initio potential energy curves for heavy rare gas dimers: Ar2, Kr2, and Xe2. J. Chem. Phys. 119 (4), 2102–2119. http://scitation.aip.org/content/aip/journal/jcp/119/4/10.1063/1.1582838.

Tkatchenko, A., Scheffler, M., 2009. Accurate molecular van der Waals interactions from ground-state electron density and free-atom reference data. Phys. Rev. Lett. 102 (7), 073005.

Tkatchenko, A., DiStasio, R.A., Car, R., Scheffler, M., 2012. Accurate and efficient method for many-body van der Waals interactions. Phys. Rev. Lett. 108, 236402. http://link.aps.org/doi/10.1103/PhysRevLett.108.236402.

Vogt, P., De Padova, P., Quaresima, C., Avila, J., Frantzeskakis, E., Asensio, M.C., Resta, A., Ealet, B.A.A., Le Lay, G., 2012. Silicene: compelling experimental evidence for graphenelike two-dimensional silicon. Phys. Rev. Lett. 108 (15), 155501. http://link.aps.org/doi/10.1103/PhysRevLett.108.155501.

Vydrov, O.A., Van Voorhis, T., 2009. Nonlocal van der Waals density functional made simple. Phys. Rev. Lett. 103 (6), 063004. http://link.aps.org/abstract/PRL/v103/e063004.

Vydrov, O.A., Van Voorhis, T., 2010. Dispersion interactions from a local polarizability model. Phys. Rev. A 81, 062708. http://link.aps.org/doi/10.1103/PhysRevA.81.062708.

Vydrov, O.A., Van Voorhis, T., 2010. Nonlocal van der Waals density functional: the simpler the better. J. Chem. Phys. 133, 244103.

Wang, S., Li, J.X., Du, Y.L., Cui, C., 2014. First-principles study on structural, electronic and elastic properties of graphene-like hexagonal Ti2C monolayer. Comput. Mater. Sci. 83, 290–293. http://www.sciencedirect.com/science/article/pii/S0927025613007088.

Włodarczyk, R., Sauer, J., Yu, X., Boscoboinik, J.A., Yang, B., Shaikhutdinov, S., Freund, H.J., 2013. Atomic structure of an ultrathin Fe-silicate film grown on a metal: a monolayer of clay? J. Am. Chem. Soc. 135 (51), 19222–19228. http://dx.doi.org/10.1021/ja408772p.

Xia, F., Wang, H., Jia, Y., 2014. Rediscovering black phosphorus as an anisotropic layered material for optoelectronics and electronics. Nat. Commun. 5, 4458. http://www.nature.com/ncomms/2014/140721/ncomms5458/full/ncomms5458.html.

Xu, M., Liang, T., Shi, M., Chen, H., 2013. Graphene-like two-dimensional materials. Chem. Rev. 113 (5), 3766–3798. http://dx.doi.org/10.1021/cr300263a.

Xu, Y., Yan, B., Zhang, H.J., Wang, J., Xu, G., Tang, P., Duan, W., Zhang, S.C., 2013. Large-gap quantum spin Hall insulators in tin films. Phys. Rev. Lett. 111 (13), 136804. http://link.aps.org/doi/10.1103/PhysRevLett.111.136804.

Yamanaka, S., Matsu-ura, H., Ishikawa, M., 1996. New deintercalation reaction of calcium from calcium disilicide. Synthesis of layered polysilane. Mater. Res. Bull. 31 (3), 307–316. http://www.sciencedirect.com/science/article/pii/0025540895001956.

Zhang, S., Yan, Z., Li, Y., Chen, Z., Zeng, H., 2015. Atomically thin arsenene and antimonene: semimetal semiconductor and indirect direct band-gap transitions. Angew. Chem. Int. Ed. Engl. 54 (10), 3112–3115. http://onlinelibrary.wiley.com/doi/10.1002/anie.201411246/abstract.

Zhao, S., Kang, W., Xue, J., 2014. Manipulation of electronic and magnetic properties of M2C (M = Hf, Nb, Sc, Ta, Ti, V, Zr) monolayer by applying mechanical strains. Appl. Phys. Lett. 104 (13), 133106. http://scitation.aip.org/content/aip/journal/apl/104/13/10.1063/1.4870515.

Zheng, H., Rivest, J.B., Miller, T.A., Sadtler, B., Lindenberg, A., Toney, M.F., Wang, L.W., Kisielowski, C., Alivisatos, A.P., 2011. Observation of transient structural-transformation dynamics in a Cu2S nanorod. Science 333 (6039), 206–209. http://www.sciencemag.org/content/333/6039/206.

Zlyomi, V., Drummond, N.D., Falko, V.I., 2013. Band structure and optical transitions in atomic layers of hexagonal gallium chalcogenides. Phys. Rev. B 87 (19), 195403. http://link.aps.org/doi/10.1103/PhysRevB.87.195403.

Zlyomi, V., Drummond, N.D., Fal'ko, V.I., 2014. Electrons and phonons in single layers of hexagonal indium chalcogenides from *ab initio* calculations. Phys. Rev. B 89 (20), 205416. http://link.aps.org/doi/10.1103/PhysRevB.89.205416.

CHAPTER TWO

Efficient Multiscale Lattice Simulations of Strained and Disordered Graphene

N. Leconte*^,1^, A. Ferreira†, J. Jung*^,2^
*University of Seoul, Seoul, South Korea
†University of York, York, United Kingdom
[2]Corresponding author: e-mail address: jeil.jung@gmail.com

Contents

1. INTRODUCTION

Graphene—a single-layer honeycomb array of carbon atoms with fascinating structural and electronic properties—has emerged as one of the most active research topics in condensed matter physics since the observation of its distinctive quantum Hall effect (Novoselov et al., 2005; Zhang et al., 2005). Part of this interest stems from the breadth of graphene's scope

[1] Present address: Department of Physics, The University of Texas at Austin, Austin, TX, United States.

Semiconductors and Semimetals, Volume 95
ISSN 0080-8784
http://dx.doi.org/10.1016/bs.semsem.2016.04.002

of research that encompasses questions of fundamental interest related to the physics of massless Dirac fermions (Semenoff, 1984), as well as questions that potentially have implications for engineering applications that exploit the material's exceptional mechanical, thermal, electrical, and optical properties. The beautiful simplicity of its atomic structure leading to a $(2+1)$-dimensional relativistic Dirac theory in the continuum limit allows to often find analogies with concepts discussed in the field of high-energy physics. The continuum approximation has been useful in many theoretical studies of the electronic structure of graphene whose properties are dominated by features of the Hamiltonian in the vicinity of the Dirac point. However, it is often desirable to develop real-space formulations that can go beyond the limitations of effective long-wavelength continuum descriptions. Direct lattice simulation approaches are particularly suitable for studying structurally disordered systems where fine-scale lattice features become important, ranging from systems with dilute isolated point defects, and extended grain boundaries, all the way up to the amorphous limit, where the periodicity of the crystal is progressively suppressed. Recent studies have also demonstrated that real-space simulations are useful for describing impurities with a rich lattice-scale structure, and for tackling systems where multiple length scales coexist to form complex patterns. The two main challenges facing real-space methods can be summarized, on the one hand, in the need of formulating compact, yet accurate lattice models representative of real disordered materials, and, on the other hand, in the need for developing efficient algorithms enabling the calculation of physical observables in sufficiently large systems.

Here we present a summary of recent research efforts aimed at achieving these two goals. The first part of this book chapter is devoted to the formulation of accurate first-principles parameterized Hamiltonians employing localized atomic orbital basis sets. The resulting tight-binding (TB) models for clean systems are further complemented by local changes in the real-space Hamiltonian terms either to include the effect of external random perturbations, such as disorder and strain fields, or to introduce regular superlattices due to larger scale patterns. To mimic the effect of realistic disorder, and profiting from the self-averaging property in macroscopic systems, these local changes can be chosen following a random distribution in the sample. The matrices that are obtained for such systems containing tens of millions of elements can be analyzed using Lanczos recursion techniques (Lanczos, 1950), giving easy access, among others, to the density of states (DOS) using continued fractions (Turek, 1988). These matrices can be fed into real-space implementations of the Kubo–Greenwood formalism

(Fan et al., 2014; Foa Torres et al., 2014; Latil et al., 2004; Leconte et al., 2011; Lherbier et al., 2008; Radchenko et al., 2013; Roche, 1999; Roche and Mayou, 1997; Roche and Saito, 2001; Roche et al., 2012; Triozon et al., 2004) and, more recently, into direct evaluations of the Kubo formula by means of large-scale polynomial methods (Ferreira, 2016; Ferreira and Mucciolo, 2015; Ferreira et al., 2011; Garcia et al., 2015). The favorable scaling of state-of-the-art real-space methods enables calculations of linear response functions in extremely large systems, which then defines the top level of our multiscale simulation approach.

In what follows, we illustrate the versatility of the multiscale approach on graphenic systems currently attracting much attention. As a common thread between the case studies, we will consider different types of random disorder that can modify the onsite energies and hopping terms in the bare Hamiltonian. The latter is ubiquitous in graphene due to the inevitable random strain associated with the naturally occurring rippling of graphene on common substrates.

The first system of interest is freestanding graphene. By means of real-space wave packet evolution, implementing the Kubo–Greenwood formalism, or direct real-space evaluation of the Kubo formula through Chebyshev polynomial expansions, one can access quantum transport regimes typically out of reach of perturbation theory. Although the diagrammatic technique for disordered electron systems provides a powerful tool to address the weakly disordered regime (Rammer, 2008), its application beyond the dilute impurity regime, where coherent multiple scattering events cannot be neglected, is in general a daunting task. Furthermore, real-space approaches provide an unbiased approach to discriminate the role of different types of disorder on charge transport. Recent literature (Couto et al., 2014) concluded that random strain should be the dominant disorder source in graphene, outweighing long-range (LR) disordered caused by, eg, charged impurities trapped between graphene and its substrate, or resonant scatterers stemming from adsorbed hydrocarbons or structural defects at the graphene's basal plane (Ferreira et al., 2011; Peres, 2010). Such claim can be put to a test by considering the effect of a random strain field, whose defining parameters are based on experimental and simulation works (Amorim et al., 2016; Fasolino et al., 2011), and analyze how its impact compares against other types of disorder for realistic concentrations.

The next case study is devoted to Bernal-stacked bilayer graphene for a system with increased complexity outlining ways for obtaining effective TB parameters based on DFT calculations that are compatible with accurate continuum models. The additional degree of freedom, due to the layer index, opens the avenue to a tunable DOS gap opening by application of an external electric field (Park et al., 2015; Rutter et al., 2011; Zhang et al., 2009), which we can easily include in our TB model. We further include a random strain field as well as long-range disorder to study their respective impact on the electronic transport properties.

We subsequently consider as the next step in complexity the Hofstadter butterfly in the energy spectrum of graphene deposited on boron nitride expanding discussions presented in the earlier work using similar techniques (Leconte et al., 2016). The prediction by Hofstadter in 1976 of the fractal butterfly structure by the interplay between the lattice periodicity and magnetic length (Hofstadter, 1976; Zak, 1964) has been observed in recent quantum Hall experiments of nearly aligned graphene (G) on hexagonal boron nitride (BN) substrates (Dean et al., 2013; Hunt et al., 2013; Ponomarenko et al., 2013). The G/BN vertical heterojunction, with a lattice mismatch between the respective crystals of about 1.7%, allows to form conveniently long nanometer scale moiré patterns to observe the fractal spectrum using moderately large magnetic fields accessible in a conventional laboratory on the order of Teslas that are strong enough to overcome the signal smearing due to disorder. The theoretical study of Hofstadter butterfly signatures of nearly aligned G/BN has been studied recently (Chen et al., 2014; Chizhova et al., 2014; Diez et al., 2014; Moon and Koshino, 2014; Slotman et al., 2015), often using phenomenological Hamiltonians based for the moiré potentials experienced by the graphene sheet. These potentials control the behavior of the band structure in the moiré Brillouin zone and determine the Landau fan diagram DOS. By building the TB and continuum Hamiltonians of incommensurable layered crystals using ab initio calculation input, used to formulate the Hamiltonians for twisted bilayer graphene and graphene on hexagonal boron nitride (Jung et al., 2014), it is possible to reduce the arbitrariness involved in fitting the multiple parameters for the Hamiltonian from educated guesses.

The DOS and the transport properties of G/BN as a function of carrier density and magnetic field using the aforementioned TB Hamiltonians provide a reliable reference calculation that is able to capture the noninteracting moiré pattern effects in the band Hamiltonian upon which the many-body effect can introduce additional modifications of the bands (Bokdam et al., 2014; Chen et al., 2014; Chizhova et al., 2014; Dean et al., 2013; Hunt et al., 2013; Ponomarenko et al., 2013; Shi et al., 2014; Song et al., 2013). Our new multiscale methodology maps the continuum moiré pattern Hamiltonians back onto effective TB models capturing the effect of the virtual strains in the graphene sheet represented in the off-diagonal real-space vector fields (Leconte et al., 2016). This method is aimed at preserving formal simplicity and accuracy in the low energy sector and can likewise be used to build the lattice Hamiltonian in other types of moiré patterns tailored through lithography and gating techniques. Extrinsic disorder plays a crucial role as well, as evidenced by its impact on the observation of satellite resistivity peaks associated with the secondary Dirac points (Martinez-Gordillo et al., 2014).

Through our multiscale approach, we highlight the importance of all the terms in the single-particle contributions to the moiré pattern Hamiltonian by comparison with the literature where subsets of our full Hamiltonian are used. Earlier models for the G/BN have focused on the role of the local mass term by sublattice-selective potential energy difference introduced by the substrate (Giovannetti et al., 2007; Kindermann et al., 2012; Sachs et al., 2011), whereas subsequent scanning probe measurements in nearly aligned G/BN revealed the relevance of the triangular local potential variations in the moiré patterns in defining the associated secondary Dirac cones. Hamiltonians that also include the off-diagonal vector fields associated to strains in the graphene sheet have been formulated based on symmetry considerations of the triangular moiré patterns together with various educated guesses for the model Hamiltonian (Wallbank et al., 2013), while specific model Hamiltonian parameters have been predicted from ab initio-based approaches where it was noted that substantial off-diagonal vector-field corrections to the Hamiltonian stem from virtual strains due to graphene electrons hopping to and back from the BN sheet even in the absence of actual lattice distortions (Jung et al., 2014). The commensuration strains that have been observed in atomic force microscopy measurements in the limit of long moiré patterns near zero twist angle (Tang et al., 2013; Woods et al., 2014) naturally give rise to a finite average band gap (Jung et al., 2015). The bond distortions due to moiré strains (Jung et al., 2015; Neek-Amal and Peeters, 2014; San-Jose et al., 2014; van Wijk et al., 2014) lead to additional modifications in the moiré superlattice electronic structure that will be discussed elsewhere.

Finally, we combine the Hamiltonian from Bernal-stacked bilayer graphene and the effects of the hexagonal boron nitride (hBN) substrate potential to understand the impact of bilayer graphene deposited on an hBN substrate. A final straightforward extension can be made by encapsulating bilayer graphene with hBN sheets on both sides that we model by applying two independent moiré patterns, one on each graphene layer which allows us to explore the interplay between the two moiré patterns by varying the relative angle between the top and bottom BN layers. The strength of our method further provides the means to consider both small/large and commensurate/incommensurate angles on the same footing.

2. MODELS

In this section we discuss about the modelization of the Hamiltonian for both clean and disordered cases and about the numerical methods to actually obtain physical observables from simulations of large-scale lattices. We start by defining the lattice TB Hamiltonians that describe the clean limit of our systems of interest and then discuss the details on the disorder models that can be incorporated in the pristine systems. We also outline the methodology to map in a TB lattice Hamiltonian the LR moiré patterns that can be captured by continuum models.

2.1 Disorder-Free Models

2.1.1 Graphene

The electronic structure models of graphene date back to mid 20th century, where graphite started to be used more extensively in industrial applications (Slonczewski and Weiss, 1958; Wallace, 1947; Weiss, 1956, 1957). The electronic structure of graphene, the building block of the three-dimensional graphite allotrope, exhibits the beautiful simplicity of a regular honeycomb array of sp_2-bonded carbon atoms providing structural stability. The electronic structure of graphene consists of π-orbital bands close to the Fermi energy originating in the hopping of the electrons between the p_z orbitals centered at the carbon atoms and σ-orbital bands associated with the sp_2 bonding network (Saito et al., 1998). The π and π^* bands that are responsible for most observable electronic properties of graphene are usually described using TB models obtained by fitting either to experiment or to theoretical first-principles bands (Jung and MacDonald, 2013; Reich et al., 2002; Saito et al., 1998; Wallace, 1947). Many qualitative features are correctly captured when only near-neighbor hopping is retained, although more accuracy can be achieved by increasing the number of distant hopping parameters. For instance, the model introduced many years ago by Wallace (1947) includes first- and second-neighbor hopping terms. Another useful model retains only nearest-neighbor hopping, but introduces an additional parameter to allow for a finite overlap between orbitals (Saito et al., 1998) localized on neighboring sites. Both improvements make it possible to account for the electron–hole asymmetry of graphene's band structure.

The graphene TB models find a convenient representation in the localized Wannier functions of the p_z orbitals. The Wannier approach provides a physically intuitive, and fully rigorous, representation of graphene's π-bands that reproduce first-principles calculation results (Jung and MacDonald, 2013). In the Wannier representation, the band Hamiltonian is succinctly represented in terms of parameters with an intuitive physical meaning as amplitudes for electron hopping from one site to another; the more physically opaque overlap parameters of some localized combination of atomic orbital theories vanish exactly because of the orthonormality of the Wannier basis set. An appropriate choice of the localized atomic orbitals provides a convenient atomic orbital basis to represent the Hamiltonian.

While a single-parameter nearest-neighbor model is able to describe a particle–hole symmetric band structure and will be the model of choice in many applications, the use of up to five nearest-neighbor hopping terms can account with good accuracy for all major features of graphene's band structure in the whole Brillouin zone including the Fermi velocity at the Dirac cone K, the particle–hole symmetry breaking at K and Γ points, and the correct band widths at M and Γ points (Jung and MacDonald, 2013; see Fig. 1). We note that the Fermi velocity measured in experiments is around $v_F \sim 10^6\,\mathrm{m/s}$ with small variations depending on the specific substrate used. This is typically 10–20% higher than the value of $v_F \sim 0.84 \times 10^6\,\mathrm{m/s}$ obtained from first-principles calculations local density approximation (LDA). This enhancement, which roots in the nonlocal many-body corrections to the band velocity (Barlas et al., 2007; Gonzalez et al., 1994; Jung and MacDonald, 2011), can be partially accounted for in the TB models discussed hereafter, by accordingly increasing the value of the off-diagonal hopping terms. The set of hopping parameters in a five-parameter model able to reproduce closely the LDA band structures is given by $t_1 = -3.00$, $t_2 = -0.225$, and $t_3 = 0.0520$ eV for the interlattice terms and $t_1' = 0.205$ and $t_2' = 0.0691$ eV for the intralattice terms. A strategy to enhance the Fermi velocity near the Dirac cone in order to reproduce experiments more accurately and minimize the modifications in the band structure in the rest of the Brillouin zone consists in tuning the value of the most distant hopping parameters in our TB model.

In the following, we present technical details for the formulation of graphene's TB model in the Wannier orbital basis and discuss the optimum

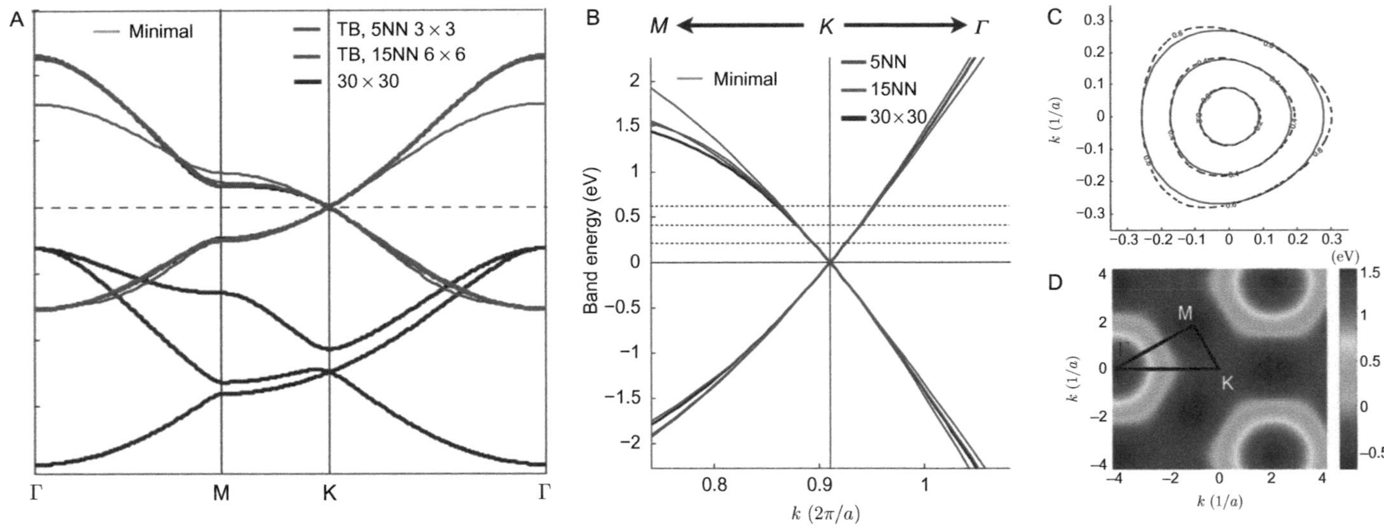

Fig. 1 Tight-binding bands of graphene in the whole Brillouin zone and near the Dirac point. (A) The band structure of graphene captured within different levels of *k*-point sampling approximation. (B) Deviation from band dispersion linearity near the Dirac point. (C) Anisotropy of the bands and (D) particle–hole asymmetry of opposite signs at different symmetry points in the Brillouin zone. *Reprinted with permission from Jung, J., MacDonald, A.H., 2013. Tight-binding model for graphene pi-bands from maximally localized Wannier functions. Phys. Rev. B 87, 195450. Copyright (2013) American Physical Society.*

parameters that accurately model the electronic structure of graphene (Jung and MacDonald, 2013). The π-band TB Hamiltonian is a 2×2 matrix that describes one π-electron per site:

$$H(\mathbf{k}) = \begin{pmatrix} H_{AA}(\mathbf{k}) & H_{AB}(\mathbf{k}) \\ H_{BA}(\mathbf{k}) & H_{BB}(\mathbf{k}) \end{pmatrix}$$

The Bloch function basis for this Hamiltonian is related to the Wannier functions by

$$|\psi_{\mathbf{k}\alpha}\rangle = \frac{1}{\sqrt{N}} \sum_{\mathbf{R}} e^{i\mathbf{k}(R+\tau_\alpha)} |\mathbf{R} + \tau_\alpha\rangle$$

where α is the sublattice index, τ_α is the position of the sublattice relative to the lattice vectors $\mathbf{R}$, and $|\mathbf{R} + \tau_\alpha\rangle$ is a Wannier function. The matrix elements of the Hamiltonian are related to the Wannier representation hopping amplitudes by

$$H_{\alpha\beta}(\mathbf{k}) = \langle \psi_{\mathbf{k}\alpha} | H | \psi_{\mathbf{k}\beta} \rangle = \frac{1}{N} \sum_{\mathbf{RR}'} e^{i\mathbf{k}(\mathbf{R}-\mathbf{R}')} t_{\alpha\beta}(\mathbf{R} - \mathbf{R}')$$

where

$$t_{\alpha\beta}(\mathbf{R} - \mathbf{R}') = \left\langle \mathbf{R} + \tau_\alpha | H | \mathbf{R}' + \tau'_\beta \right\rangle$$

represents tunneling from β to α sublattice sites located, respectively, at $\mathbf{R}' + \tau'_\beta$ and $\mathbf{R} + \tau_\alpha$. It follows from inversion symmetry that the site potentials at sublattices A and B are equivalent $H_{AA}(\mathbf{k}) = H_{BB}(\mathbf{k})$.

The Hamiltonian matrix elements can be expressed as a sum over neighbor indices n:

$$H_{AB}(\mathbf{k}) = \sum_n t_n f_n(\mathbf{k}) \quad \text{or} \quad H_{AA}(\mathbf{k}) = \sum_n t'_n g_n(\mathbf{k})$$

where $t_n = t_{ABn}$ and $t'_n = t'_{AAn}$ are the common hopping of members of the set of nth neighbors for a given sublattice and $f_n(\mathbf{k})$ and g_n are the corresponding structure factors obtained by summing phase factors $\exp(i\mathbf{k} \cdot \mathbf{R})$ over this set distinguishing neighbor groups that are off-diagonal in the sublattice from those that are diagonal. We have chosen a coordinate system in which the honeycomb's Bravais lattice has primitive vectors $\mathbf{a}_1 = a(1, 0)$, $\mathbf{a}_2 = a\left(\frac{1}{2}, \frac{\sqrt{3}}{2}\right)$, where $a = 2.46\,\text{Å}$ is the lattice constant of graphene. The reciprocal lattice vectors are then $\mathbf{b}_1 = \frac{4\pi}{\sqrt{3}a}\left(\frac{\sqrt{3}}{2}, -\frac{1}{2}\right)$, $\mathbf{b}_2 = \frac{4\pi}{\sqrt{3}a}(0, 1)$. We choose $\tau_\alpha = (0, 0)$ and $\tau_\beta = \left(0, a/\sqrt{3}\right)$. The associated structure factors $f_n(\mathbf{k})$ near the band-crossing Brillouin-zone corner points $\mathbf{k}_D = (4\pi/3a, 0)$ corresponding to the nearest five hopping terms grouped as inter- or intrasublattice hopping terms are:

$$
\begin{aligned}
n=1,\quad & e^{i\left(k_y a/\sqrt{3}\right)} + 2e^{-i\left(k_y a/2\sqrt{3}\right)} \cos\left(\frac{k_x a}{2}\right) \\
n=2,\quad & e^{-i\left(k_y 2a/\sqrt{3}\right)} + 2e^{i\left(k_y a/\sqrt{3}\right)} \cos(k_x a) \\
n=3,\quad & 2e^{i\left(k_y 5a/2\sqrt{3}\right)} \cos\left(\frac{k_x a}{2}\right) + 2e^{-i\left(k_y a/2\sqrt{3}\right)} \cos\left(\frac{k_x 3a}{2}\right) \\
& +2e^{-i\left(k_y 2a/\sqrt{3}\right)} \cos(k_x a)
\end{aligned}
$$

For the off-diagonal elements for which the Dirac velocity and trigonal warping are specified at low energies through $H_{\alpha\beta}(\mathbf{k}_D + \mathbf{k}) \simeq C_{\alpha\beta 1} k e^{-i\theta_{\mathbf{k}}} + C_{\alpha\beta 2} k^2 e^{i2\theta_{\mathbf{k}}}$ involving sums of f_n terms for matrix elements connecting sites $\alpha\beta = AB$. The explicit form of the expansion coefficients third nearest intersublattice contribution is

$$
\begin{aligned}
C_{\alpha\beta 1} &= \frac{\sqrt{3}a}{2}(-t_1 + 2t_2 + t_3) \\
C_{\alpha\beta 2} &= \frac{a^2}{8}(t_1 + 4t_2 - 13t_3)
\end{aligned}
$$

where we have abbreviated the notation by denoting $t_n = t_{\alpha\beta n}$ removing the $\alpha\beta$ subscripts. The $g_n(\mathbf{k})$ structure factors for the intrasublattice contribution

$$
\begin{aligned}
n=1,\quad & 2\cos(k_x a) + 4\cos\left(\frac{k_y\sqrt{3}a}{2}\right)\cos\left(\frac{k_x a}{2}\right) \\
n=2,\quad & 2\cos\left(k_y\sqrt{3}a\right) + 2\cos\left(\frac{k_y\sqrt{3}a}{2}\right)\cos\left(\frac{k_x 3a}{2}\right)
\end{aligned}
$$

for the diagonal elements responsible for particle–hole symmetry breaking lead to

$$
H_{\alpha\beta}(\mathbf{k}_D + \mathbf{k}) \simeq C'_{\alpha\beta 0} + C'_{\alpha\beta 2} k^2
$$

and the value for their constants is given by

$$C'_{\alpha\beta 0} = -3t'_1 + 6t'_2$$
$$C'_{\alpha\beta 2} = \frac{3a^2}{4}\left(t'_1 - 6t'_2\right)$$

for matrix elements involving sums of g_n terms that connect the sites $\alpha\beta = AA, BB$ for intrasublattice hopping parameters. For inversion symmetric systems where $H_{AA}(\mathbf{k}) = H_{BB}(\mathbf{k})$ the π-band energies of graphene are given by $E_{\pm}(\mathbf{k}) = |H_{AA}(\mathbf{k})| \pm |H_{AB}(\mathbf{k})|$. It follows that the band velocity at the Dirac point is determined by $|H_{AB}(\mathbf{k})|$, ie, by the intersublattice hopping contribution to the Hamiltonian, and that particle–hole symmetry is broken whenever $|H_{AA}(\mathbf{k})| \neq 0$, ie, when there is an intrasublattice contribution. For all intersublattice hopping processes, $f_n(\mathbf{k})$ vanishes at the Brillouin-zone corner $\mathbf{k}_D = (4\pi/3a, 0)$ and has a leading correction proportional to $q\exp\left(-i\theta_{\vec{q}}\right)$. Here, $\vec{q}$ is a wave vector measured from the Brillouin-zone corner. The intrasublattice processes on the other hand have no linear in q terms and are isotropic to second order in q. The subleading term in the expansion of the intersublattice terms behaves like $q^2\exp\left(2i\theta_{\vec{q}}\right)$.

2.1.2 Bilayer Graphene

The energetically favored stacking sequence for bilayer graphene is the Bernal or AB stacking configuration where only one of the carbon atoms in the unit cell is aligned on top of each other. Modeling of a bilayer graphene uses the Hamiltonian of a graphene single layer as building block, and the description of the interlayer coupling becomes the key element in defining the Hamiltonian. A π-band model of bilayer graphene contains four p_z orbitals centered on the A, B sites in the bottom layer and A' and B' sites in the top layer. We fix the lattice constants at the experimental values of graphite, for which the in-plane lattice constant is $a = 2.46$ Å and the interlayer separation is $c = 3.35$ Å, and label the Bernal (AB) stacked honeycomb lattice sites so that the A' site on the top layer is directly above the B site in the bottom layer. Our definitions of the in-plane lattice vectors and Bloch functions and the Hamiltonian of each individual layer remain similar to the discussions presented above for single-layer graphene.

The π-band Hamiltonian for bilayer graphene is a $\mathbf{k}$-dependent four-dimensional matrix:

$$H(\mathbf{k}) = \begin{pmatrix} H_B(\mathbf{k}) & H_{BT}(\mathbf{k}) \\ H_{TB}(\mathbf{k}) & H_T(\mathbf{k}) \end{pmatrix}$$

where $H_B(\mathbf{k})$ and $H_T(\mathbf{k})$ are two-dimensional matrices that describe intra- and intersublattice hopping within the bottom and top graphene layers.

The Hamiltonians of the bottom and top graphene sheets are

$$H_B(\mathbf{k}) = \begin{pmatrix} H_{AA}(\mathbf{k}) & H_{AB}(\mathbf{k}) \\ H_{BA}(\mathbf{k}) & H_{BB}(\mathbf{k}) \end{pmatrix}, \quad H_T(\mathbf{k}) = \begin{pmatrix} H_{A'A'}(\mathbf{k}) & H_{A'B'}(\mathbf{k}) \\ H_{B'A'}(\mathbf{k}) & H_{B'B'}(\mathbf{k}) \end{pmatrix}$$

Both intralayer and interlayer hopping processes can be classified into symmetry-equivalent groups identical to that of single-layer graphene. The intralayer Hamiltonian can be written as a sum over groups of the product of a hopping strength for the group and a structure factors defined above for single-layer graphene, while the hopping parameters are essentially the same as the isolated layers:

$$H_{A(')B(')}(\mathbf{k}) = \sum_n t_{A(')B(')n} f_n(\mathbf{k}), \quad H_{B(')A(')}(\mathbf{k}) = \sum_n t_{B(')A(')n} f_n^*(\mathbf{k})$$
$$H_{A(')A(')}(\mathbf{k}) = \sum_n t'_{A(')A(')n} g_n(\mathbf{k}), \quad H_{B(')B(')}(\mathbf{k}) = \sum_n t'_{B(')B(')n} g_n(\mathbf{k})$$

Inversion symmetry leads to the following relations between the hopping amplitudes in top and bottom layers:

$$t_{ABn} = t_{B'A'n}, \quad t'_{AAn} = t'_{B'B'n}, \quad t'_{BBn} = t'_{A'A'n}$$

Coupling between the layers is described by the $H_{BT}(\mathbf{k}) = H^*_{TB}(\mathbf{k})$, where

$$H_{BT}(\mathbf{k}) = \begin{pmatrix} H_{AA'}(\mathbf{k}) & H_{AB'}(\mathbf{k}) \\ H_{BA'}(\mathbf{k}) & H_{BB'}(\mathbf{k}) \end{pmatrix}$$

where the different interlayer coupling terms

$$H_{AA'}(\mathbf{k}) = \sum_n t_{AA'n} f_n(\mathbf{k}), \quad H_{AB'}(\mathbf{k}) = \sum_n t_{AB'n} f_n^*(\mathbf{k})$$
$$H_{BA'}(\mathbf{k}) = \sum_n t'_{BA'n} g_n(\mathbf{k}), \quad H_{BB'}(\mathbf{k}) = \sum_n t_{BB'n} f_n(\mathbf{k})$$

share structure factors similar to those for the interlayer matrix elements. The nearest-neighbor hopping terms are the most prominent, but the smaller terms do play a role in defining the details of the electronic structure because the number of important interlayer hopping processes is larger than in Slonczewski–Weiss–McClure (SWM)-type phenomenological models (Slonczewski and Weiss, 1958).

The continuum approximation of the bands near the Dirac point is captured by the four-dimensional π-band Hamiltonian in the form (Jung and MacDonald, 2014)

$$H(\mathbf{k}) = \begin{pmatrix} G_{AA}(\mathbf{k}) & F_{AB}(\mathbf{k}) & F_{AA'}(\mathbf{k}) & F^*_{AB'}(\mathbf{k}) \\ & G_{BB}(\mathbf{k}) & G_{BA'}(\mathbf{k}) & [F_{AA'}(\mathbf{k})] \\ \vdots & \ddots & [G_{BB}(\mathbf{k})] & [F_{AB}(\mathbf{k})] \\ & \cdots & & [G_{AA}(\mathbf{k})] \end{pmatrix}$$

where the Hamiltonian matrix elements labeled by the letters F and G emphasize that they consist of sums of the f_n and g_n structure factors, respectively. The matrix elements surrounded by square brackets are equivalent to other matrix elements by the symmetry relations $F_{BB'}(\mathbf{k}) = F_{AA'}(\mathbf{k})$, $F_{A'B'}(\mathbf{k}) = F_{AB}(\mathbf{k})$, $G_{A'A'}(\mathbf{k}) = G_{BB}(\mathbf{k})$, and $G_{B'B'}(\mathbf{k}) = G_{AA}(\mathbf{k})$. We discuss all the inequivalent Hamiltonian matrix elements below. Note that $F_{\alpha\beta}(\mathbf{k}_D) = 0$, that the intralayer nonzero values of $G_{\alpha\beta}(\mathbf{k}_D)$ can be interpreted as site energy shifts, and that the interlayer nonzero values of $G_{\alpha\beta}(\mathbf{k}_D)$ can be interpreted as interlayer tunneling amplitudes. The nonzero values of $G_{\alpha\beta}(\mathbf{k}_D)$ play the most essential role in defining the low-energy physics of bilayers. We choose our zero of energy so that $G_{AA}(\mathbf{k}_D) = G_{B'B'}(\mathbf{k}_D) = 0$. For $\alpha\beta = AB, AA', BA'$, and wave vectors near $\mathbf{k}_D = (4\pi/3a, 0)$ we write $F_{\alpha\beta}(\mathbf{k}_D + \mathbf{q}) \simeq C_{\alpha\beta 1} q e^{-i\theta_\mathbf{q}} + C_{\alpha\beta 2} q^2 e^{i2\theta_\mathbf{q}}$. The expansion of the $F_{\alpha\beta}$ functions near $(-4\pi/3a, 0)$ differs by an overall sign and complex conjugation. For $\alpha\beta = AA, BB, BA'$ we write $G_{\alpha\beta}(\mathbf{k}_D + \mathbf{q}) \simeq C'_{\alpha\beta 0} + C'_{\alpha\beta 2} q^2$. When the small q expansions are used for the matrix elements, the electronic structure near the Dirac point is accurately reproduced.

A simplified model that is reminiscent of the SWM model for graphite (Slonczewski and Weiss, 1958) can be constructed by identifying the shortest range hopping model that can capture the correct zeroth order terms in the expansion of $G_{\alpha\beta}(\mathbf{k}_D + \mathbf{q})$ and the correct first-order terms in the expansion of $F_{\alpha\beta}(\mathbf{k}_D + \mathbf{q})$. Given a choice for the zero of energy, this leaves in a model with five independent parameters:

$$\begin{aligned} H_{AA}(\mathbf{k}) &= H_{BB'}(\mathbf{k}) = t'_{AA0} \\ H_{BB}(\mathbf{k}) &= H_{A'A'}(\mathbf{k}) = t'_{BB0} \\ H_{BA'}(\mathbf{k}) &= t'_{BA'0} \end{aligned}$$

and

$$H_{AB}(\mathbf{k}) = H_{A'B'}(\mathbf{k}) = t_{AB1}f_1(\mathbf{k})$$
$$H_{AA'}(\mathbf{k}) = H_{BB'}(\mathbf{k}) = t_{AA'1}f_1(\mathbf{k})$$
$$H_{AB'}(\mathbf{k}) = t_{AB'1}f_1^*(\mathbf{k})$$

Here $f_1(\mathbf{k}) = \exp\left(ik_y a/\sqrt{3}\right) + 2\exp\left(-ik_y a/2\sqrt{3}\right)\cos(k_x a/2)$ and we use the correct linear dispersion coefficient associated with the hopping term $t_{\alpha\beta 1} = -2C_{\alpha\beta,1}/\sqrt{3}a$. The effective TB parameters in this model rely on folding all hopping parameters down to a model with only near-neighbor and onsite or purely vertical type hopping, but have the same continuum model limit as the full TB model. This model achieves accuracy near the Dirac point, is more accurate across the full Brillouin zone than the continuum model, and thus serves as a useful approximation for the models of bilayer graphene in a lattice. Differences in the conventions for the hopping parameter signs in the early literature (McCann and Falko, 2006; Partoens and Peeters, 2006) can lead to opposite particle–hole asymmetry features or to relative rotation of 60 degrees of the triangle vertices forming the Fermi points near charge neutrality.

The accuracy of effective TB models far from the Dirac point can be improved by increasing the number of hopping parameters and associated structure factors. In single-layer graphene, we saw that a five-parameter model with two intersublattice f_n-type structure factors and three intrasublattice g_n-type structure factors is able to capture the relevant features of the band structure throughout the whole Brillouin zone. One useful recipe to systematically increase the number of parameters in the effective model is to increase the number of structure factors n, and hence the range of the effective TB model, while maintaining correct values for the zeroth- and first-order $\vec{k}\cdot\vec{p}$ expansion coefficients and also correct values for the strongest shortest range hopping parameters which have a dominant influence on the π-band width. A compromise between simplicity and accuracy can be achieved with an $n=2$ truncation in the expansions of both the F and G functions where we use the correct near-neighbor hopping terms for the shortest hops and correct the more distant $n=2$ hopping amplitudes using the relations

$$t_{\alpha\beta 2} = C_{\alpha\beta 1}/\sqrt{3}a + t_{\alpha\beta 1}/2$$
$$t'_{\alpha\beta 2} = \frac{1}{6}\left(C'_{\alpha\beta 0} - t'_{\alpha\beta 0} + 3t'_{\alpha\beta 1}\right)$$

to recover the correct $\vec{k}\cdot\vec{p}$ expansion coefficients near the Dirac point.

The optimum parameter list in a nearest intersublattice hopping model that captures accurately the band structure in the vicinity of the K-point is given by

$$t_1 = t'_{AB} = 0.36\,\text{eV}, \quad t_0 = t_{AB,0} = -2.6\,\text{eV}, \quad t_3 = t_{AB',1} = 0.28\,\text{eV},$$
$$t_4 = t_{AA',1} = 0.138\,\text{eV}, \quad \Delta' = t'_{BB,0} = 0.015\,\text{eV}$$

The corresponding Hamiltonian is able to capture accurately the features of the band structure near the Dirac point (see Fig. 2).

2.1.3 Graphene on Boron Nitride

Graphene on hexagonal boron nitride (hBN) is an important system for device applications of graphene-based field effect transistors where the atomic scale smoothness of the hBN substrate allows to achieve electron mobilities of graphene comparable to those of suspended samples (Dean et al., 2010). The dominantly *van der Waals* nature of the interlayer interaction between the graphene sheet and the hBN substrate, albeit weaker than usual covalent bonds, is strong enough to give rise to measurable effects in scanning probe microscopy or transport experiments. Especially interesting are features associated with the long range moiré superlattice patterns arising from the lattice constant mismatch or due to angle misalignment between the layered crystals. In the case of G and BN layers vertical heterojunctions, the finite lattice constant mismatch leads to an incommensurate crystal structure even in the absence of a finite twist angle posing a challenge in the theoretical study of the system. Early calculation approaches of incommensurable crystal structures relied on large commensurate supercell calculations that closely resemble the incommensurable geometry (Chu et al., 2013; Landgraf et al., 2013; Shallcross et al., 2008, 2010, 2013; Wang et al., 2012a) and have led many groups to use simplified TB models (Bistritzer and MacDonald, 2010; Kindermann and First, 2011; Kindermann et al., 2012; Lopes dos Santos et al., 2012; Mele, 2011; Mele, 2012; Trambly de Laissardiere et al., 2010) or to approximate the real nonperiodic structures with periodic ones by increasing the period as much as possible in density functional theory or TB simulations. Alternatively, the so-called moiré band models, which reintroduce the validity of the Bloch theorem by applying it on the periodicity of the moiré pattern, were proven as effective means for obtaining model Hamiltonians for the relevant low-energy sector (Bistritzer and MacDonald, 2011; de Gail et al., 2011; Gonzalez, 2013; Lopes dos Santos et al., 2012; Moon and Koshino, 2013; San-Jose and Prada, 2013;

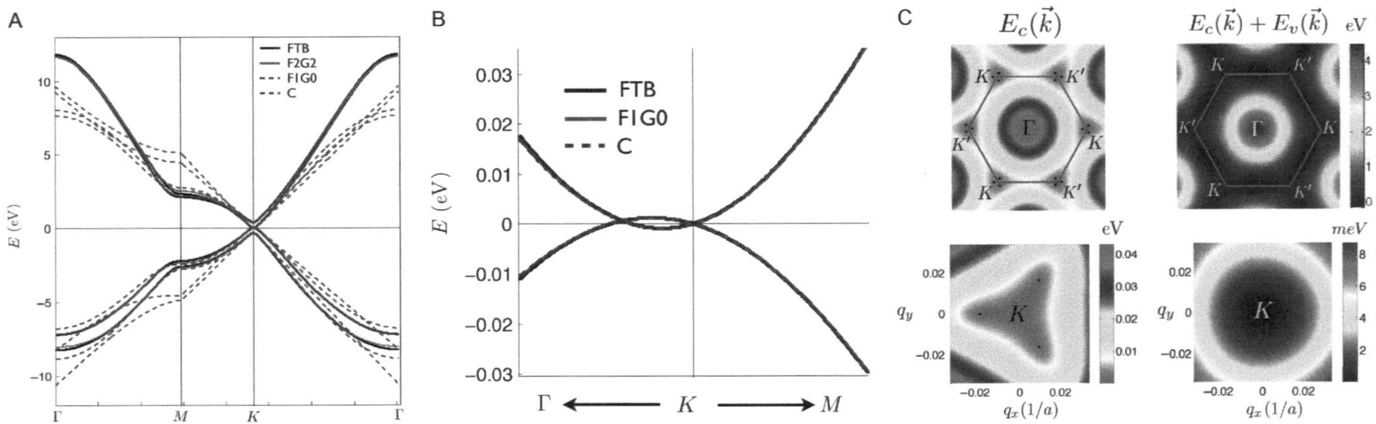

Fig. 2 (A) Comparison of tight-binding bands of bilayer graphene within different levels of approximation. (B) Comparison of the band structure near the Dirac point. (C) Trigonal distortion of the bands and particle–hole symmetry breaking. *Reprinted with permission from Jung J., MacDonald A. H., 2014. Accurate tight-binding models for the π bands of bilayer graphene. Phys. Rev. B 89, 035405. Copyright (2014) American Physical Society.*

San-Jose et al., 2012; Stauber et al., 2013; Tabert and Nicol, 2013). Here we summarize a recently developed (Jung et al., 2014) systematic approach to link short-period stacking order variations to model long-period moiré pattern systems using input from commensurate density functional theory simulations to formulate a model beyond the two-center approximation assumed in simplified theories. The goal is to obtain an accurate lattice Hamiltonian

$$H_{\text{lat}} = \langle ls\mathbf{L}|H|l's'\mathbf{L}'\rangle$$

where the l, l' label the layer, s, s' are the sublattice indices, and $\mathbf{L}, \mathbf{L}'$ are the lattice vectors specifying the position of a given atomic site. The short-period Hamiltonian of the commensurately translated geometry is given by

$$\langle ls\mathbf{L}|H(\mathbf{d})|l's'\mathbf{L}'\rangle$$

that can be expressed using the Wannier representation of the Bloch band Hamiltonian $H_{ls,l's'}(\mathbf{k}:\mathbf{d})$. The local variation of the stacking configuration in real space due to a finite rotation or a lattice constant mismatch is captured within the small-angle approximation through

$$\mathbf{d}(\mathbf{L}) = \varepsilon\mathbf{L} + \theta\hat{z} \times \mathbf{L}$$

where the parameter $\varepsilon \equiv \alpha - 1$ defines the relative lattice constant mismatch, where α is the factor by which the top layer needs to be expanded or compressed to have the same lattice constant as the bottom layer and θ is the relative twist angle between the layers. This association between the relative shift between the small commensurate unit cell displacements $\mathbf{d} = (d_x, d_y)$ with the real-space position vector leads to a lattice moiré superlattice Hamiltonian that depends on both $\mathbf{L}$ and $\mathbf{L}'$ simultaneously rather than only on their relative distance. Because the Wannier representation of the Bloch band Hamiltonian incorporates all the relevant distant hopping terms, it allows for a formally rigorous microscopic representation of the moiré patterns in the lattice Hamiltonian as a function of position. Considering an effective Hamiltonian of graphene where we integrate out the hBN whose band edges are a few eV away from the charge neutrality of graphene, we can obtain the following Hamiltonian of the moiré pattern broken down into three different sublattice pseudospin terms of graphene's Hamiltonian:

$$H^M(\mathbf{d}) = H_0(\mathbf{d}) + H_z(\mathbf{d}) + H_{AB}(\mathbf{d})$$

where

$$H_0(\mathbf{d}) = 2C_0[f(\mathbf{d})\exp i\varphi_0]$$

represents a periodic modulation of the moiré pattern site potential,

$$H_z(\mathbf{d}) = 2C_z[f(\mathbf{d})\exp i\varphi_z]$$

is a mass or local gap term representing the sublattice-dependent site potential difference, and

$$\begin{aligned} H_{AB}(\mathbf{d}) = 2C_{AB}\left[\cos\left(\frac{\sqrt{3}}{2}G_1 d_x\right)\cos\left(\frac{G_1 d_y}{2} - \varphi_{AB}\right) + \sin\left(\frac{G_1 d_y}{2} - \varphi_{AB} - \frac{\pi}{6}\right)\right] \\ + 2C_{AB}\sin\left(G_1 d_y + \varphi_{AB} - \frac{\pi}{6}\right) \\ + i2C_{AB}\left[\sin\left(\frac{\sqrt{3}}{2}G_1 d_x\right)\cos\left(\frac{G_1 d_y}{2} - \varphi_{AB}\right) - \sin\left(\frac{G_1 d_y}{2} - \varphi_{AB} - \frac{\pi}{6}\right)\right] \end{aligned}$$

is an off-diagonal *virtual strain* term which mimics the asymmetric hopping between boron–carbon and nitrogen–carbon. Fig. 3 represents the real-space map of these three pseudospin Hamiltonian terms.

We note that the contribution of the off-diagonal virtual strain term has a magnitude that is comparable with the other terms and remains significant even in the absence of real strains due to bond distortions in the graphene sheet.

The function $f(\mathbf{d})$ is given by $f(\mathbf{d}) = \exp\left[-iG_1 d_y\right] + 2\exp\left[i\frac{G_1 d_y}{2}\right]\cos\left[\frac{\sqrt{3}}{2}G_1 d_x\right]$ with $G_1 = 4\pi/\left(\sqrt{3}a\right)$

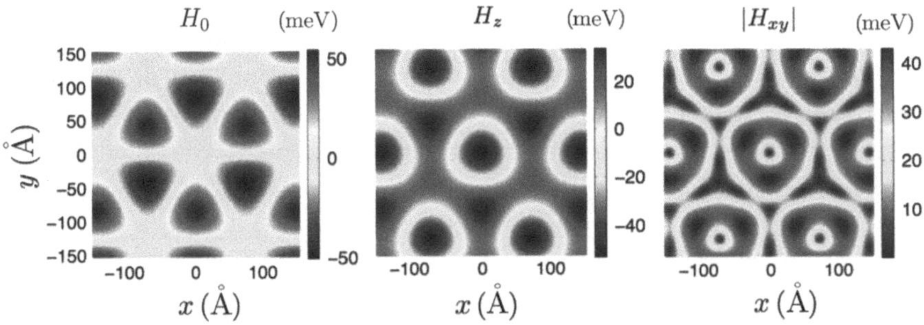

Fig. 3 Real-space map of moiré pattern potentials generated in graphene by the hexagonal boron nitride substrate where we can distinguish the potential fluctuation, the local mass variation, and virtual strain terms due to unequal hopping to neighboring atoms. *Reprinted with permission from Jung J., Raoux A., Qiao Z.H., MacDonald A.H., 2014. Ab initio theory of moiré superlattice bands in layered two-dimensional materials. Phys. Rev. B 89, 205414. Copyright (2014) American Physical Society.*

and the different parameters that define the moiré pattern Hamiltonian of graphene on hexagonal boron nitride for the *rigid model* are

$$
\begin{aligned}
C_0 &= -10.13\,\mathrm{meV}, & \phi_0 &= 86.53\,\mathrm{degree}\\
C_z &= -9.01\,\mathrm{meV}, & \phi_z &= 8.43\,\mathrm{degree}\\
C_{AB} &= 11.34\,\mathrm{meV}, & \phi_{AB} &= 19.60\,\mathrm{degree}
\end{aligned}
$$

or alternatively the equivalent parameters represented in the sublattice basis

$$
\begin{aligned}
C_{AA} &= -14.88\,\mathrm{meV}, & \phi_{AA} &= 50.19\,\mathrm{degree}\\
C_{BB} &= 12.09\,\mathrm{meV}, & \phi_{BB} &= -46.64\,\mathrm{degree}
\end{aligned}
$$

The lattice relaxation effects due to interlayer interaction can play a role in modifying the electronic structure, with the most notorious effect being the opening of a band gap due to the generation of a nonzero average mass term H_z (Jung et al., 2015; San-Jose et al., 2014). However, the rigid model is still expected to be capable of capturing fairly accurately many qualitative features of the electronic bands and the above model can serve as a realistic reference model Hamiltonian for the moiré pattern map.

We briefly outline here how to map the continuum model on a TB site model. By substituting with the expression for the small-angle approximation, the position ($\mathbf{L}$)-dependent $H_0(\mathbf{L})$ and $H_z(\mathbf{L})$ onsite terms are readily found. For the virtual strain term, we proceed as follows (also applicable for other types of strain). Using the atom positions for the first neighbors (in B-sublattice) of the atom at $(0, 0)$ (in A-sublattice) as

$$
\tau_1 = \left(0, \frac{1}{\sqrt{3}}\right), \quad \tau_{2,3} = \left(\pm\frac{1}{2}, -\frac{1}{2\sqrt{3}}\right)
$$

we find

$$
H_{AB} = t_1 \exp\frac{ik_y a}{\sqrt{3}} + t_2 \exp\frac{-ik_y a}{2\sqrt{3}} \exp\frac{-ik_x a}{2} + t_3 \exp\frac{-ik_y a}{2\sqrt{3}} \exp\frac{ik_x a}{2}
$$

which, with the unperturbed hopping Hamiltonian H^0_{AB}, gives

$$
\begin{aligned}
H_{AB} = H^0_{AB} &+ \delta_1 \exp\left(\frac{ik_y a}{\sqrt{3}}\right) + \delta_2 \exp\left(\frac{-ik_y a}{2\sqrt{3}}\right) \exp\left(\frac{-ik_x a}{2}\right)\\
&+ \delta_3 \exp\left(\frac{-ik_y a}{2\sqrt{3}}\right) \exp\left(\frac{ik_x a}{2}\right)
\end{aligned}
$$

where $t_i = -t_0 + \delta_i$.

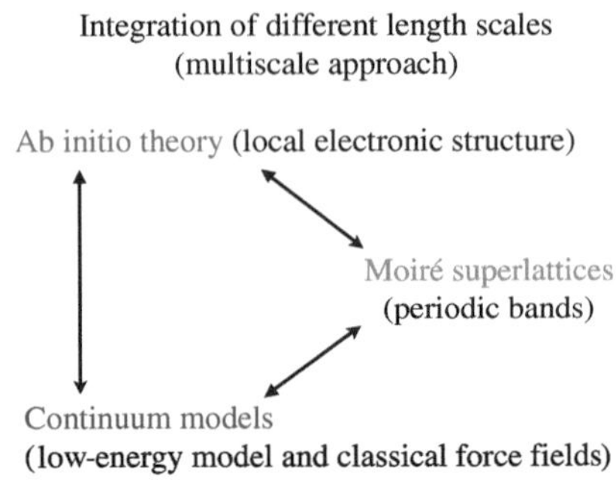

Fig. 4 Schematic representation of the multiscale approach employed in our calculations where the ab initio methods provide information on the local physical effects, the moiré bands theory is able to capture the physics of the nanometer scale superlattices, while this physics can be cast into a continuum model, which in turn can be used to map an effective lattice Hamiltonian.

At $K=(4\pi/3a, 0)$, one gets

$$H_{AB}=H_{AB}^{0}+\delta_1-\frac{\delta_2+\delta_3}{2}+i\frac{\sqrt{3}}{2}(\delta_3-\delta_2)$$

Setting $\delta_1+\delta_2+\delta_3=0$, $\Re e(H_{AB})=A$, and $\Im m(H_{AB})=B$, after some simple algebra, the corrections to the hopping terms are finally given by

$$\delta_1=\frac{2}{3}A,\quad \delta_2=-\frac{A+\sqrt{3}B}{3},\quad \delta_3=\frac{-A+\sqrt{3}B}{3}$$

This practical approach for building effective lattice Hamiltonians from accurate continuum approximations establishes a multiscale approach where the physical properties proper to different length scales mutually feed each other and can be captured on the same footing (see Fig. 4).

These considerations for building the Hamiltonian for graphene on hBN can straightforwardly be extended to model bilayer (BL) graphene on hBN and encapsulated G or BLG, by applying the moiré potential to one or both sides of the system. As a first approximation, we will only consider modifications in the orbital terms of the carbon atoms in the layer directly interacting with a neighboring hBN layer neglecting possible distant interactions between hBN and the carbon layer shielded by the neighboring layer.

2.2 Disorder Models

The effect of disorder can routinely be introduced in the different systems above, by locally modifying the Hamiltonian terms in the real-space representation. For the sake of generality in this contribution, we consider a variety of typical generic models that have proven their worth in the literature. The impact of more advanced chemical modifications could also be extracted from precise ab initio calculations and included in our multiscale methodology whenever warranted for more exact quantitative comparison with experiment (Leconte et al., 2010, 2014; Soriano et al., 2011).

Random strain (RS) fields stemming from rippling in the otherwise flat surface can be modeled by the Gaussian function $F(x) = \exp\left(-x^2/2\lambda^2\right)$ that enters the correlation function (Burgos et al., 2015)

$$\langle h(r)h(r')\rangle = h_{\text{rms}}^2 F\left(\frac{r-r'}{\lambda}\right)$$

where $h(r)$ defines the height profile of each atom at position r. We use realistic values for the correlation length λ and typical height h_{rms}, based on the experiment and the theory (Amorim et al., 2016; Fasolino et al., 2011). The chosen respective values are, unless stated otherwise, 8 nm and 0.7 Å. The randomness is assured by stochastic picking of ripple centers in the sample. From $h(r)$, we can extract the deformation tensor using

$$\begin{aligned} u_{xx} &= \frac{\partial u_x}{\partial x} + \frac{1}{2}\left(\frac{\partial h}{\partial x}\right)^2 \\ u_{yy} &= \frac{\partial u_y}{\partial y} + \frac{1}{2}\left(\frac{\partial h}{\partial y}\right)^2 \\ u_{xy} &= \frac{1}{2}\left(\frac{\partial h}{\partial x}\frac{\partial h}{\partial y}\right) \end{aligned}$$

For simplicity, we neglect the in-plane deformations given by $\vec{u}$ and only consider the $h(\vec{r})$ contributions (Burgos et al., 2015). The bond lengths are then modified as

$$d_{ij} = \left(a_0^2 + u_{xx}x_{ij}^2 + u_{yy}y_{ij}^2 + 2_{xy}x_{ij}y_{ij}\right)/a_0$$

where a_0 is the equilibrium bond length equal to 0.142 nm. Finally, the hopping parameters are then changed as

$$t_{ij} = t_0 \exp\left[-\beta\left(d_{ij}/a_0 - 1\right)\right]$$

where $\beta = \partial\log(t)/\partial d(a)|_{a=a_0} = 3.37$.

An alternative approach to include random strain would be to perform structural relaxation of the system of interest, eg, using molecular dynamics or Monte Carlo simulations, and to extract the hopping terms between each site using an environment-dependent TB model (Tang et al., 1996). This model goes beyond the traditional two-center approximations and entails parameters that are dependent on the binding environment. By including a screening function that mimics the electron accumulation between two atoms, and by incorporating a scaling function that modifies the distance between two atoms based on their effective coordination numbers, accurate hopping terms can be extracted that in inherently include the effect of strain. We provide this expression for reference:

$$t(r_{ij}) = \alpha_1 R_{ij}^{-\alpha_2} \exp\left[-\alpha_3 R_{ij}^{\alpha_4}\right](1 - S_{ij})$$

where the screening function is modeled as

$$S_{ij} = \frac{\exp(\xi_{ij}) - \exp(-\xi_{ij})}{\exp(\xi_{ij}) + \exp(-\xi_{ij})}$$

with

$$\xi_{ij} = \beta_1 \sum_l \exp\left[-\beta_2 \left(\frac{r_{il} + r_{jl}}{r_{ij}}\right)^{\beta_3}\right]$$

The effective interatomic distance R_{ij} is given by

$$R_{ij} = r_{ij} 1 + \frac{\delta}{2}\left[\left(\frac{g_i - g_0}{g_0}\right) + \left(\frac{g_j - g_0}{g_0}\right)\right]$$

where the angular insensitive expression for the coordination number in Tang et al. (1996), $g_i = 3.17678$ for graphitic materials, was optimized to include the angle dependency in Lee et al. (2006). The α_i and β_i parameters are obtained through fitting procedures and are summarized in Tang et al. (1996).

The effect of *LR disorder* can be introduced by modifying the diagonal TB parameters using the site potential given by Roche et al. (2012)

$$\sum_i \epsilon_i e^{-|r_0 - r_i|^2/(2\lambda^2)}$$

where ϵ_i is taken at random from $\epsilon_i \in [-W/2, W/2]t_0$, where we vary the value of W, and where we set $\lambda = \sqrt{3}a$ for a fixed distribution of impurity centers at position i of 0.125%. As long as the value of W does not exceed 1, the system will not experience a significant localization effect (for the considered timescales); ie, the system remains in the diffusive regime, subject to classical scattering events.

For *ultra-long-range* (*ULR*) *impurities* (Wong et al., 2015), the same Gaussian model is used as for LR impurities, but with a different set of parameters, namely a concentration of 0.0003%, a correlation length of 20 nm, and a strength given by $W = 0.05$.

For the generic *Anderson* (*AND*) *disorder* model, the onsite energies are modified from their unperturbed value by a random correction given by (Roche et al., 2012)

$$\delta\epsilon \in [-W/2, W/2]t_0$$

Finally, for *sublattice-selective disorder*, we use a simple toy model that shifts the onsite energy of a dilute random concentration of sublattice-specific atoms, with a value given by $\delta\epsilon = At_0$, where A controls the strength of the local potential. A (nonreconstructed) vacancy would correspond to the limiting case of $A \to \infty$.

3. LARGE-SCALE SIMULATION METHODOLOGIES

In the first part of this section, we describe the computational details of the Lanczos recursion method for the evaluation of the DOS and the real-space implementation of the Kubo–Greenwood formalism to obtain the conductivity from the wave packet time evolution. In the second part, we describe two powerful alternatives to the time evolution method exploring large-scale polynomial methods that enable the accurate determination of linear response functions in disordered lattices containing billions of sites. Fig. 5 is an illustration of hydrogen-doped graphene on hexagonal boron nitride that can be studied through direct quantum mechanical calculations using the methods we describe below.

3.1 Wave Packet Evolution and Lanczos Recursion Technique

All the models described above are, *in fine*, fully captured by modifications of the onsite energies of and hopping terms between single p_z-orbitals per atom. This opens the avenue to large-scale simulations, where the number of the atoms is the defining scaling parameter, giving access to the DOS using Lanczos recursion techniques and quantum simulations of

Fig. 5 Illustration of a realistic system consisting of hydrogen-doped graphene on hexagonal boron nitride that can be studied through direct real-space quantum mechanical calculations containing billions of atoms for mesoscopic size simulations.

the scattering time and conductivity utilizing a real-space implementation of the Kubo–Greenwood formalism (Fan et al., 2014; Foa Torres et al., 2014; Latil et al., 2004; Leconte et al., 2011; Lherbier et al., 2008; Radchenko et al., 2013; Roche, 1999; Roche and Mayou, 1997; Roche and Saito, 2001; Roche et al., 2012; Triozon et al., 2004).

We briefly lay out the cornerstones of this formalism. The implementation is based on the trace expressions of the Kubo–Greenwood formula (Mott, 1967):

$$\begin{aligned}&\Re[\sigma(\omega)]\\&=\frac{2\pi e^2\hbar}{V}\int_{-\infty}^{\infty}dE\frac{f(E)-f(E+\hbar\omega)}{\hbar\omega}\mathrm{Tr}\left[\hat{V}_x\delta\left(E-\hat{H}\right)\hat{V}_x\delta\left(E+\hbar\omega-\hat{H}\right)\right]\end{aligned}$$

By using,

$$\delta\left(E+\hbar\omega-\hat{H}\right)=\frac{1}{2\pi\hbar}\int_{-\infty}^{\infty}dte^{i\left(E+\hbar\omega-\hat{H}\right)t/\hbar}$$

the above expression becomes

$$\begin{aligned}&\Re(\sigma(\omega))\\&=\frac{e^2}{V}\int_{-\infty}^{\infty}dte^{i\omega t}\int_{-\infty}^{\infty}dE\frac{f(E)-f(E+\hbar\omega)}{\hbar\omega}\mathrm{Tr}\left[\hat{V}_x(0)\delta\left(E-\hat{H}\right)\hat{V}_x(t)\right]\end{aligned}$$

Defining

$$n(E) = \frac{1}{V}\mathrm{Tr}\left[\delta\left(E - \hat{H}\right)\right]$$

the DOS per unit volume, and using the average of an operator on the states at energy E,

$$\left\langle \hat{A} \right\rangle_E = \frac{\mathrm{Tr}\left[\delta\left(E - \hat{H}\right)\hat{A}\right]}{\mathrm{Tr}\left[\delta\left(E - \hat{H}\right)\right]}$$

the real part of the conductivity is given by

$$\Re(\sigma(\omega)) = e^2 \int_{-\infty}^{\infty} dt e^{i\omega t} \int_{-\infty}^{\infty} dE \frac{f(E) - f(E + \hbar\omega)}{\hbar\omega} n(E) \left\langle \hat{V}_x(t)\hat{V}_x(0) \right\rangle_E$$

For the dc conductivity, take $\omega \to 0$, so that

$$\lim_{\omega \to 0} \frac{f(E) - f(E + \hbar\omega)}{\hbar\omega} = \delta(E - E_F), \quad \text{allowing to find}$$

$$\sigma_{\mathrm{dc}} = e^2 n(E_F) \int_{-\infty}^{\infty} dt \left\langle \hat{V}_x(t)\hat{V}_x(0) \right\rangle_{E_F}$$

Defining the velocity autocorrelation function of the states at energy E as

$$C(E, t) = \left[\hat{V}_x(t)\hat{V}_x(0) + \hat{V}_x(0)\hat{V}_x(t)\right]_E$$

and the mean quadratic displacement as

$$\Delta X^2(E, t) = \left\langle \left(\hat{X}(t) - \hat{X}(0)\right)^2 \right\rangle_E$$

and considering the fact that

$$\frac{d^2}{dt^2}\Delta X^2(E, t) = C(E, t)$$

the conductivity finally becomes

$$\sigma_{\mathrm{dc}} = e^2 n(E_F) \lim_{t \to \infty} \frac{d}{dt} \Delta X^2(E_F, t)$$

A wave-packet evolution implementation allows following this time-dependent mean quadratic displacement. We rewrite the mean quadratic displacement using the time evolution operator $\hat{U}(t) = e^{-i\hat{H}t/\hbar}$. Some simple manipulations finally give

$$\Delta X^2(E,t)=\frac{\mathrm{Tr}\left[\left[\hat{X},\hat{U}\right]^{\dagger}\delta\left(E-\hat{H}\right)\left[\hat{X},\hat{U}(t)\right]\right]}{\mathrm{Tr}\left[\delta\left(E-\hat{H}\right)\right]}$$

Instead of calculating the sum over wave packets initially localized on each orbital of the system, which would be very time consuming, one can replace the trace by an initial state with a random phase on each orbital of the system:

$$\mathrm{Tr}\left[\delta\left(E-\hat{H}\right)\right]=\sum_{j=1}^{M}\left\langle\varphi_j\middle|\delta\left(E-\hat{H}\right)\middle|\varphi_j\right\rangle=\frac{M}{N_{\mathrm{RP}}}\sum_{i=1}^{N_{\mathrm{RP}}}\left\langle\varphi_{\mathrm{RP}}^{i}\middle|\delta\left(E-\hat{H}\right)\middle|\varphi_{\mathrm{RP}}^{i}\right\rangle$$

with

$$|\varphi_{\mathrm{RP}}\rangle=\frac{1}{\sqrt{M}}\sum_{j=1}^{M}e^{i2\pi\theta_j}\left|\varphi_j\right\rangle$$

where θ_j is a random number between 0 and 1. If the system is large enough, the average over a few random phases is usually sufficient to capture the physical quantities of interest (Iitaka and Ebisuzaki, 2004). This technique is referred to as the stochastic trace evaluation in the remainder of the chapter.

Using this approximation, the mean quadratic displacement becomes

$$\Delta X^2(E,t)=\frac{\left\langle\varphi_{\mathrm{RP}}'(t)\middle|\delta\left(E-\hat{H}\right)\middle|\varphi_{\mathrm{RP}}'(t)\right\rangle}{\left\langle\varphi_{\mathrm{RP}}\middle|\delta\left(E-\hat{H}\right)\middle|\varphi_{\mathrm{RP}}\right\rangle}$$

where $\varphi_{\mathrm{RP}}'(t)=\left[\hat{X},\hat{U}(t)\right]|\varphi_{\mathrm{RP}}\rangle$. The latter is obtained using a Chebyshev polynomial expansion (Roche et al. 2013).

The main interest of this last expression for the mean quadratic displacement is that both the numerator and the denominator can be calculated using Lanczos recursion techniques with a continued fraction:

$$\left\langle\varphi_1\middle|\frac{1}{E+i\eta-\hat{H}'}\middle|\varphi_1\right\rangle=\cfrac{1}{E+i\eta-a_1+\cfrac{b_1^2}{\ddots\cfrac{}{\cdots E+i\eta+a_{N-1}-\cfrac{b_{N-1}^2}{E+i\eta-a_N-b_N^2\times\mathrm{Term}}}}}$$

$\hat{H}'$, containing the recursion coefficients, is obtained by tridiagonalizing the TB Hamiltonian in a Lanczos basis. We note that, for this diagonal conductivity, only the recursion coefficients have to be saved, while the wavefunctions at each time step can be discarded after usage. The initial Hamiltonian often contains millions of elements, while this approximated Hamiltonian usually contains a few thousands of elements. The complexity of the numerical evaluation of $\delta(E-\hat{H})$ is generally very favorable as compared to exact diagonalizations. The exact scaling of the numerical procedure depends on the desired resolution η, as well as the energy-level statistics of the system of interest. By construction, the total number of operations is given by $N \times D \times R$, where N is the number of matrix–vector multiplications, D is the dimension of the Hamiltonian matrix, and R is the number of random vectors. In practice, the number of recursive steps required to attain convergence depends on the system size. For instance, near the Dirac point of pristine graphene, the linear spectrum results in a level spacing given by $\Delta\varepsilon \propto 1/\sqrt{D}$. Since optimal resolution requires $\eta \approx \Delta\varepsilon$, the total number of recursive steps is $N \propto \sqrt{D}$. For very large sparse matrices, a few random vectors are often enough to perform stochastic traces, and hence the overall scaling in the calculation of the spectral operator is $D^{3/2}$.

From the mean quadratic displacement, one can estimate several observables, such as the diffusion coefficient, which is defined as

$$D(t) = \frac{2\Delta X^2(t)}{t}$$

the mean free path (extracted from the maximum in the diffusion coefficient):

$$\ell_e(E) = \frac{D^{\max}(E)}{2v(E)}$$

the semiclassical conductivity:

$$\sigma_{SC}(E) = \frac{1}{4} e^2 \rho(E) D^{\max}(E)$$

the mobility:

$$\mu(E) = \frac{\sigma_{SC}(E)}{ne}$$

and the scattering time:

$$\tau(E) = \frac{\ell_e(E)}{v(E)}$$

where $v(E)$ is the charge carrier velocity at energy E, e is the electronic charge, n is the number of charge carriers, and $\rho(E)$ is the DOS.

This methodology was recently generalized to evaluate a transverse conductivity expression within the same Kubo linear response theory (Ortmann et al., 2015):

$$\sigma_{xy}(E, t) = -\frac{2}{V}\int_0^{\infty} dt e^{-\eta t/\hbar} \int_{-\infty}^{\infty} dE' f\left(E' - E\right)$$
$$\Re e\left[\left\langle \phi_{\mathrm{RP}} \middle| \delta\left(E' - \hat{H}\right)\hat{j}_y \frac{1}{E' - \hat{H} + i\eta} \hat{j}_x(t) \middle| \phi_{\mathrm{RP}} \right\rangle\right]$$

with ϕ_{RP} representing the same random phase wavefunction as used for the stochastic trace evaluation in the longitudinal conductivity, V the volume of the system, the current operator $\hat{j}_x$, f the Fermi function, and $\eta \rightarrow 0$ a small parameter required for numerical convergence. Alternatively, a direct Chebyshev polynomial expansion of the Kubo–Bastin formula by means of kernel polynomial methods can also be employed to evaluate transverse conductivities (Garcia et al., 2015).

In contrast with the evaluation of longitudinal response functions, the transverse conductivity calculation requires saving and evolving wavefunctions at each time step, which results in a considerable increase in the workload. To assure reasonable computation times, a highly parallelized implementation has been developed to take the advantage from high-performance computing (HPC) infrastructures with thousands of processors.

The limitations of semiclassical approaches become apparent away from the dilute regime, where coherent multiple scattering events become important (Ferreira et al, 2011; Klos and Zozoulenko, 2010; Leconte et al., 2011; Mucciolo and Lewenkopf, 2010; Peres, 2010; Radchenko et al., 2013), and it is thus advisable to use quantum methods to get accurate numerical estimates. Our Lanczos recursion and the time evolution methods give quantitatively accurate results that have been contrasted with analytical results in certain limits, including for instance the minimum value of longitudinal conductivity in graphene and weak localization corrections (Leconte et al., 2011; Lherbier et al., 2012). We provide further evidence by contrasting experimental Hofstadter fan diagrams with our simulated one in Section 4.

For the superlattice systems, we work with moiré unit cells containing 6050 atoms, ie, $55 \times 55 \times 2$ atoms. This moiré unit cell is then repeated 60 times both in the x and in the y-direction, containing more than 20 million atoms (factor 2 for bilayer systems), reaching near micrometer scale. For best energy resolution, 4000 recursion coefficients are required. One random phase configuration in the stochastic trace evaluation is usually sufficient to get fine DOS vs energy and magnetic field maps. We average over six random phase configurations because integration of the DOS (vs energy), to get the charge carrier concentration dependency, is quite sensitive to noise. As for the dynamic part of the simulation, a time step of $15\hbar/t_0$ can be accurately expanded using 100 Chebyshev polynomials.

Yet, powerful alternatives to Lanczos and time evolution methods exist to evaluate the Kubo conductivity, such as the kernel polynomial method (KPM) and Chebyshev polynomial Green function (CPGF) approaches. For instance, these methods have been successfully applied to capture the physics of the critical point at zero energy (Ferreira and Mucciolo, 2015; Gattenlohner et al., 2014; Ostrovsky et al., 2006, 2010). We describe and illustrate these methodologies with a case study in the following section. Future work plans to feed the TB Hamiltonians from the present work into such large-scale polynomial evaluations of response functions.

3.2 Chebyshev Polynomial Expansion Methods

The use of Chebyshev polynomials to obtain spectral functions is a powerful approach for accurate calculation of observables in very large lattices. Chebyshev polynomial expansion-based approaches have the advantage that the numerical convergence can be controlled by increasing the number of polynomial moments and through the judicious choice of the kernels avoiding ad hoc numerical broadening that can limit the accuracy of the calculations. Here we review the KPM (Weisse et al., 2006) that uses kernel convolution during the calculation of the moments associated to the target function, and the recently introduced CPGF expansion scheme that allows treating giant lattices with many billions of atoms (Ferreira, 2016; Ferreira and Mucciolo, 2015). We show how these methods can be used for calculating the DOS and conductivity of graphene-based systems and comment on the future outlook of this type of approaches.

3.2.1 The Kernel Polynomial Method

The starting point of Chebyshev polynomial-based methods is the representation of the target function of interest $f(x)$ in terms of a complete sum of orthogonal polynomials $p_n(x)$

$$f(x) = \sum_{n=0}^{\infty} \alpha_n p_n(x)$$

Among the family of orthogonal polynomials, Chebyshev polynomials are a popular choice owing to their optimal convergence properties and relation to the Fourier transform (Boyd, 2001; Weisse et al., 2006).

For Chebyshev polynomials of the first kind given by $T_n(x) = \cos(n \arccos(x))$, we rescale the spectrum into $x \in [-1:1]$ and use expansion coefficients defined by the following scalar product:

$$\alpha_n \equiv \langle T_n | f \rangle = \int_{-1}^{1} \frac{dx}{\pi\sqrt{1-x^2}} T_n(x) f(x)$$

The Chebyshev polynomials $\{T_n(x)\}_{n \in N_0}$ satisfy the recursion relations $T_0(x) = 1$, $T_1(x) = x$, and $T_{n+1}(x) = 2xT_n(x) - T_{n-1}(x)$. They obey the orthogonality relations

$$\int_{\mathcal{J}} dx \omega(x) T_n(x) T_m(x) = \frac{1 + \delta_{n,0}}{2} \delta_{n,m}$$

where $\omega(x) = 1/\left(\pi\sqrt{1-x^2}\right)$, thus forming a complete set in the domain $\mathcal{J}$. For a function $f(x)$ and $x \in \mathcal{J}$ one can write the expansion

$$f(x) = \omega(x) \sum_{n=0}^{\infty} \frac{2\mu_n}{1 + \delta_{n,0}} T_n(x)$$

where $\mu_n = \int_{\mathcal{J}} dx f(x) T_n(x)$. Upon truncation of the expansion, the Chebyshev polynomials distribute errors uniformly, providing a superior polynomial expansion with uniform resolution $\delta x \propto 1/N$, where N is the highest polynomial order used (Boyd, 2001). Fig. 6 illustrates the accuracy of the polynomial expansion method to describe functions with sharp features, such as a delta or a step function, and the effects of vacancies in biased bilayer graphene.

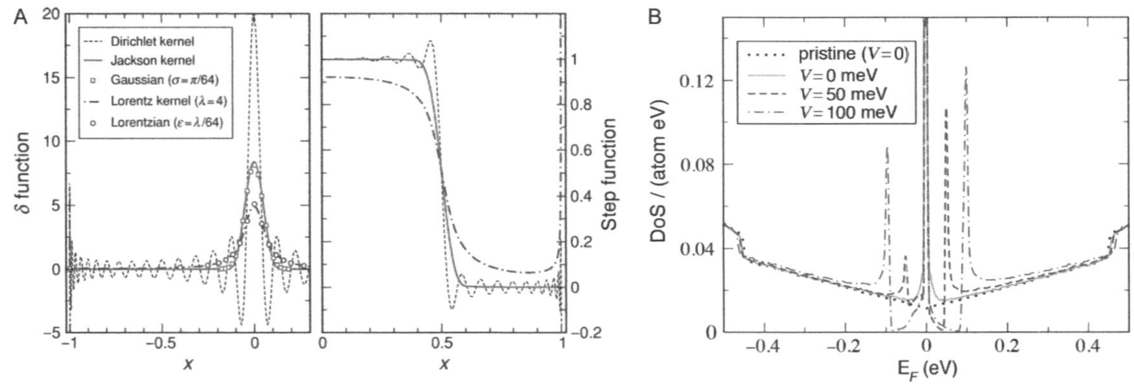

Fig. 6 (A) Expansion of $\delta(x)$ (*left*) and the step function (*right*) for different kernels. The Jackson kernel shows a good convergence, whereas the Lorentz kernel leads to Lorentz-broadened approximations. (B) Average DOS of a real-size bilayer graphene system with $D = 2 \times 14,142^2$ carbon atoms at selected positive bias values for 0.05% vacancy concentration. $N = 20,000$ Chebyshev moments have been used, resulting in a superior meV resolution. The DOS of pristine unbiased system is shown as a guide to the eye. Other parameters: $t = 2.7$ eV and $t_{\perp} = 0.45$ eV. *Reprinted with permission from Weisse, A., Wellein, G., Alvermann, A., Fehske, H., 2006. The kernel polynomial method. Rev. Mod. Phys 78, 2 75; Stabile, A.A., Ferreira, A., Li, J., Peres, N.M.R., Zhu, J., 2015. Electrically tunable resonant scattering in fluorinated bilayer graphene. Phys. Rev. B 92, 121411(R). Copyright (2006, 2015) American Physical Society.*

One of the main observable quantities of interest in solid-state physics is the DOS formally given by $\rho(E)=\frac{1}{D}\text{Tr}\delta(E-\hat{H})$, where $\hat{H}$ is the Hamiltonian and D is the dimension of the Hilbert space (ie, number of lattice sites and orbitals). The application of the KPM in the calculation of the average DOS results in

$$\rho(\epsilon)=\frac{2}{D\pi\sqrt{1-\epsilon^2}}\sum_{n=0}^{\infty}\frac{1}{1+\delta_{n,0}}\mu_n T_n(\epsilon)$$

where $\epsilon\in[-1:1]$ is the rescaled energy, and the Chebyshev moments are given by $\mu_n=\text{Tr}\,T_n\left(\widetilde{H}\right)$, where $\widetilde{H}=\hat{H}/W$ for Hamiltonians with symmetric spectrum (here, W is the half-bandwidth). In the KPM approach, the infinite sum in the evaluation of the DOS is truncated and a kernel $\{g_n\}_{n=0\ldots N-1}$ is introduced to damp the Gibbs oscillations (Weisse et al., 2006):

$$\rho_{\text{KPM}}(\epsilon)=\frac{2}{D\pi\sqrt{1-\epsilon^2}}\sum_{n=0}^{N-1}\frac{g_n}{1+\delta_{n,0}}\mu_n T_n(\epsilon)$$

where N depends on the desired resolution η, usually $N\propto W/\eta$ so larger values of N are required for better resolution. This resembles the numerical broadening parameter that exists in the time evolution method. The Jackson kernel performs very well close to the Dirac point of graphene (Ferreira et al., 2011):

$$g_n=\frac{(N-n+1)\cos\left(\frac{\pi n}{N+1}\right)+\cot\left(\frac{\pi}{N+1}\right)\sin\left(\frac{\pi n}{N+1}\right)}{N+1}$$

The computation of the moments μ_n makes use of the same stochastic trace evaluation technique as explained in Section 3.1. We provide a few additional details about this technique at the end of this section.

Another quantity of interest is the dc conductivity that can be obtained from the application of the Kubo formula at zero temperature (Mahan, 2000)

$$\sigma_{\text{dc}}(E)=\frac{\pi g_s\hbar e^2}{\Omega}\text{Tr}\left[\hat{v}_x\delta(E-\hat{H})\hat{v}_x\delta(E-\hat{H})\right]$$

where $\hat{v}_x=(i/\hbar)\left[\hat{H},\hat{x}\right]$ is the xth component of the velocity operator, $g_s=2$ is a spin degeneracy factor, and Ω is the area of the system. The conductivity $\sigma_{\text{dc}}(E)$ can be recast in terms of Chebyshev polynomial's KPM expansion after rescaling the energy

$$\sigma_{\text{KPM}}(\epsilon) = \frac{2g_s\hbar e^2}{\pi\Omega(1-\epsilon^2)} \sum_{n,m=0}^{N-1} \Delta_{nm}\left[T_{n+m}(\epsilon) + T_{|n-m|}(\epsilon)\right]$$

where the expansion coefficients

$$\Delta_{nm} \equiv \frac{g_n}{1+\delta_{n,0}} \frac{g_m}{1+\delta_{m,0}} \text{Tr}\left[\tilde{v}_x T_n\left(\tilde{H}\right)\tilde{v}_x T_m\left(\tilde{H}\right)\right]$$

can be computed with the stochastic trace evaluation and the Lorentz kernel

$$g_n = \frac{\sinh\left[\tilde{\eta}(N-n)\right]}{\sinh\left(\tilde{\eta}N\right)}$$

to approximate the delta functions in the Kubo conductivity by Lorentzians with resolution $\eta \equiv \tilde{\eta}\, W$. The calculation of conductivities is more demanding than the calculation of the DOS (linear in N) due to the higher cost in the calculation of the matrix expansion coefficients (quadratic in N).

3.2.2 The Chebyshev Polynomial Green's Function Method

An alternative implementation that bypasses the use of kernels employs a exact Chebyshev polynomial expansion of disordered Green functions to calculate the spectral properties and response functions of large systems (Ferreira, 2016; Ferreira and Mucciolo, 2015). Below, we provide a short description of their main properties and a brief derivation of the CPGF expansion. In what follows, z is a rescaled complex energy variable, $z := \epsilon + i\lambda$ with $\lambda > 0$. Let $\{\epsilon_m\}$ and $\{|m\rangle\}$ be the eigenvalues and eigenvectors of the Hamiltonian $\hat{h}$. In order to find an exact expansion of the lattice Green function,

$$\hat{\mathcal{G}}(\epsilon + i\lambda) = \sum_m \frac{|m\rangle\langle m|}{\epsilon + i\lambda - \epsilon_m}$$

in terms of Chebyshev polynomials, we make use of the identity (Tal-Ezer and Kosloff, 1984)

$$e^{-ixz} = \sum_{n=0}^{\infty} \frac{2i^{-n}}{1+\delta_{n,0}} J_n(z) T_n(x), \quad |x| \leq 1$$

where $J_n(z)$ is the Bessel function of order n in order to obtain

$$\hat{\mathcal{G}}(\epsilon + i\lambda) = \frac{1}{i}\int_0^\infty dt e^{i(\epsilon + i\lambda)t}\left[\sum_{n=0}^{\infty}\frac{2i^{-n}}{1+\delta_{n,0}}J_n(t)\mathcal{T}_n(\hat{h})\right]$$

where $\{\mathcal{T}_n(\hat{h})\}$ are operators defined by the matrix version of the Chebyshev recursion relations, that is,

$$\mathcal{T}_0(\hat{h}) = \mathbb{I}_D,\ \ \mathcal{T}_1(\hat{h}) = \hat{h},\ \ \mathcal{T}_{n+1}(\hat{h}) = 2\hat{h}\cdot\mathcal{T}_n(\hat{h}) - \mathcal{T}_{n-1}(\hat{h})$$

with D denotes the Hilbert space dimension. The Laplace transform of the Bessel function has a well-known solution (Gradshteyn and Ryzhik, 2007)

$$\int_0^\infty dt e^{-st}J_n(t) = \frac{1}{\sqrt{1+s^2}}\left(\sqrt{1+s^2} - s\right)^n$$

Using this expression, after the analytic continuation $s \longrightarrow -iz$ and after some algebra

$$\hat{\mathcal{G}}(\epsilon + i\lambda) = \sum_{n=0}^{\infty} g_n(\epsilon + i\lambda)\mathcal{T}_n(\hat{h}),\ \ g_n(z) \equiv \frac{2i^{-1}}{1+\delta_{n,0}}\frac{\left(z - i\sqrt{1-z^2}\right)^n}{\sqrt{1-z^2}}$$

The Green functions are versatile objects that relate to many different operators of interest in physics. In the following we illustrate the use of the exact representation of the Green functions in practical calculations. The first quantity we study is the thermodynamic DOS that is formally given by $\rho(\epsilon) = \lim_{\lambda\to 0}\lim_{D\to\infty} \nu(\epsilon, \lambda)$ where we defined

$$\nu(\epsilon, \lambda) = -\frac{1}{\pi D}\overline{\mathrm{Tr}\mathrm{Im}\hat{\mathcal{G}}(\epsilon + i\lambda)}$$

Here bar denotes disorder averaging. Using the expansion of the Green function in terms of Chebyshev polynomials we obtain

$$\nu(\epsilon, \lambda) = -\frac{1}{\pi D}\sum_{n=0}^{\infty}\mathrm{Im}[g_n(\epsilon + i\lambda)]\mu_n$$

Similar to the KPM (Weisse et al., 2006), the calculation of the DOS amounts to the determination of the Chebyshev moments. This scheme is very convenient as $\{\mu_n\}$ can be efficiently calculated even in giant systems containing billions of atoms (Ferreira, 2016; Ferreira and Mucciolo, 2015). Once the moments are determined, the smeared DOS in the entire parameter space (ϵ, λ) can be quickly retrieved from the sum involving μ_n and g_n.

In practice, one requires only a finite set of moments, which defines the order-N approximation to the target function, ie,

$$\nu_N(\epsilon, \lambda) = -\frac{1}{\pi D}\sum_{n=0}^{N-1} \mathrm{Im}[g_n(\epsilon + i\lambda)]\mu_n$$

For moderate λ highly accurate results can be obtained for modest N, but the thermodynamic limit $\lambda \to \Delta\epsilon \to 0$ requires large N. The two main advantages of Chebyshev expansion-based approaches are the numerical stability offered by the recursive properties of the Chebyshev polynomials that can be used in the calculation of Chebyshev moments $\{\mu_n\}$ and the unambiguous definition of the resolution defined through the broadening parameter $\eta = \lambda W$ appearing in the definition of the Green's function.

The convergence rate of the exact expansion for the DOS depends crucially on the smoothness of the target function, while the presence of sharp resonances in the DOS requires a large number of moments (Ferreira et al., 2011). As a rule of thumb, the total number of Chebyshev moments N determines the resolution $\delta\epsilon_N$ according to $\delta\epsilon_N \approx 1/N$ (Weisse et al., 2006). For instance, to probe features with small width η an accurate calculation requires $\delta\epsilon_N \lesssim \eta$ and hence many Chebyshev moments to achieve convergence (Ferreira et al., 2011; Weisse et al., 2006).

The CPGF formulation can be readily used to calculate the dc conductivity for which we use the finite-size Kubo formula at zero temperature,

$$\sigma(\epsilon, \lambda) = \frac{2\hbar e^2}{\pi\Omega}\mathrm{Tr}\left[\overline{\mathrm{Im}\hat{\mathcal{G}}(\epsilon + i\lambda)\hat{v}_x \mathrm{Im}\hat{\mathcal{G}}(\epsilon + i\lambda)\hat{v}_x}\right]$$

Here, the broadening parameter λ defines a timescale $\tau_i \propto 1/\lambda$ for phase coherence in the system (Imry, 2002). Using the expansion in terms of the Green's function we find

$$\sigma_N(\epsilon, \lambda) = \frac{2\hbar e^2}{\pi\Omega}\sum_{n,m=0}^{N-1} \mathrm{Im}[g_n(\epsilon + i\lambda)]\mathrm{Im}[g_m(\epsilon + i\lambda)]\mathcal{V}_{nm}$$

where

$$\mathcal{V}_{nm} = \mathrm{Tr}\left[\hat{v}_x \mathcal{T}_n\left(\hat{h}\right)\hat{v}_x \mathcal{T}_m\left(\hat{h}\right)\right]$$

Like in the KPM, the calculation of dc conductivity is computationally more demanding than the DOS due to the presence of a double sum during the evaluation of the trace. The number of moments is now N^2, which can severely limit the resolutions and/or system sizes attainable. However, this limitation can be overcome if enough memory exists to store the random vectors used for a stochastic evaluation of the moments.

3.2.3 Efficient Calculation of Chebyshev Moments

The complexity of a trace operation is D^2. However, for large sparse matrices, such as those appearing in effective TB models, the trace can be replaced by a stochastic average. For instance, for the DOS calculation we can use the expression

$$\mu_n \approx \frac{1}{R}\sum_{r=1}^{R} \langle r|\mathcal{T}_n(\hat{h})|r\rangle$$

where $|r\rangle = \sum_{i=1}^{D} \xi_i|i\rangle$ are complex random vectors with coefficients satisfying $\langle\langle\xi_i\rangle\rangle = 0$ and $\left\langle\left\langle\xi_i^*\xi_j\right\rangle\right\rangle = \delta_{ij}$ (real vectors may be used for spin rotational and time reversal symmetric Hamiltonians) (Iitaka and Ebisuzaki, 2004). The number of operations required to compute each moment is $\mathcal{O}(D \times R)$, where D is the number of sites and R is the number of random phases.

It is often assumed that the error in this approximation for the moment has the very favorable scaling $\mathcal{O}\left(1/\sqrt{RD}\right)$ (Weisse et al., 2006). However, the sparsity of the matrix $\mathcal{T}_n(\hat{h})$ decreases with increasing number of moments n and a larger R (or a larger system size D) is needed to obtain a stochastic trace evaluation with good precision (Ferreira, 2016; Ferreira and Mucciolo, 2015). In practice, for very large systems, with $D \approx 10^9$, a single random vector is often enough to obtain errors below 1% for n up to 10,000.

We now overview the recursive method that allows us to efficiently calculate Chebyshev moments. For concreteness, we describe the calculation of conductivity moments, ie,

$$\mathcal{V}_{nm} = \frac{1}{R}\sum_{r=1}^{R} \langle r|\hat{v}_x\mathcal{T}_n(\hat{h})\hat{v}_x\mathcal{T}_m(\hat{h})|r\rangle$$

The DOS moments are computed with a similar scheme. Suppose we start with a random vector $|r\rangle$. Then, using the recursion relations we obtain

$$\mathcal{T}_{m+1}(\hat{h})|r\rangle = 2\hat{h}\mathcal{T}_m(\hat{h})|r\rangle - \mathcal{T}_{m-1}(\hat{h})|r\rangle$$

which inspires us to write

$$|r\rangle_{m+1} = 2\hat{h}|r\rangle_m - |r\rangle_{m-1}$$

where

$$|r\rangle_m = \mathcal{T}_m(\hat{h})|r\rangle$$

In the above, $|r\rangle_0 = |r\rangle$. In fact, the best way to proceed is to define a second, auxiliary truncated basis $\{|\bar{r}\rangle\}$, $\bar{r} = 1, .., R$, with $|\bar{r}\rangle = \hat{v}_x|r\rangle$. Then, we can apply the Chebyshev recursion to write

$$|\bar{r}\rangle_{n+1} = 2\hat{h}|\bar{r}\rangle_n - |\bar{r}\rangle_{n-1}$$

The idea now is to implement a recursive calculation for each pair of random vectors, $\{|r\rangle\}$ and $\{|\bar{r}\rangle\}$, to generate two sequences $\{|r\rangle_0, \ldots, |r\rangle_{N-1}\}$ and $\{|\bar{r}\rangle_0, \ldots, |\bar{r}\rangle_{N-1}\}$, since they can be used to directly compute the stochastic averages

$$\mathcal{V}_{nm}(r) = {}_n\langle\bar{r}|\hat{v}_x|r\rangle_m$$

needed for the calculation of the conductivity moments, ie,

$$\mathcal{V}_{nm} = \frac{1}{R}\sum_{r=1}^{R}\mathcal{V}_{nm}(r)$$

If large amounts of RAM are available, one can recursively compute $\{|r\rangle_n\}$ and $\{|\bar{r}\rangle_m\}$ for all $n, m = 0, \ldots, N-1$, store them, and then evaluate the coefficients $\mathcal{V}_{nm}(r)$ for each r considering that there is no need to store $\{|r\rangle_n\}$ and $\{|\bar{r}\rangle_m\}$ for more than a given r at a time.

We now show how to evaluate efficiently the matrix elements $\mathcal{V}_{nm}(r) = {}_n\langle\bar{r}|\hat{v}_x|r\rangle_m$ using a site representation for the random vector. Let

$$\psi_n^{(r)}(x_k, y_k) = \langle x_k, y_k|r\rangle_n$$
$$\phi_m^{(r)}(x_k, y_k) = \langle x_k, y_k|\bar{r}\rangle_m$$

with $k = 1, \ldots, D$, where (x_k, y_k) are lattice site coordinates.

Then,

$$\mathcal{V}_{nm}(r) = \sum_{k,k'=1}^{D}\left[\phi_m^{(r)}(x_k, y_k)\right]^*\psi_n^{(r)}(x_{k'}, y_{k'})\langle x_k, y_k|\hat{v}_x|x_{k'}, y_{k'}\rangle$$

We can write

$$\langle x_k, y_k|\hat{v}_x|x_{k'}, y_{k'}\rangle = i(x_{k'} - x_k)\langle x_k, y_k|\hat{h}|x_{k'}, y_{k'}\rangle$$

When only nearest-neighbor hopping is allowed, there is a substantial reduction in the number of terms required to compute the matrix element:

$$\begin{aligned}\mathcal{V}_{nm}(r) = i\sum_{k=1}^{D}\left[\phi_m^{(r)}(x_k, y_k)\right]^*\sum_{\tau}\tau_x\psi_n\left(x_k + \tau_x, yk + \tau_y\right)\\ \times\langle x_k, y_k|\hat{h}|x_k + \tau_x, y_k + \tau_y\rangle\end{aligned}$$

where the number of lattice vectors $\boldsymbol{\tau}$ depends on the topology of the problem. The number of computational steps is thus precise $D \times z$, where z is a coordination number, which is much lower than D^2. In most cases of interest, the Hamiltonian matrix element is just a constant hopping amplitude $-t_s$, in which case we have

$$\mathcal{V}_{nm}(r) = -it_s\sum_{k=1}^{D}\left[\phi_m^{(r)}(x_k, y_k)\right]^*\sum_{\boldsymbol{\tau}}\tau_x\psi_n\left(x_k + \tau_x, y_k + \tau_y\right)$$

Notice that t_s is a dimensionless hopping amplitude since, by construction, $||\pm\hat{h}|| \leq 1$ (for graphene with vacancy defects, $t_s \equiv t/W$, where t is the carbon–carbon hopping integral and W is half-bandwidth). Clearly, there is no need to store the entire $D \times D$ Hamiltonian matrix, a connectivity table with information about neighbor coordinates $\{n_1(x_k, y_k), \ldots, n_z(x_k, y_k)\}_k$ suffices. This shows that the current scheme is just limited by the memory required to store the amplitudes $\{\{\psi_n^{(r)}\}_n^r, \{\phi_m^{(r)}\}_m^r\}$ needed to compute the overlap $\mathcal{V}_{nm}(r)$ for any two vectors $\{|r\rangle_n, |\bar{r}\rangle_m\}$.

The calculation can be made substantially more efficient if we are just interested in evaluating the conductivity in a small rectangular parametric grid $\left\{\{\epsilon_p\} \times \left\{\eta_q\right\}\right\}$, $1 \leq p, q \leq p_{\max}, q_{\max}$. The Chebyshev moments $\mathcal{V}_{nm}$ contain more information than any such grid since they allow one to retrieve the complete spectral conductivity. Recall that η is only limited by the number of Chebyshev iterations, min $\eta \propto N^{-1}$, and hence can be made arbitrarily small by increasing N. The conductivity for each point in the grid can be calculated efficiently using the *single-energy algorithm* outlined in the main text. The idea is to write the conductivity $\sigma_N(\epsilon_p, \eta_q)$ for each pair $\{\epsilon_p, \eta_q\}$ as

$$\sigma_N\left(\epsilon_p, \eta_q\right) = \frac{2\hbar e^2}{\pi \Omega R} \sum_{r=1}^{R} \left\langle \varphi_-^{(r)} \epsilon_p, \eta_q) | \varphi_+^{(r)} \left(\epsilon_p, \eta_q\right) \right\rangle$$

where

$$|\varphi_+^{(r)}\left(\epsilon_p, \eta_q\right)\rangle = \sum_{n=0}^{N-1} \mathrm{Im}\left[g_n\left(\epsilon_p + i\eta_q\right)\right] \hat{v}_x |r_n\rangle$$

and

$$|\varphi_-^{(r)}\left(\epsilon_p, \eta_q\right)\rangle = \sum_{n=0}^{N-1} \mathrm{Im}\left[g_n\left(\epsilon_p + i\eta_q\right)\right] |\bar{r}_n\rangle$$

The conductivity can now be computed iteratively with only a few vectors stored in memory (instead of $2 \times N$ vectors). The substantial reduction in memory allocation allows to treat very large TB systems, in excess of a billion atoms ($D = 3.6 \times 10^9$), with high resolution; see the following section.

3.2.4 Graphene with Random Vacancies as Benchmark Example

Here we provide the full numerical details of the KPM and CPGF methods applied for single graphene with vacancies as a stringent test of the Chebyshev polynomial expansion-based calculations.

The DOS of a macroscopic large honeycomb lattice ($N_x = N_y = 60,000$; periodic boundary conditions) with dilute randomly distributed vacancies (concentration $n_i = 0.4\%$) has been calculated using the CPGF method and numerical implementations as described above. The N-order approximation to the DOS is given by

$$\nu_N(E, \eta) = -\frac{1}{R} \sum_{n=0}^{N-1} \sum_{r=1}^{R} \frac{\mathrm{Im}[g_n(\epsilon + i\lambda)]}{\pi D W} \left\langle \overline{r_n} | r_n \right\rangle$$

where $W = 3t$ is graphene's half-bandwidth and $|r_n\rangle$ are the random vectors, $E = \epsilon W$, and $\eta = \lambda W$. The initial random vector used in the Chebyshev recursion reads as $|r_0\rangle = \sum_{i=1}^{D} x_i |i\rangle$, where $\{x_i\}$ are generated from a uniform distribution on the interval $\left[-\sqrt{3}, \sqrt{3}\right]$. In such a large Hilbert space ($D = N_x \times N_y = 3.6 \times 10^9$), self-averaging guarantees that a single random vector $R=1$ and one disorder realization suffice to obtain accurate results even for fine resolutions, that is, large N.

The accuracy of the stochastic evaluation of $\nu_N(\epsilon, \lambda)$ is illustrated with a few examples in Table 1. The superior precision, better than 0.1% for zero-energy modes (ZEMs), is a consequence of the size of the system simulated. We note that at larger values of the resolution parameter $\lambda(\eta)$, the data precision improves because convergence is achieved at smaller values of N (see below).

We now assess the convergence of the N-order approximation. N must be sufficiently large such that $\nu_N(\epsilon, \lambda)$ is well converged (say to 1% accuracy or better) for the smallest desired resolution λ. In Fig. 7A, we show the variation of the DOS of ZEMs $\nu_N(0, \lambda)$ with N. The calculations highlight the need for many thousands of Chebyshev iterations when the spectrum is probed with fine resolutions (ie, a few meV). Similar conclusions hold for other energies (not shown). For comparison we show the KPM approximation to the DOS using a Lorentz kernel (Ferreira et al., 2011; Weisse et al., 2006). Despite being accurate in the limit $N \to \infty$, the KPM convergence rate is manifestly poorer in this case.

Having established the convergence and accuracy of the CPGF method in the case of graphene with vacancies, we show the fully converged DOS for 1 meV resolution in Fig. 7B. A single-system realization and a random vector were employed. As an independent error estimator we use the electron–hole asymmetry degree, ie, $|\nu_\infty(\epsilon,\lambda) - \nu_\infty(-\epsilon,\lambda)|/\nu_\infty(\epsilon,\lambda)$. The magnitude of the error and its dependencies with the Fermi energy are consistent with the earlier statistical analysis.

In order to illustrate the divergent behavior of the DOS at $E=0$ we show in the right inset of Fig. 7B a plot of $E\nu(E,\eta)$ at several values of the resolution. According to the standard nonlinear sigma model picture

Table 1 The Estimation of the Data Precision Obtained Through Three Independent System Realizations of Disorder and Initial Random Vector Labeled S_1, S_2, and S_3 at Several Energies for a Resolution $\eta = \lambda W$ of 1 meV

	ZEM	$E=0.05$eV	$E=0.10$eV	$E=0.20$eV
S_1	1.0782	2.1507×10^{-2}	1.6777×10^{-1}	1.11326×10^{-2}
S_2	1.0786	2.1495×10^{-2}	1.6764×10^{-1}	1.11259×10^{-2}
S_3	1.0784	2.1501×10^{-2}	1.6705×10^{-1}	1.11310×10^{-2}
$\max|\nu_{S_i} - \bar{\nu}|/\bar{\nu}$	$\approx$0.02%	$\approx$0.03%	$\approx$0.30%	$\approx$0.35%

The relative maximum deviation from the average is shown in the last row.

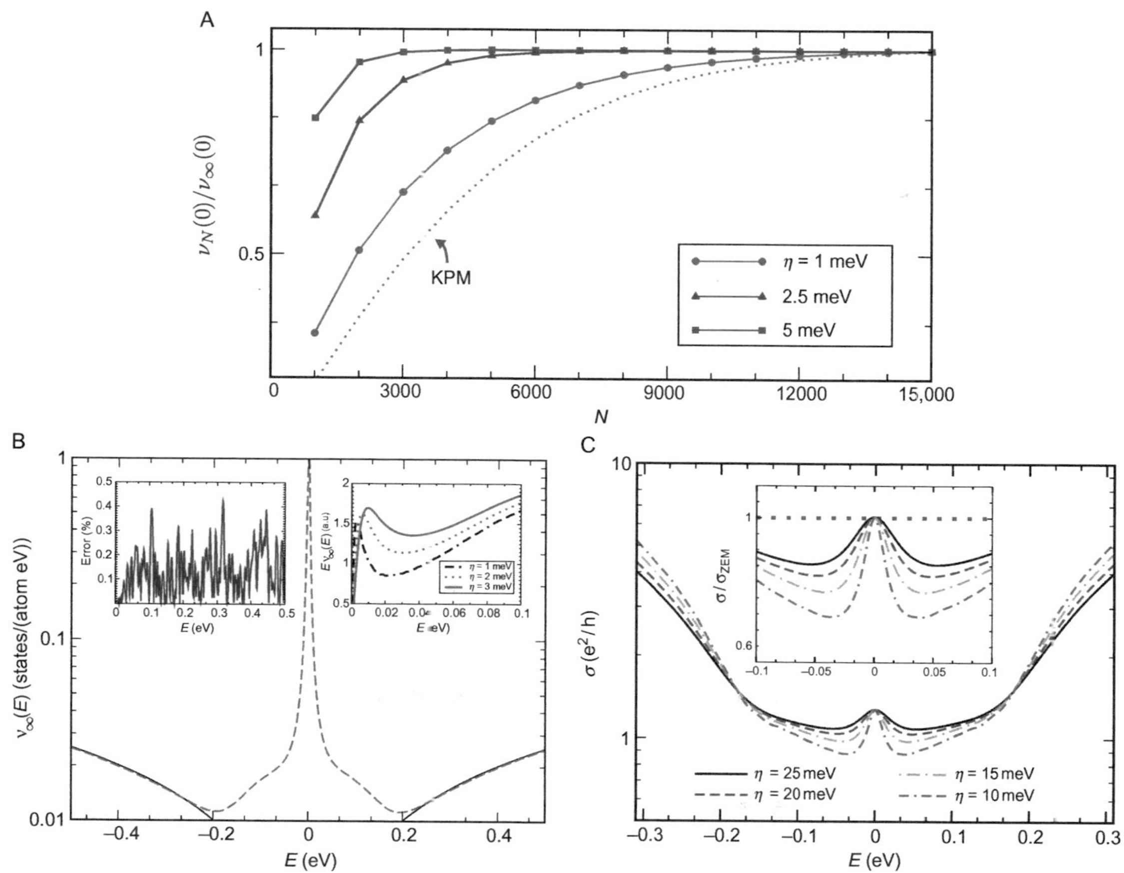

Fig. 7 See legend on next page.

(Gade, 1993; Gade and Wegner, 1991), the thermodynamic DOS behaves as $\nu(E,0) \to |E|^{-1}\exp\left[-|\ln|E||^{-1/2}\right]$ as $|E| \to 0$, whereas Häfner and coworkers observed a stronger singularity $\nu(E,0) \to |E|^{-1}|\ln|E||^{-x}$ with $2 > x \geq 1$ (Häfner et al., 2014) in consistency with a recent prediction (Gattenlohner et al., 2014). Our results are qualitatively consistent with the results reported by $E\nu(E,\eta) \to 0$ Häfner et al. (2014), however the extraction of the exact behavior of the DOS approaching the Dirac point would require a more detailed study beyond the scope of this chapter. A more detailed analysis would be needed to reveal the exact dependence as obtained in the CPGF.

The conductivity calculation relies on the knowledge of individual Chebyshev moments $\mathcal{V}_{nm} = \mathrm{Tr}\left[\hat{v}_x \mathcal{T}_n(\hat{h})\hat{v}_x \mathcal{T}_m(\hat{h})\right]$. An efficient numerical implementation requires enough memory to store $2 \times N$ vectors of dimension D which in practice limits the attainable D and/or N. To boost the size of the simulations, we implemented the Chebyshev recursive method in large memory machines with sequences $\{|r\rangle_0, \ldots, |r\rangle_{N-1}\}$ and $\{|\overline{r}\rangle_0, \ldots, |\overline{r}\rangle_{N-1}\}$ for a quick evaluation of the Chebyshev moments through optimized linear algebra subroutines. The results reported in this section are for a honeycomb lattice with $D = 3200 \times 3200$ sites and a vacancy concentration $n_i = 0.4\%$, for which we computed $N = 8000$ Chebyshev iterations (corresponding to $N^2 = 6.4 \times 10^7$ moments in the expansion of $\sigma_N(E,\eta)$). The resulting $N \times N$ matrix is subsequently used to evaluate $\sigma_N(E,\eta)$ on a fine grid. D is large enough so that the thermodynamic limit $\Omega \to \infty$ can be safely extrapolated.

Fig. 7 (A) Convergence of the N-order approximation to the DOS of ZEMs at selected values of resolution (broadening) parameter $\eta \equiv \lambda W$ with $W = 8.1$ eV. A single realization of a disordered system with $N_x = N_y = 60{,}000$ and 0.4% vacancy concentration has been considered. The limiting value $\nu_{N\to\infty}(\epsilon,\lambda)$ has been estimated—with precision better than 1%—from the value of $\nu_N(\epsilon,\lambda)$ at $N = 15{,}000$. The KPM result is shown (*dotted line*) for comparison. (B) DOS of graphene with vacancy defects ($n_i = 0.4\%$) as a function of Fermi energy (*green* (*gray* in the print version) *dashed line*). The resolution of the calculation is $\eta = 1$ meV. A logarithmic scale has been chosen to highlight the singular behavior of $\nu(\epsilon,\lambda)$ as $\epsilon \to 0$. The *solid black line* shows the DOS of pristine graphene as a guide to the eye. The insets show the estimated error as a function of Fermi energy (*left*) and a close look at the DOS singularity at $E = 0$ (*right*). The energy grid contains 1000 points. (C) The numerical calculations are able to reproduce the critical delocalization leading to a finite conductivity at the Dirac point of graphene. *Reprinted with permission from Ferreira, A., Mucciolo, E.R., 2015. Critical delocalization of chiral zero energy modes in graphene. Phys. Rev. Lett. 115, 106601. Copyright (2015) American Physical Society.*

In Fig. 7C we show the conductivity near the Dirac point of disordered graphene. Remarkably, the stochastic trace required thousands of random vectors to converge $\sigma_N(E,\eta)$ to a good precision. Moreover, statistical fluctuations in the STE become important at small η, which requires more random vector realizations. The error in $\sigma_N(E,\lambda)$ is estimated to be in the range 0.1–1% as observed from the achieved degree of electron–hole symmetry $\sigma(E,\eta)=\sigma(-E,\eta)$. The convergence study of the N-order approximation for ZEMs shows that for poor resolutions of ≈ 20 meV a few thousand Chebyshev iterations are sufficient, whereas probing resolutions ≈ 1 meV is manifestly more demanding. Our results show that the accurate determination of the spectral properties of disordered graphene is highly demanding, requiring very large N, especially near singular regions of the spectrum. Similar challenges were reported in Häfner et al. (2014) where a time-domain stochastic method was used to extract the DOS. Finally, we note that the calculations are not sensitive to the system dimension beyond a certain characteristic system size, as long as the mean-level spacing is the smallest energy scale $\delta\epsilon \lesssim \lambda$. This makes the CPGF a convenient tool to extract the thermodynamic limit.

4. RESULTS

In this section we illustrate the flexibility and wide applicability of our methodology by discussing the application of the real-space Lanczos recursion-based study on a variety of graphene-based layered systems.

4.1 Graphene

4.1.1 Strain vs Long-Range Disorder

In this section, we weigh in on the controversy regarding the dominant disorder contribution, balancing on the one hand, the effect of random strain, inevitably present in freestanding graphene, and, in a lesser extent, in graphene deposited on a substrate, and, on the other hand, LR disorder contributions due to, for instance, charged impurities trapped between graphene and its substrate. The random strain field has been introduced in Section 2.2, with parameters $\lambda = 8$ nm and $h_{\text{rms}} = 0.7$ Å. Note that this corresponds to about 0.2% of 15 Å wide ripples. Due to the fact that these parameters are sample dependent, and because of the generic strain model, we can only go so far in our conclusions based on the results in Fig. 8.

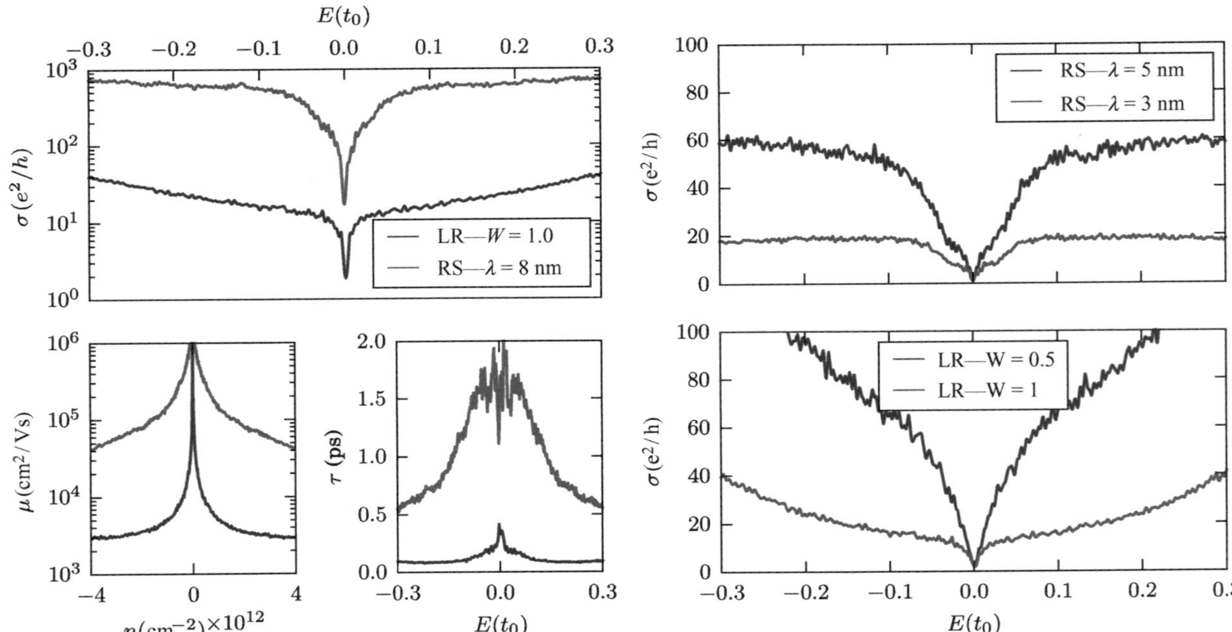

Fig. 8 Conductivity, mobility, and scattering times for SL graphene. *Left panels*: Comparison between long-range scatterers with a correlation length of 0.426 nm, and random strain profiles with correlation length equal to 8 nm. *Right panels*: Comparison between long-range scatterers with a correlation length of 0.426 nm and varying amplitude (*lower panel*), and random strain profiles with varying correlation lengths (*upper panel*).

For these parameters, the extracted ultra-high mobilities (>100,000 $cm^2/V\,s$) and conductivities are comparable to those from experiments on exfoliated graphene (Dean et al., 2010) or CVD graphene deposited on copper (Banszerus et al., 2015). Actually, for the calculated times, the lowest energy charge carriers are not completely diffusive yet, so the reported values should be considered as lower bounds. Strain may thus very well be the dominant scattering for ultraclean samples (Couto et al., 2014).

For lower quality samples (~20,000 $cm^2/V\,s$), smaller random strain correlation length ($\lambda \sim 3\,nm$), or additional LR disorder (eg, $W = 1$) are required to bring the mobility down to the desired values.

The right panels in Fig. 8 depict the relative strength of LR disorder vs RS. The upper right panel shows that in the regime where RS becomes stronger for decreasing λ, the effect is comparable, in order of magnitude, to the one induced by the conventional LR disorder model in Roche et al. (2012). For instance, at low energies, 0.125% of LR ($\lambda < 0.5\,nm$) with $W = 0.5$ (1) induces diffusive scattering comparable to RS with $\lambda = 5$ (3) nm. These latter values are on the low end side of the spectrum of experimentally reported correlation lengths for ripples, suggesting that ripples alone can often not explain the scattering mechanism in low mobility samples. We note that at higher energies ($> \pm 0.1\,t_0$), RS disorder has a relatively stronger impact than at low energy, compared to LR Coulomb scatterers.

We come back to the discussion about the dominant disorder source, when discussing the case of G/BN in Section 4.2. Indeed, recently, ULR impurities ($\lambda = 20\,nm$, not to be confused with the LR impurities considered here, with $\lambda = 0.426\,nm$) have also been proposed as an important disorder source (Wong et al., 2015).

We finally note that the evaluation of τ at ~0 energy becomes numerically challenging with our method (due to the very low DOS), causing the numerical instabilities in SL graphene, mostly. Numerical interpolation would be a reasonable workaround for this problem, as explained in de Jamblinne de Meux et al. (2015).

4.2 Graphene on hBN

4.2.1 Hofstadter and Wannier Diagrams

We remind that the Hamiltonian described in Section 2.1 contains three terms, namely the periodic moiré modulation H_0, the periodic mass term H_Z, and the virtual strain term H_{AB}. In Figs. 9 and 10, we map out the DOS vs energy and magnetic field to obtain the Hofstadter butterfly. The

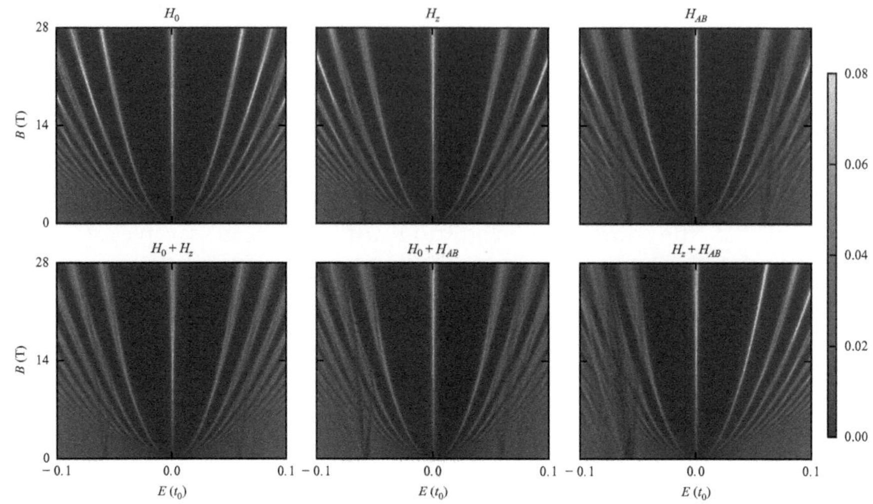

Fig. 9 *B*-Field-dependent DOS map for graphene on hBN, for different subsets of the full continuum model Hamiltonian where we combine the potential fluctuations, the local mass, and the off-diagonal strain contributions.

different panels of Fig. 7 are based on the subsets of these three terms, while Fig. 10 includes all three of them. To observe the experimental electron–hole asymmetry and the gap opening at negative energy, a mixing of terms with the virtual strain term is required. If one were to observe a gap using the mass-term contribution only, the value of C_Z would have to be at least twice as large (not shown here).

The full Hamiltonian maps in Fig. 10 capture well most of the single-particle features in experiment. The lower panel is obtained from the upper panel by integration of the DOS, to replace the energy dependency to a charge carrier density. The complex diamond-like fractal structure at negative energies depicts an intricate structure of secondary features. For visual reference, we draw a selection of lines in the Wannier representation of the data (lower panel). Such lines can be represented by a Diophantine equation, for which the slope t is directly linked to the transverse conductivity quantization by $\sigma_{xy} = t\frac{e^2}{h}$. Our detailed Wannier maps are expected to provide sound reference for experimentalists performing high magnetic field dissipationless conductivity measurements.

The opening of the experimental gap at $n/n_0 = -4$ at zero field, and subsequent closing at around six roots in the inequivalent K and K' contributions (Moon and Koshino, 2014). The transport behavior of the charge

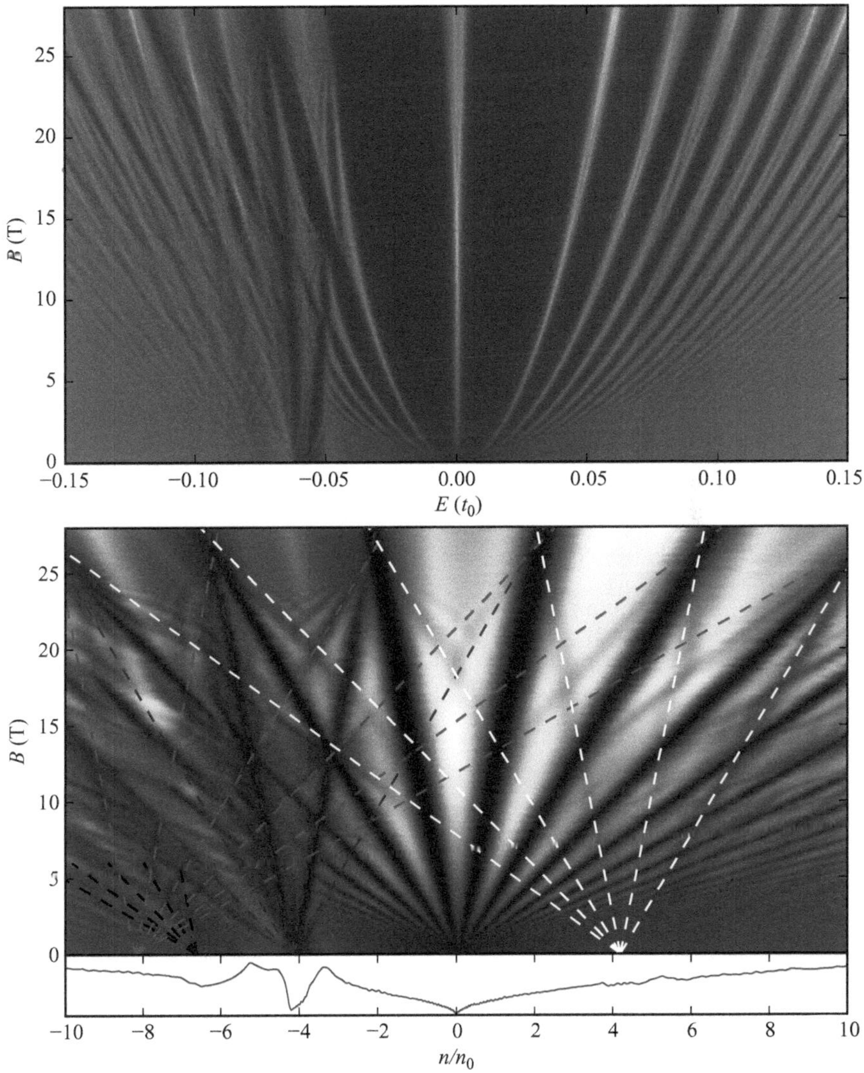

Fig. 10 *B*-Field-dependent DOS map (*upper panel*) and Wannier diagram (*lower panel*) for the full continuum model Hamiltonian. Selected number of *Diophantine lines* are superimposed for reference.

carriers at the secondary level is predicted to follow the Schrödinger equation, rather than complying to the Dirac equation, unlike the primary charge carriers at zero energy (Chizhova et al., 2014). The linear energy-dependent behavior in the upper panel seems to confirm this, even if slight distortions are apparent. Additional features develop at filling factors below $n/n_0 = -4$, which, by comparison with Fig. 7, are induced by the virtual strain term (most prominent when mixed with the mass term).

A fundamental limitation of periodic supercell simulations restricts the accessible twist angle values to well-defined commensurate numbers. Formally, this is no exception for the present method. However, because our actual supercells contain a repetition of 60 moiré unit cells in both the x and y-directions, the periodic mismatch arising at the boundaries becomes increasingly negligible in our bulk picture with increasing system size. Convergence studies (not shown here) confirm this. We illustrate this capability by giving some zero field and finite magnetic field DOS in Figs. 9 and 10, using the small-angle approximation. Large angles are also accessible, but require the complete expression given in Jung et al. (2014). We will illustrate such large angles in Section 4.5.

4.2.2 Transport Properties

The origin of electron scattering in graphene transport has been attributed mainly to the effects of Coulomb impurity scattering (Adam et al., 2007; Ando, 2006; Nomura and MacDonald, 2006; Sarma et al., 2011) although random strain fields have also been suggested as sources of scattering (Couto et al., 2014). In ultraclean devices such as suspended graphene or ultra clean graphene on G/BN several scattering mechanisms could sensitively impact the mobility. One scenario that we wish to explore through our numerical simulation is to examine the role of the charge scatterers in G/BN that are related to ULR disorder with very large correlation lengths on the order of 20 nm observed in recent measurements (Wong et al., 2015). In order to verify the feasibility of such scenario, we compare in Fig. 11E–G the impact of our disorder models on the dynamic observables of the system. The random strain field behavior is also provided for reference.

Our calculation on ULR correlated onsite disorder (light blue (light gray in the print version) line) for $\lambda = 20$ and $W = 0.05$ follows almost exactly the curves of RS disorder. Because of this, microscopic characterization of the sample might be required to differentiate between these two possible scenarios in experiment.

Finally, the general shape for the low disorder case agrees with DaSilva et al. (2015), where a fixed mobility of 5000 cm^2/V s is used. The relatively higher conductivity for low energies in our curves is rationalized by the increasing mobility toward the Dirac point. Our results again suggest that many-body effects are *not* necessarily required to explain the hole side

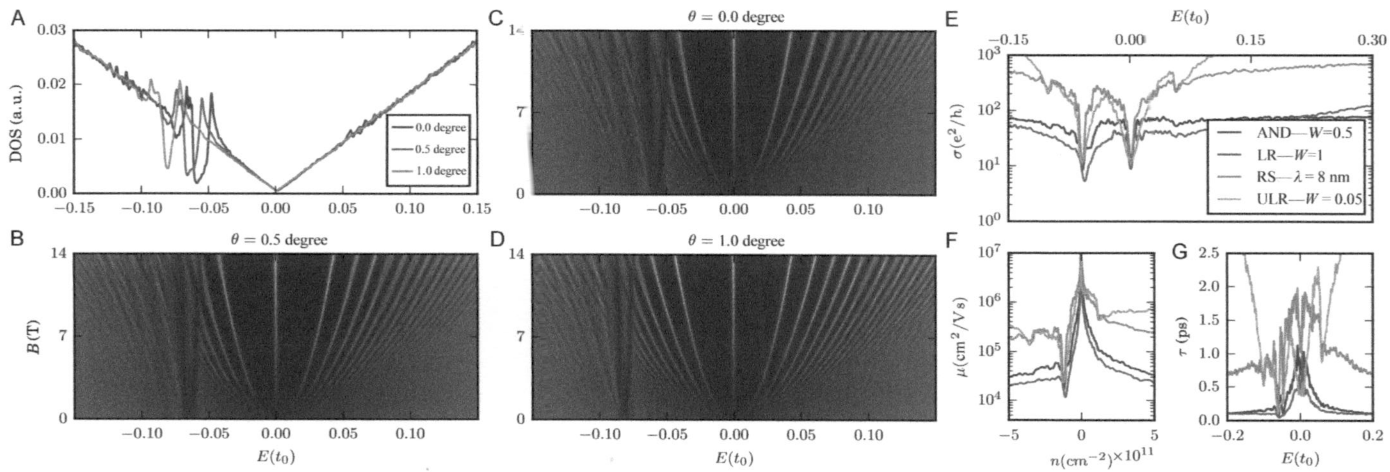

Fig. 11 (A) Density of states at zero magnetic field for different small twist angles in G/BN. (B–D) *B*-Field-dependent DOS plot map for twisted graphene on hexagonal boron nitride systems. (E)–(G) Conductivity, mobility, and scattering times for SL graphene. Comparison between Anderson disorder, long-range scatterers with a correlation length of 0.426 nm, random strain profiles with a correlation length equal to 8 nm, and ultra-long-range scatterers with a correlation length of 20 nm.

minimum of conductivity, as opposed to the claim in the claims in Slotman et al. (2015), where a simplified model is used.

4.2.3 Sublattice Asymmetric Disorder

To further illustrate the freedom that this kind of computer experiments provide us with, we calculate the Hofstadter butterfly (see Fig. 12) for the sublattice-selective functionalization disorder introduced in Section 2.2. What started as a toy in the theorist's playground (Leconte et al., 2011; Lherbier et al., 2013) has since been observed in experiments of sublattice-selective nitrogen substitution (Lv et al., 2012; Telychko et al., 2014; Wang et al., 2012b; Zabet-Khosousi et al., 2014; Zhao et al., 2011), molybdenum substitution (Wan et al., 2013) and hydrogen functionalization (Lin et al., 2015). Here, we combine our realistic model for the G/BN system with the dilute sublattice-specific modification of onsite energies, as introduced in the model and methodology section. The parameters of choice for the latter are $A = 0.05$ for a concentration of 0.01% of shifted onsite energies. In the lower panel of Fig. 12, we allow for variations in the value of A.

As expected from the zero-field literature (Lherbier et al., 2013), a gap opens at the Dirac point, which scales with concentration and onsite energy shift. The consequence of this gap opening in the finite magnetic field regime is a splitting of the zero LL, as observed in the upper panel. The

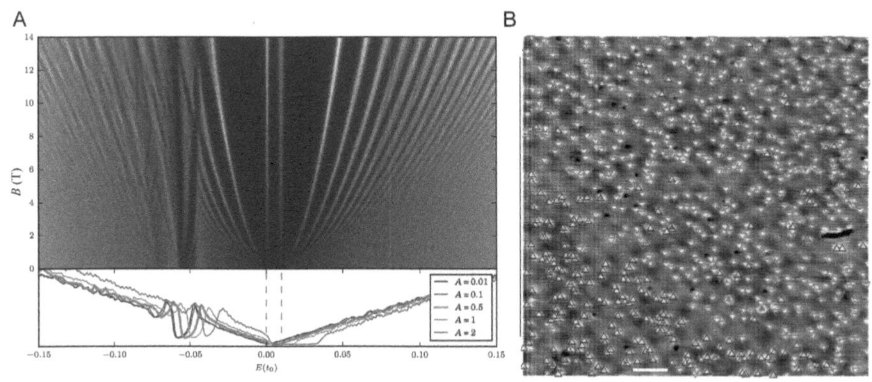

Fig. 12 (A) *B*-Field-dependent DOS map for G/BN system with sublattice-selective onsite disorder, onsite shift of random distribution of sites (*upper panel*). Zero-field DOS for different values of A (*lower panel*). (B) Highly efficient experimental sublattice-selective nitrogen-doping in graphene. *Reprinted with permission from Zabet-Khosousi, A., Zhao, L., Palova, L., Hybertsen, M.S., Reichman, D.R., Pasupathy, A.N., Flynn, G.W., 2014. Segregation of sublattice domains in nitrogen doped graphene. J. Am. Chem. Soc. 136, 1391–1397. Copyright (2014) American Chemical Society.*

dashed lines in the lower panel indicate the band gap edges and correspond to the position of two LL0-type features in the upper panel. The reason for the LL0 splitting roots in degeneracy lifting of the valley degree of freedom, which is directly related to the global sublattice symmetry breaking. Additionally, our simulations suggest that controlled sublattice chemical modification of this moiré system at the atomic level, by introducing onsite potential breaking disorders such as hydrogen or vacancies, can lead to a downward shift in the energy position of the superlattice Dirac features.

4.3 Bilayer Graphene

For (Bernal-stacked) BL graphene, we differentiate between the zero and finite electric field cases. The application of an electric field gives access to a tunable band-gap opening (Park et al., 2015; Rutter et al., 2011; Zhang et al., 2009). We first provide the magnetic B-field-dependent DOS maps and turn after that to the transport signatures of these systems.

As expected, Landau levels develop, on the one hand, from the levels in the electronic band structure that touch at $E=0$ and, on the other hand, from the set of bands at higher energy (see dips around $\pm 0.12t_0$ in the zero-field DOS; see Fig. 13A).

The applied electric field (bottom panel) opens a low-energy gap and a splitting occurs in the Landau levels that initially developed from the second set of bands. LLs are eightfold degenerate, due to the additional layer degree of freedom. The degeneracy of this layer degree of freedom is lifted by the electric field, leading to the split levels. It seems that the first of these levels remains degenerate, while the higher levels come in pairs.

The transport observables for the case without applied electric field are presented in Fig. 13B. We induce scattering by introducing RS, AND, or LR potential disorders. The latter two bring the system into the diffusive regime and reduce the conductivity from the hundreds of thousands of $cm^2/V\,s$ characteristic for ultraclean samples, to values smaller by one or two orders of magnitudes, typical for dirty samples.

By comparing the scattering times for the AND scatterers of BL graphene with the same curves for SL graphene at low energy, we see a downward trend toward the zero energy for the former, while an upward trend is observed for the latter. This opposite behavior of these elastic scattering times between BL and SL graphene agrees with the experimental data (Monteverde et al., 2010) on the transport and the elastic scattering time

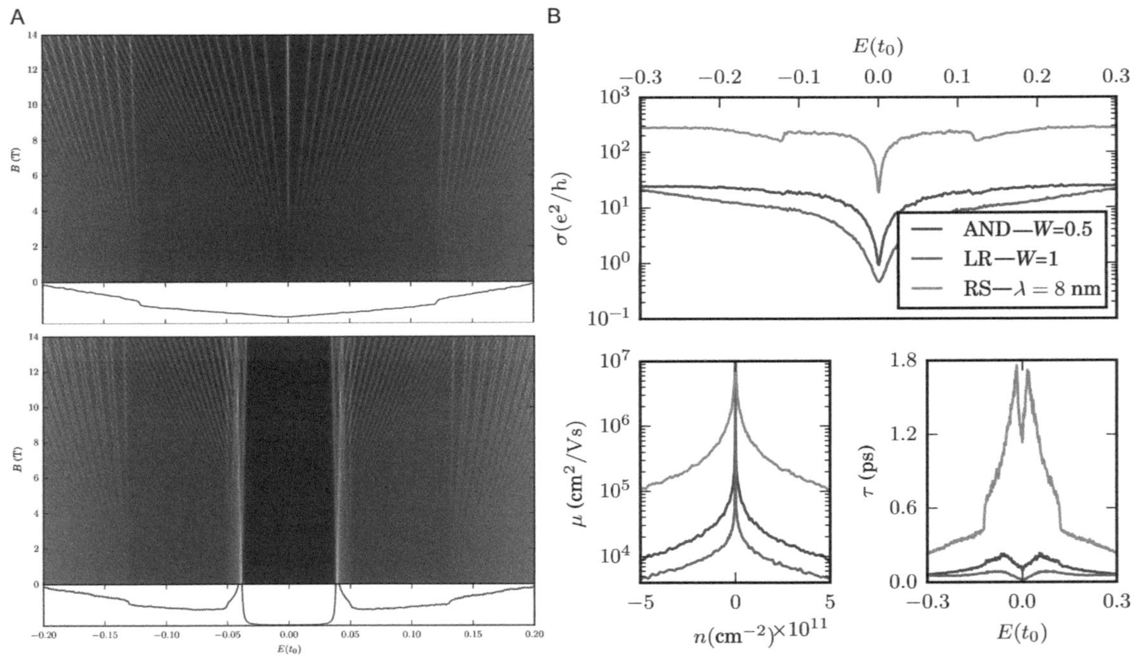

Fig. 13 (A) *B*-Field-dependent DOS map for Bernal-stacked bilayer graphene, $E=0$ (*top*) and $E=0.06t_0$ (*bottom*). *Lower subpanels* depict the zero-field DOS. (B) Conductivity, mobility, and scattering times for BL graphene. Comparison between LR scatterers and RS profiles.

(as well as their ratio). A more in-depth analysis of resonant scattering in SL and BL graphene is provided in Ferreira et al. (2011).

4.4 Bilayer Graphene on hBN

By depositing the bilayer graphene on hBN, superlattice features appear (see left panel of Fig. 14). Compared to the SL system, new sets of butterfly features develop at very low energy, whose energy position agrees with Moon and Koshino (2014), and correspond to filling factors $n/n_0 = \pm 2$, in addition to the single-layer features at $n/n_0 = \pm 4$. Yet again, the intensity of these features is more pronounced on the hole side than on the electron side. The transport observables in the right panels indicate that these secondary low-energy features are robust up to moderate disorder.

The gap opening due to electric field (see Fig. 14) slightly shifts the energy dependence of all characteristic features to higher energies. For instance, the superlattice features at high energy have been shifted from below to above $-0.15t_0$. At the same time, due to the full DOS depletion induced by the electric field, the new characteristic features at low energy disappear. Only a slight asymmetry exists between DOS features of the holes and the electrons at the edges of the bands. From the transport properties (not shown here), we conclude that this asymmetry is negligible.

4.5 Encapsulated Bilayer Graphene

Encapsulating graphene is a means to screen the system from external perturbations, provided the additional layer does not induce significant changes to the envisioned properties of, in this case, BL graphene. In this section, we thus investigate the impact of this additional layer of hBN. We consider different orientation angles of the top hBN layer, from small to large angles. In Fig. 15, we first consider the same angle orientation for the top and bottom layer of hBN. In the Hofstadter map (left panel), we see that the secondary features are not the same as the ones in Fig. 14. This is because the Moiré pattern of the top layer is shifted by a distance of one interatomic distance compared to the one applied to the bottom layer.

In Fig. 16, we consider the small rotation angles of 0.5 degree and 1 degree. The position of the secondary low-energy features is affected by the angle value, and large rotation angles are required to scale out the secondary features from the top BN layer. Indeed, we checked that the B-field dependent DOS and the dynamics for a 20 degree angle in the top layer of

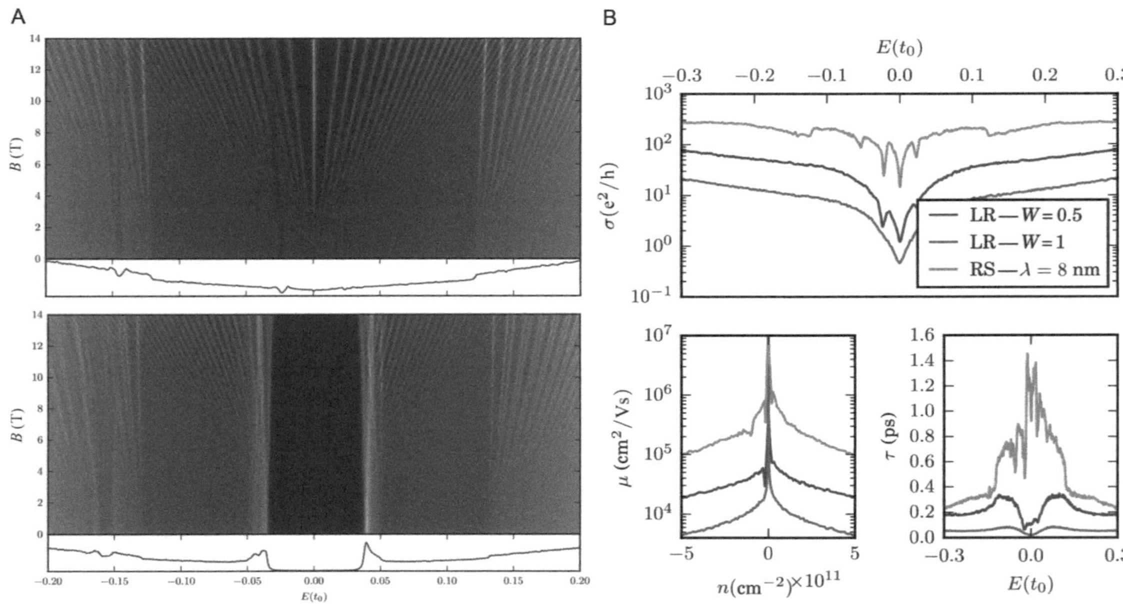

Fig. 14 BL graphene on hBN. (A) *B*-Field-dependent DOS map, without and with *E*-field. (B) Conductivity, mobility, and scattering times. Comparison between LR scatterers and RS disorder.

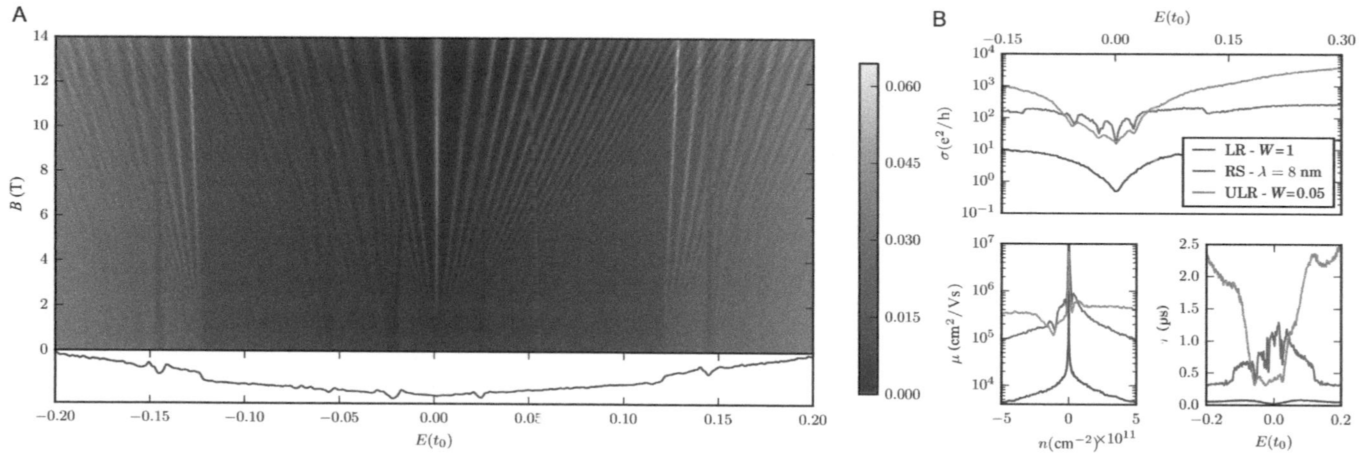

Fig. 15 Encapsulated BL graphene, no twist angle for top hBN layer. (A) *B*-Field-dependent DOS map, without *E*-field. (B) Conductivity, mobility, and scattering times. Comparison between LR scatterers, ULR scatterers, and RS disorder.

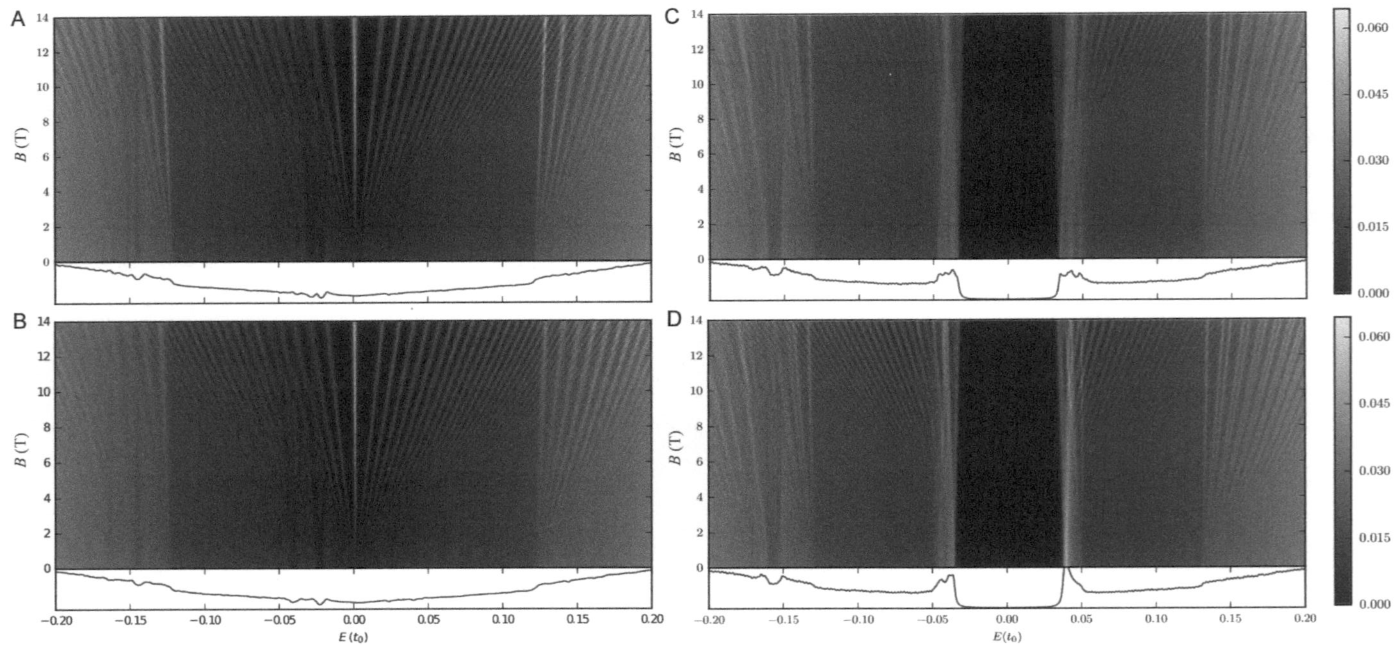

Fig. 16 Encapsulated BL graphene. *B*-Field-dependent DOS map, no *E*-field, for 0.5 degree (A) and 1 degree (B) twist angle for the top hBN layer. Maps with *E*-field, 0.5 degree (C) and 20 degree (D) twist angle.

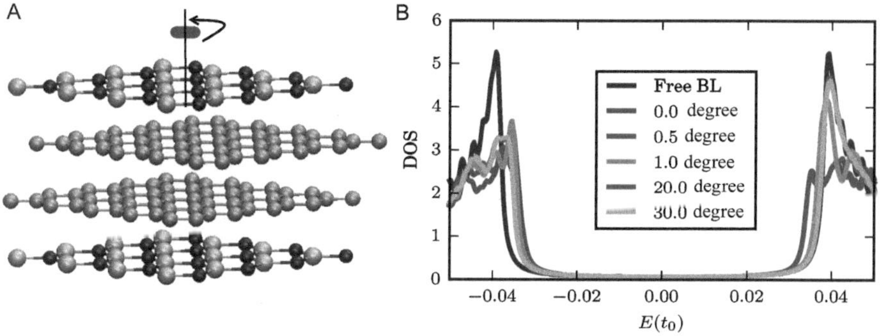

Fig. 17 (A) Sketch of encapsulated BL graphene. (B) Zero-field DOS for this system under *E*-field, for different twist angles for the top hBN layer.

hBN (not shown here) are the same as the ones presented in Fig. 15 in the absence of the secondary layer.

We further investigate the angle dependency whenever an electric field is applied. Similar to BL on hBN, the gap opening dominates the low-energy physics. Consequently, controlling the orientation of the top layer of hBN becomes of secondary concern; the *B*-field-dependent DOS maps at low energies for small angles are practically the same (illustrated for 0.5 degree and 20 degree angle in Fig. 16). Indeed, the lowest energy super-periodic features due to the top and bottom layer of hBN have disappeared for all angles.

Finally, Fig. 17 focuses on the size of the gap. Only a very small reduction of its amplitude due to the interplay between the mass contributions from the electric field and the moiré pattern is observed.

In conclusion, when adding a top layer of hBN layer to shield the system from external contamination, the transport physics are not significantly affected whenever an electric field is applied. In the absence of this electric field, large angles should be preferred to minimize impact from the top layer.

5. CONCLUSION

We have presented a methodology to investigate the electronic and transport properties of graphene-based two-dimensional materials through the large-scale direct real-space modeling of the lattice Hamiltonian and calculation of observables of interest through numerically efficient methods. This approach has the advantage of being able to capture on the same footing the effects of realistic disorder and crystal incommensurability often found in artificial two-dimensional materials without further assumptions on the model. We present two key advancements, on the one hand, the methodology to map accurate effective lattice Hamiltonians and, on the other hand, the implementation of efficient numerical tools to perform large-scale simulation of systems containing up to billions of atoms. The specific systems targeted in our study were graphene-based systems subject to the effects of disorder, strains, and moiré pattern potentials due to coupling with hexagonal boron nitride substrates, all of which are systems of current interest in the scientific community. The methods to formulate the accurate TB Hamiltonians of commensurate and incommensurate 2D materials heterojunction used in this work rely on the use of effective hopping parameters that can correctly reproduce the Fermi velocity of a continuum Hamiltonian, and are also able to capture the site potential variations, as well as the strain fields arising due to unequal hopping strengths to neighboring carbon atoms. These contributions have been informed using first-principles calculation methods to obtain a model both for commensurate and for incommensurate pristine systems with predictive accuracy. We have further considered the role of interlayer coupling with an external hBN substrate that introduces a moiré pattern that is not necessarily commensurate with the crystal.

The purpose of this chapter was to introduce an efficient multiscale approach that, one the one hand, captures accurately the physics at play in layered materials and, on the other hand, implements a numerically efficient method to evaluate both the static and dynamic transport properties of these materials. We aimed at providing a modular approach to easily model increasingly complex systems which we apply effectively in the example cases of graphene and bilayer graphene subject to the effects of disorder, strains, and coupling with the hBN substrate. Our analysis has allowed us to address, among others, the following questions:

(i) Regarding the controversy of the dominant scattering source in graphene, either freestanding or deposited on a substrate, we confirm that certain random scattering profiles can induce the required scattering to explain the experimental ultra-high mobilities (as long as the correlation length is not too small). At the same time, one cannot rule out the fact that ULR disorder (corresponding to charged impurities trapped because of the substrate) may sometimes be the dominant scattering source as well. Indeed, the two conductivity profiles are often very similar.

(ii) Regarding the Hofstadter butterfly in SL graphene deposited on hBN, we give a clear image on the experimental features that are captured by single-particle physics, and that do not necessarily require to invoke a many-body interpretation. These features include secondary gap opening and activated higher energy secondary features.

(iii) We illustrate the rich physics that may occur when experimentalists manage to induce sublattice-selective doping in Gr/BN superlattice models in the same way as it has been achieved for freestanding graphene.

(iv) We quantify the negligible impact of adding a top layer on top of the BL/BN system on its transport properties. Consequently, this layer can safely be used as a buffer layer to protect the system from external contamination.

(v) We quantify the negligible impact of the moiré pattern on the width of the gate-tunable band-gap opening in bilayer graphene.

The steady increase of computing power available for mid-size scientific computing facilities is a fact that is enormously reshaping the flavor of research. From the analysis provided in this work, we can conclude that it is already routinely feasible to carry out a direct computational study of experimentally relevant large-scale systems for capturing the quantum mechanical effects in nano-, meso-, and macroscale devices. Keeping in mind the projected increase in computing power during the years to come, we can anticipate a more widespread use of this type of simulation techniques in the future. The methods outlined here are expected to be applicable for a wide variety of systems and experiments that are currently being studied in the field on 2D materials beyond graphene.

REFERENCES

Adam, S., Hwang, E.H., Galitski, V.M., Sarma, S. Das, 2007. A self-consistent theory of graphene transport. Proc. Natl. Acad. Sci. U.S.A. 104, 18392.

Amorim, B., Cortijo, A., de Juan, F., Grushin, A.G., Guinea, F., Gutiérez-Rubio, A., Ochoa, H., Parente, V., Roldan, R., San-José, P., Schiefele, J., Sturla, M., Vozmediano, M.A.H., 2016. Novel effects of strains in graphene and other two dimensional materials. Phys. Rep. 617, 1–54.

Ando, T., 2006. Screening effect and impurity scattering in monolayer graphene. J. Phys. Soc. Jpn. 75, 074716.

Banszerus, L., Schmitz, M., Engels, S., Dauber, J., Oellers, M., Haupt, F., Watanabe, K., Taniguchi, T., Beschoten, B., Stampfer, C., 2015. Ultrahigh-mobility graphene devices from chemical vapor deposition on reusable copper. Sci. Adv. 1, e1500222.

Barlas, Y., Pereg-Barnea, T., Polini, M., Asgari, R., MacDonald, A.H., 2007. Chirality and correlations in graphene. Phys. Rev. Lett. 98, 236601.

Bistritzer, R., MacDonald, A.H., 2010. Transport between twisted graphene layers. Phys. Rev. B 81, 245412.

Bistritzer, R., MacDonald, A.H., 2011. Moire bands in twisted double-layer graphene. PNAS 108, 12233.

Bokdam, M., Amlaki, T., Brocks, G., Kelly, P.J., 2014. Band gaps in incommensurable graphene on hexagonal boron nitride. Phys. Rev. B 89, 201404.
Boyd, J.P., 2001. Chebyshev and Fourier Spectral Methods, second revised ed. Dover Publications, New York.
Burgos, R., Warnes, J., Lima, L.R.F., Lewenkopf, C., 2015. Effects of a random gauge field on the conductivity of graphene sheets with disordered ripples. Phys. Rev. B 91, 115403.
Chen, Z.G., Shi, Z., Yang, W., Lu, X., Lai, Y., Yan, H., Wang, F., Zhang, G., Li, Z., 2014. Observation of an intrinsic bandgap and Landau level renormalization in graphene/boron-nitride heterostructures. Nat. Commun. 5, 4461.
Chizhova, L.A., Libisch, F., Burgdorfer, J., 2014. Graphene quantum dot on boron nitride: Dirac cone replica and Hofstadter butterfly. Phys. Rev. B 90, 165404.
Chu, Z.D., He, W.Y., He, L., 2013. Coexistence of van Hove singularities and superlattice Dirac points in a slightly twisted graphene bilayer. Phys. Rev. B 87, 155419.
Couto, N.J.G., Costanzo, D., Engels, S., Ki, D.-K., Watanabe, K., Taniguchi, T., Stampfer, C., Guinea, F., Morpurgo, A.F., 2014. Random strain fluctuations as dominant disorder source for high-quality on-substrate graphene devices. Phys. Rev. X 4, 041019.
DaSilva, A.M., Jung, J., Adam, S., MacDonald, A.H., 2015. Transport and particle-hole asymmetry in graphene on boron nitride. Phys. Rev. B 91, 245422.
de Gail, R., Goerbig, M.O., Guinea, F., Montambaux, G., Castro Neto, A.H., 2011. Topologically protected zero modes in twisted bilayer graphene. Phys. Rev. B 84, 045436.
de Jamblinne de Meux, A., Leconte, N., Charlier, J.-C., Lherbier, A., 2015. Velocity renormalization and Dirac cone multiplication in graphene superlattices with various barrier-edge geometries. Phys. Rev. B 91, 235139.
Dean, C.R., Young, A.F., Meric, I., Lee, C., Wang, L., Sorgenfrei, S., Watanabe, K., Taniguchi, T., Kim, P., Shepard, K.L., Hone, J., 2010. Boron nitride substrates for high-quality graphene electronics. Nat. Nanotechnol. 5, 722–726.
Dean, C.R., Wang, L., Maher, P., Forsythe, C., Ghahari, F., Gao, Y., Katoch, J., Ishigami, M., Moon, P., Koshino, M., Taniguchi, T., Watanabe, K., Shepard, K.L., Hone, J., Kim, P., 2013. Hofstadter's butterfly and the fractal quantum Hall effect in moiré superlattices. Nature 497, 598.
Diez, M., Dahlhaus, J.P., Wimmer, M., Beenakker, C.W.J., 2014. Emergence of massless Dirac fermions in graphene's Hofstadter butterfly at switches of the quantum hall phase connectivity. Phys. Rev. Lett. 112, 196602.
Fan, Z., Uppstu, A., Siro, T., Harju, A., 2014. Efficient linear-scaling quantum transport calculations on graphics processing units and applications on electron transport in graphene. Comp. Phys. Commun. 185, 28–39.
Fasolino, A., Los, J.H., Katsnelson, M.I., 2011. Intrinsic ripples in graphene. Nat. Mater. 6, 858–861.
Ferreira, A., 2016. Unpublished.
Ferreira, A., Mucciolo, E.R., 2015. Critical delocalization of chiral zero energy modes in graphene. Phys. Rev. Lett. 115, 106601.
Ferreira, A., Viana-Gomes, J., Nilsson, J., Mucciolo, E.R., Peres, N.M.R., Castro Neto, A.H., 2011. Unified description of the dc conductivity of monolayer and bilayer graphene at finite densities based on resonant scatterers. Phys. Rev. B 83, 165402.
Foa Torres, L.E.F., Roche, S., Charlier, J.C., 2014. Introduction to Graphene-Based Nanomaterials. Cambridge University Press, Cambridge.
Gade, R., 1993. Anderson localization for sublattice models. Nucl. Phys. B 398, 499.
Gade, R., Wegner, F., 1991. The n = 0 replica limit of U(n) and U(n)SO(n) models. Nucl. Phys. B 360, 213.
Garcia, J.H., Covaci, L., Rappoport, T.R., 2015. Real-space calculation of the conductivity tensor for disordered topological matter. Phys. Rev. Lett. 114, 116602.

Gattenlohner, S., Hannes, W.-R., Ostrovsky, P.M., Gorny, I.V., Mirlin, A.D., Titov, M., 2014. Quantum hall criticality and localization in graphene with short-range impurities at the Dirac point. Phys. Rev. Lett. 112, 026802.

Giovannetti, G., Khomyakov, P.A., Brocks, G., Kelly, P.J., van den Brink, J., 2007. Substrate-induced band gap in graphene on hexagonal boron nitride: ab initio density functional calculations. Phys. Rev. B 76, 073103.

Gonzalez, J., 2013. Magnetic and Kohn-Luttinger instabilities near a Van Hove singularity: monolayer versus twisted bilayer graphene. Phys. Rev. B 88, 125434

Gonzalez, J., Guinea, F., Vozmediano, M.A.H., 1994. Non-Fermi liquid behavior of electrons in the half-filled honeycomb lattice (a renormalization group approach). Nucl. Phys. B 424, 595.

Gradshteyn, I.S., Ryzhik, I.M., 2007. Table of Integrals, Series, and Products, seventh ed. Elsevier, Academic Press, Boston.

Häfner, V., Schindler, J., Weik, N., Mayer, T., Balakrishnan, S., Narayanan, R., Bera, S., Evers, F., 2014. Density of states in graphene with vacancies: midgap power law and frozen multifractality. Phys. Rev. Lett. 113, 186802.

Hofstadter, D.R., 1976. Energy levels and wave functions of Bloch electrons in rational and irrational magnetic fields. Phys. Rev. B 14, 2239.

Hunt, B., Sanchez-Yamagishi, J.D., Young, A.F., Yankowitz, M., LeRoy, B.J., Watanabe, K., Taniguchi, T., Moon, P., Koshino, M., Jarillo-Herrero, P., Ashoori, R.C., 2013. Science 340, 1427.

Iitaka, T., Ebisuzaki, T., 2004. Random phase vector for calculating the trace of a large matrix. Phys. Rev. E 69, 057701.

Imry, Y., 2002. Introduction to Mesoscopic Physics, second ed. Oxford University Press, Oxford.

Jung, J., MacDonald, A.H., 2011. Enhancement of nonlocal exchange near isolated band crossings in graphene. Phys. Rev. B 84, 085446.

Jung, J., MacDonald, A.H., 2013. Tight-binding model for graphene pi-bands from maximally localized Wannier functions. Phys. Rev. B 87, 195450.

Jung, J., MacDonald, A.H., 2014. Accurate tight-binding models for the π bands of bilayer graphene. Phys. Rev. B 89, 035405.

Jung, J., Raoux, A., Qiao, Z.H., MacDonald, A.H., 2014. Ab initio theory of moiré superlattice bands in layered two-dimensional materials. Phys. Rev. B 89, 205414.

Jung, J., DaSilva, A.M., MacDonald, A.H., Adam, S., 2015. Origin of band gaps in graphene on hexagonal boron nitride. Nat. Commun. 6, 6308.

Kindermann, M., First, P.N., 2011. Local sublattice-symmetry breaking in rotationally faulted multilayer graphene. Phys. Rev. B 83, 045425.

Kindermann, M., Uchoa, B., Miller, D.L., 2012. Zero-energy modes and gate-tunable gap in graphene on hexagonal boron nitride. Phys. Rev. B 86, 115415.

Klos, J.W., Zozoulenko, I.V., 2010. Effect of short- and long-range scattering on the conductivity of graphene: Boltzmann approach vs tight-binding calculations. Phys. Rev. B 82, 081414.

Lanczos, C., 1950. An iteration method for the solution of the eigenvalue problem of linear differential and integral operators. J. Res. Natl. Bur. Stand. 45, 255.

Landgraf, W., Shallcross, S., Türschmann, K., Weckbecker, D., Pankratov, O., 2013. Electronic structure of twisted graphene flakes. Phys. Rev. B 87, 075433.

Latil, S., Roche, S., Mayou, D., Charlier, J.-C., 2004. Mesoscopic transport in chemically doped carbon nanotubes. Phys. Rev. Lett. 92, 256805.

Leconte, N., Moser, J., Ordejon, P., Tao, H., Lherbier, A., Bachtold, A., Alsina, F., Sotomayor Torres, C.M., Charlier, J.-C., Roche, S., 2010. Damaging graphene with ozone treatment: a chemically tunable metal–insulator transition. ACS Nano 4, 4033–4038.

Leconte, N., Lherbier, A., Varchon, F., Ordejon, P., Roche, S., Charlier, J.-C., 2011. Quantum transport in chemically modified two-dimensional graphene: from minimal conductivity to Anderson localization. Phys. Rev. B 84, 235420.
Leconte, N., Ortmann, F., Cresti, A., Charlier, J.-C., Roche, S., 2014. Quantum transport in chemically functionalized graphene at high magnetic field: defect-induced critical states and breakdown of electron–hole symmetry. 2D Mater. 1, 021001.
Leconte, N., Martinez-Gordillo, R., MacDonald, A.H., Jung, J., 2016. Hofstadter butterfly and quantum transport in graphene on hexagonal boron nitride from multiscale lattice simulations, submitted for publication.
Lee, G.D., Wang, C.Z., Yoon, E., Hwang, N.M., Ho, K.M., 2006. Vacancy defects and the formation of local haeckelite structures in graphene from tight-binding molecular dynamics. Phys. Rev. B 74, 245411.
Lherbier, A., Biel, B., Niquet, Y.-M., Roche, S., 2008. Transport length scales in disordered graphene-based materials: strong localization regimes and dimensionality effects. Phys. Rev. Lett. 100, 036803.
Lherbier, A., Dubois, S.M.M., Declerck, X., Niquet, Y.M., Roche, S., Charlier, J.C., 2012. Transport properties of graphene containing structural defects. Phys. Rev. B 86, 075402.
Lherbier, A., Botello-Mendez, A.R., Charier, J.C., 2013. Electronic and transport properties of unbalanced sublattice N-doping in graphene. Nano Lett. 13, 1446–1450.
Lin, C., Feng, Y., Xiao, Y., Drr, M., Huang, X., Xu, X., Zhao, R., Wang, E., Li, X.-Z., Hu, Z., 2015. Direct observation of ordered configurations of hydrogen adatoms on graphene. Nano Lett. 15, 903–908.
Lopes dos Santos, J.M.B., Peres, N.M.R., Castro Neto, A.H., 2012. Continuum model of the twisted graphene bilayer. Phys. Rev. B 86, 155449.
Lv, R., Li, Q., Botello-Mendez, A.R., Hayashi, T., Wang, B., Berkdemir, A., Hao, Q., Elias, A.L., Cruz-Silva, R., Gutierrez, H.R., 2012. Nitrogen-doped graphene: beyond single substitution and enhanced molecular sensing. Sci. Rep. 2, 586.
Mahan, G.D., 2000. Many Particle Physics, third ed., Springer Science @ Business Media.
Martinez-Gordillo, R., Roche, S., Ortmann, F., Pruneda, M., 2014. Transport fingerprints at graphene superlattice Dirac points induced by a boron nitride substrate. Phys. Rev. B 89, 161401.
McCann, E., Falko, V.I., 2006. Landau-level degeneracy and quantum hall effect in a graphite bilayer. Phys. Rev. Lett. 96, 086805.
Mele, E.J., 2011. Band symmetries and singularities in twisted multilayer graphene. Phys. Rev. B 84, 235439.
Mele, E.J., 2012. Interlayer coupling in rotationally faulted multilayer graphenes. J. Phys. D Appl. Phys. 45, 154004.
Monteverde, M., Ojeda-Aristizabal, C., Weil, R., Bennaceur, K., Ferrier, M., Gueron, S., Glattli, C., Bouchiat, H., Fuchs, J.N., Maslov, D.L., 2010. Transport and elastic scattering times as probes of the nature of impurity scattering in single-layer and bilayer graphene. Phys. Rev. Lett. 104, 126801.
Moon, P., Koshino, M., 2013. Optical absorption in twisted bilayer graphene. Phys. Rev. B 87, 205404.
Moon, P., Koshino, M., 2014. Electronic properties of graphene/hexagonal-boron-nitride moiré superlattice. Phys. Rev. B 90, 155406.
Mott, N.F., 1967. Electrons in disordered structures. Adv. Phys. 16, 49.
Mucciolo, E.R., Lewenkopf, C.H., 2010. Disorder and electronic transport in graphene. J. Phys. Condens. Matter 22, 273201.
Neek-Amal, M., Peeters, F.M., 2014. Graphene on hexagonal lattice substrate: stress and pseudo-magnetic field. Appl. Phys. Lett. 104, 173106.
Nomura, K., MacDonald, A.H., 2006. Quantum transport of massless Dirac fermions. Phys. Rev. Lett. 98, 076602.

Novoselov, K.S., Geim, A.K., Morozov, S.V., Jiang, D., Katsnelson, M.I., Grigorieva, I.V., Dubonos, S.V., Firsov, A.A., 2005. Two-dimensional gas of massless Dirac fermions in graphene. Nature 438, 197–200.

Ortmann, F., Leconte, N., Roche, S., 2015. Efficient linear scaling approach for computing the Kubo Hall conductivity. Phys. Rev. B 91, 165117.

Ostrovsky, P.M., Gornyi, I.V., Mirlin, A.D., 2006. Electron transport in disordered graphene. Phys. Rev. B 74, 235443.

Ostrovsky, P.M., Titov, M., Bera, S., Gornyi, I.V., Mirlin, A.D., 2010. Diffusion and criticality in undoped graphene with resonant scatterers. Phys. Rev. Lett. 105, 266803.

Park, C., Ryou, J., Hong, S., Sumpter, B.G., Kim, G., Yoon, M., 2015. Electronic properties of bilayer graphene strongly coupled to interlayer stacking and an external electric field. Phys. Rev. Lett. 115, 015502.

Partoens, B., Peeters, F.M., 2006. From graphene to graphite: electronic structure around the *K* point. Phys. Rev. B 74, 075404.

Peres, N.M.R., 2010. Colloquium: the transport properties of graphene: an introduction. Rev. Mod. Phys. 82, 2673.

Ponomarenko, L.A., Gorbachev, R.V., Yu, G.L., Elias, D.C., Jalil, R., Patel, A.A., Mishchenko, A., Mayorov, A.S., Woods, C.R., Wallbank, J.R., Mucha-Kruczynski, M., Piot, B.A., Potemski, M., Grigorieva, I.V., Novoselov, K.S., Guinea, F., Fal'ko, V.I., Geim, A.K., 2013. Cloning of Dirac fermions in graphene superlattices. Nature 497, 594.

Radchenko, T.M., Shylau, A.A., Zozoulenko, I.V., Ferreira, A., 2013. Effect of charged line defects on conductivity in graphene: numerical Kubo and analytical Boltzmann approaches. Phys. Rev. B 87, 195448.

Rammer, J., 2008. Quantum Transport Theory (Frontiers in Physics). Westview Press, Boulder.

Reich, S., Maultzsch, J., Thomsen, C., Ordejón, P., 2002. Tight-binding description of graphene. Phys. Rev. B 66, 035412.

Roche, S., 1999. Quantum transport by means of O(N) real-space methods. Phys. Rev. B 59, 2284.

Roche, S., Mayou, D., 1997. Conductivity of quasiperiodic systems: a numerical study. Phys. Rev. Lett. 79, 2518.

Roche, S., Saito, R., 2001. Magnetoresistance of carbon nanotubes: from molecular to mesoscopic fingerprints. Phys. Rev. Lett. 87, 246803.

Roche, S., Leconte, N., Ortmann, F., Lherbier, A., Soriano, D., Charlier, J.C., 2012. Quantum transport in disordered graphene: a theoretical perspective. Solid State Commun. 152, 1404.

Rutter, G.M., Jung, S., Klimov, N.N., Newell, D.B., Zhitenev, N.B., Stroscio, J.A., 2011. Microscopic polarization in bilayer graphene. Nat. Phys. 7, 649.

Sachs, B., Wehling, T.O., Katsnelson, M.I., Lichtenstein, A.I., 2011. Adhesion and electronic structure of graphene on hexagonal boron nitride substrates. Phys. Rev. B 84, 195414.

Saito, R., Dresselhaus, D., Dresselhaus, M.S., 1998. Physical Properties of Carbon Nanotubes. Imperial College Press, London.

San-Jose, P., Prada, E., 2013. Helical networks in twisted bilayer graphene under interlayer bias. Phys. Rev. B 88, 121408.

San-Jose, P., Gonzalez, J., Guinea, F., 2012. Non-Abelian gauge potentials in graphene bilayers. Phys. Rev. Lett. 108, 216802.

San-Jose, P., Gutiérrez-Rubio, A., Sturla, M., Guinea, F., 2014. Electronic structure of spontaneously strained graphene on hexagonal boron nitride. Phys. Rev. B 90, 115152.

Sarma, S. Das, Adam, S., Hwang, E.H., Rossi, E., 2011. Electronic transport in two-dimensional graphene. Rev. Mod. Phys. 83, 407.

Semenoff, G.W., 1984. Condensed-matter simulation of a three-dimensional anomaly. Phys. Rev. Lett. 53, 2449.

Shallcross, S., Sharma, S., Pankratov, O.A., 2008. Quantum interference at the twist boundary in graphene. Phys. Rev. Lett. 101, 056803.

Shallcross, S., Sharma, S., Kandelaki, E., Pankratov, O.A., 2010. Electronic structure of turbostratic graphene. Phys. Rev. B 81, 165105.

Shallcross, S., Sharma, S., Pankratov, O.A., 2013. Emergent momentum scale localization, and van Hove singularities in the graphene twist bilayer. Phys. Rev. B 87, 245403.

Shi, Z., Jin, C., Yang, W., Ju, L., Horng, J., Lu, X., Bechtel, H.A., Martin, M.C., Fu, D., Wu, J., Watanabe, K., Taniguchi, T., Zhang, Y., Bai, X., Wang, E., Zhang, G., Wang, F., 2014. Gate-dependent pseudospin mixing in graphene/boron nitride moiré superlattices. Nat. Phys. 10, 743.

Slonczewski, J.C., Weiss, P.R., 1958. Band structure of graphite. Phys. Rev. 109, 272.

Slotman, G.J., van Wijk, M.M., Zhao, P.-L., Fasolino, A., Katsnelson, M.I., Yuan, S., 2015. Effect of structural relaxation on the electronic structure of graphene on hexagonal boron nitride. Phys. Rev. Lett. 115, 186801, arXiv 1507.03146.

Song, J.C.W., Shytov, A.V., Levitov, L.S., 2013. Electron interactions and gap opening in graphene superlattices. Phys. Rev. Lett. 111, 266801.

Soriano, D., Leconte, N., Ordejón, P., Charlier, J.-C., Palacios, J.J., Roche, S., 2011. Magnetoresistance and magnetic ordering fingerprints in hydrogenated graphene. Phys. Rev. Lett. 107, 016602.

Stauber, T., San-Jose, P., Brey, L., 2013. Optical conductivity Drude weight and plasmons in twisted graphene bilayers. New. J. Phys. 15, 113050.

Tabert, C.J., Nicol, E.J., 2013. Optical conductivity of twisted bilayer graphene. Phys. Rev. B 87, 121402.

Tal-Ezer, H., Kosloff, R., 1984. An accurate and efficient scheme for propagating the time dependent Schrödinger equation. J. Chem. Phys. 81, 3967.

Tang, M.S., Wang, C.Z., Chan, C.T., Ho, K.M., 1996. Environment-dependent tight-binding potential model. Phys. Rev. B 53, 979.

Tang, S., Wang, H., Zhang, Y., Li, A., Xie, H., Liu, X., Liu, L., Li, T., Huang, F., Xie, X., Jiang, M., 2013. Precisely aligned graphene grown on hexagonal boron nitride by catalyst free chemical vapor deposition. Sci. Rep. 3, 02666.

Telychko, M., Mutombo, P., Ondracek, M., Hapala, P., Bocquet, F.C., Kolorenc, J., Vondracek, M., Jelinek, P., Svec, M., 2014. Achieving high-quality single-atom nitrogen doping of graphene/sic (0001) by ion implantation and subsequent thermal stabilization. ACS Nano 8, 7318–7324.

Trambly de Laissardiere, G., Mayou, D., Magaud, L., 2010. Localization of Dirac electrons in rotated graphene bilayers. Nano Lett. 10, 804.

Triozon, F., Roche, S., Rubio, A., Mayou, D., 2004. Electrical transport in carbon nanotubes: role of disorder and helical symmetries. Phys. Rev. B 69, 121410.

Turek, I., 1988. A maximum-entropy approach to the density of states within the recursion method. J. Phys. C Solid State Phys. 21, 17.

van Wijk, M.M., Schuring, A., Katsnelson, M.I., Fasolino, A., 2014. Moiré patterns as a probe of interplanar interactions for graphene on h-BN. Phys. Rev. Lett. 113, 135504.

Wallace, P.R., 1947. The band theory of graphite. Phys. Rev. 71, 622.

Wallbank, J.R., Patel, A.A., Mucha-Kruczynski, M., Geim, A.K., Fal'ko, V.I., 2013. Generic miniband structure of graphene on a hexagonal substrate. Phys. Rev. B 87, 245408.

Wan, W., Li, H., Huang, H., Wong, S.L., Lv, L., Gao, Y., Wee, A.T.S., 2013. Incorporating isolated molybdenum (mo) atoms into bilayer epitaxial graphene on 4 h-sic (0001). ACS Nano 8, 970–976.

Wang, Z.F., Liu, F., Chou, M.Y., 2012a. Fractal Landau-level spectra in twisted bilayer graphene. Nano Lett. 12, 3833.

Wang, H., Maiyalagan, T., Wang, X., 2012b. Review on recent progress in nitrogen-doped graphene: synthesis, characterization, and its potential applications. ACS Cat. 2, 781–794.
Weiss, J.W., 1956. Diamagnetism of graphite. Phys. Rev. 104, 666.
Weiss, J.W., 1957. Band structure of graphite and de Haas-van Alphen effect. Phys. Rev. 108, 612.
Weisse, A., Wellein, G., Alvermann, A., Fehske, H., 2006. The kernel polynomial method. Rev. Mod. Phys. 78, 275.
Wong, D., Velasco Jr., J., Ju, L., Lee, J., Kahn, S., Tsai, H.Z., Germany, C., Taniguchi, T., Watanabe, K., Zettl, A., Wang, F., Crommie, M.F., 2015. Characterization and manipulation of individual defects in insulating hexagonal boron nitride using scanning tunnelling microscopy. Nat. Nanotechnol. 10, 949.
Woods, C.R., Britnell, L., Eckmann, A., Ma, R.S., Lu, J.C., Guo, H.M., Lin, X., Yu, G.L., Cao, Y., Gorbachev, R.V., Kretinin, A.V., Park, J., Ponomarenko, L.A., Katsnelson, M.I., Gornostyrev, Y.N., Watanabe, K., Taniguchi, T., Casiraghi, C., Gao, H.J., Geim, A.K., Novoselov, K.S., 2014. Commensurate/incommensurate transition in graphene on hexagonal boron nitride. Nat. Phys. 10, 451.
Zabet-Khosousi, A., Zhao, L., Palova, L., Hybertsen, M.S., Reichman, D.R., Pasupathy, A.N., Flynn, G.W., 2014. Segregation of sublattice domains in nitrogen doped graphene. J. Am. Chem. Soc. 136, 1391–1397.
Zak, J., 1964. Magnetic translation group. Phys. Rev. 134, A1602.
Zhang, Y., Tan, Y.-W., Stormer, H.L., Kim, P., 2005. Experimental observation of the quantum Hall effect and Berry's phase in graphene. Nature 438, 201–204.
Zhang, Y., Tang, T.T., Girit, C., Hao, Z., Martin, M.C., Zettl, A., Crommie, M.F., Shen, Y.R., Wang, F., 2009. Direct observation of a widely tunable bandgap in bilayer graphene. Nature 459, 820.
Zhao, L., He, R., Rim, K.T., Schiros, T., Kim, K.S., Zhou, H., Gutiérrez, C., Chockalingam, S., Arguello, C.J., Palova, L., 2011. Visualizing individual nitrogen dopants in monolayer graphene. Science 333, 999–1003.

CHAPTER THREE

2D Boron Nitride: Synthesis and Applications

G.R. Bhimanapati*,†, N.R. Glavin‡, J.A. Robinson*,†,1
*The Center for 2-Dimensional and Layered Materials, The Pennsylvania State University, University Park, PA, United States
†NSF I/UCRC Center for Atomically Thin Multifunctional Coatings, The Pennsylvania State University, University Park, PA, United States
‡Air Force Research Laboratory, Wright Patterson AFB, Dayton, OH, United States
[1]Corresponding author: e-mail address: jrobinson@psu.edu

Contents

1. INTRODUCTION

Boron nitride (BN) is a chemical compound that is isoelectronic and isostructural to carbon with equal composition of boron and nitrogen atoms. The first synthesis of boron nitride was in 1842 by Balmain (1842) using

Semiconductors and Semimetals, Volume 95
ISSN 0080-8784
http://dx.doi.org/10.1016/bs.semsem.2016.04.004

molten boric acid and potassium cyanide; however, stabilizing the material to form powders was a challenge until recently (Anon, 1958). Similar to carbon, boron nitride is produced in amorphous and crystalline forms. In its crystalline form, boron nitride exists in three major allotropes: hexagonal boron nitride (h-BN) resembling graphite (Fig. 1A), sphalerite boron nitride (β-BN) resembling cubic diamond, and wurtzite boron nitride (ɣ-BN) resembling hexagonal diamond form (Pakdel et al., 2014). Unlike the carbon fullerenes (C60 bulky balls), BN fullerenes have mostly squares or octagons instead of pentagons to avoid the thermodynamically unfavorable B–B and N–N bonding (Fig. 1D). Similar to 1D carbon nanotubes (CNTs), BN nanotubes (BNNTs) (Fig. 1C) also exist which are isoelectric to CNTs in terms of chirality, tube diameters, and number of walls. Out of these different phases, h-BN is the most common stable form of BN and most of the interest started after the isolation of graphene sheets in 2004 (Bhimanapati et al., 2015). h-BN is also a layered structure and within each layer, the boron and nitrogen atoms are bound by strong covalent bonds in-plane and each layer is held together by van der Waals forces. A single layer of h-BN is typically referred as a BN nanosheet or BNNS. This nomenclature is only true for h-BN sheets whose aspect ratio is small. For higher aspect ratio materials where the typical widths will be <50 nm, they are referred as BN nanoribbons or BNNR's.

Although h-BN has a similar structure to graphene, it is a wide bandgap material with an intrinsic band gap (E_g) of 5.9 eV when compared to the highly conductive graphene. It is also thermally conductive, which is attractive for many electronic applications since it can be used as electrically

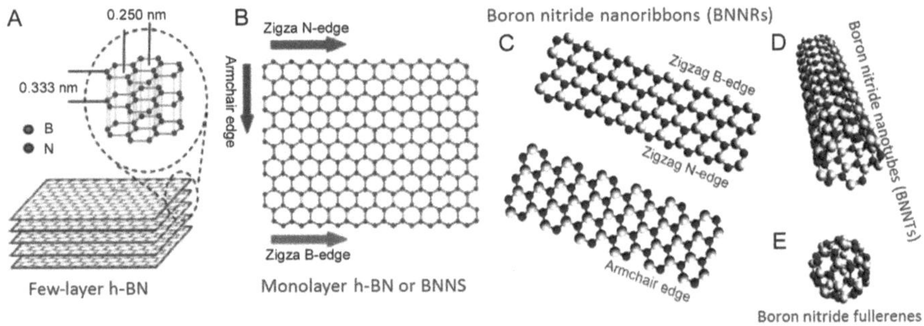

Fig. 1 Boron nitride allotropes: (A) few-layer hBN, (B) monolayer h-BN nanosheet or BNNS, (C) boron nitride nanoribbons (BNNRs) with different edge termination leading to zigzag B-edge and armchair edge structures, (D) boron nitride nanotubes (BNNT), and (E) BN fullerenes (Lin and Connell, 2012; Pakdel et al., 2014).

insulating filler material for polymer or ceramic composites (Ghosh et al., 2008; Yurdakul et al., 2012), thermal radiators (Song et al., 2012; Zhi et al., 2009), field emitters (Chen et al., 2008a), and UV emitters (Golberg et al., 2010; Zhi et al., 2005). In addition to its insulating properties, h-BN is chemically inert in a wide variety of acids, solvents, and oxidizers. h-BN is insoluble in the usual acids but is soluble in alkaline molten salts and nitrides such as LiOH, KOH, and Li_3BN_2. Owing to its high chemical resistance and thermal stability, h-BN is an attractive material for use as a chemically inert coating in hazardous environments (Liu et al., 2013; Rafiee et al., 2013; Zhi et al., 2009).

2. STRUCTURE AND PROPERTIES OF 2D BORON NITRIDE

2.1 Structual Properties

The physical form of h-BN is a white slippery powder, analogous to graphite. The flake size of commercially available h-BN ranges from hundreds of nanometers to tens of microns (information obtained from Sigma Aldrich, Alfa Aesar). Hence, the BNNS sheets obtained by exfoliation of these crystals are often limited to the maximum lateral sizes (few tens of microns) of the initial starting material (Bhimanapati et al., 2014). Individual h-BN monolayers or BNNSs are a honeycomb structure with alternating boron and nitrogen atoms. The bond type for B–N is covalent and bond length is 1.45 Å. The distance between the centers of neighboring borazine rings is 2.5 Å. The edge structure for the BNNS can be armchair or zigzag, similar to graphene. The former is a B or N-edged structure, whereas the latter is a BN pair-edged structure. The crystal structure for h-BN is hexagonal with $P6_3/mmc$ space group (Fig. 1A), lattice constants $a=b=0.2504$ nm, $c=0.6661$ nm, bond angles $\alpha=\beta=90°$, $\gamma=120°$. The partially ionic structure of BN in h-BN reduces covalency and electrical conductivity and unlike graphite, favors the AA′ stacking (Constantinescu et al., 2013). Typically, this is the most energetically favorable stacking observed in BN, where the electron-deficient B atoms are directly above or below the electron-rich N atoms in the adjacent layer (Marom et al., 2010; Warner et al., 2010). The most common way to determine the number of layers in h-BN is by looking at the transmission electron microscopy (TEM) images of the folded edges (Bresnehan et al., 2014; Han et al., 2008; Kim et al., 2012a; Lin et al., 2010a; Pacilé et al., 2008; Shi et al., 2010; Song et al., 2010; Zhi et al., 2009). Fig. 2A and B shows the high-resolution TEM images, which typically gives the number of layers along with more

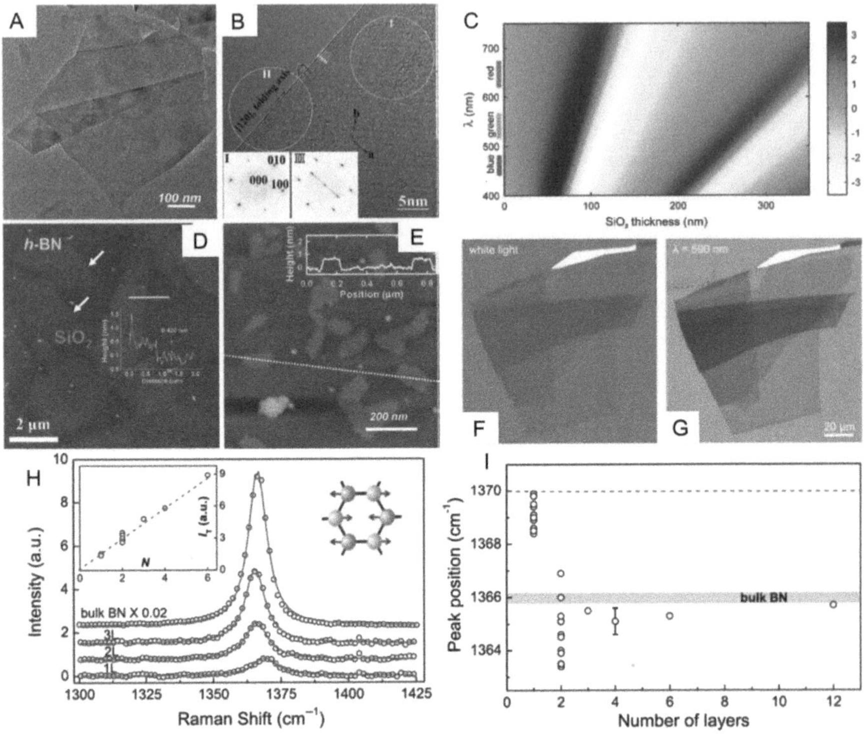

Fig. 2 (A) Low-magnification TEM image of an exfoliated BNNS. (B) HR-TEM image showing the number of layers at the folded edges (Lin et al., 2010a). (C and D) AFM height map of CVD grown BNNS showing the monolayer thickness of 0.4 nm. (E) Chemically exfoliated BNNS showing a height of 1 nm. (F) White light microscopy image of exfoliated h-BN flake showing the optical contrast on 90 nm SiO_2/Si. (G) Improved optical image of h-BN flake using a 590 nm light source showing the different contrast between layers (Gorbachev et al., 2011). (H) Raman spectra for BNNS to few-layer h-BN comparing the difference in the peak width vs number of layers. (I) Peak position variation as observed for monolayer to bulk BNNS (Gorbachev et al., 2011).

atomic lattice information. Another simple technique to determine the thickness of h-BN is by using atomic force microscopy (AFM) (Bhimanapati et al., 2014; Gorbachev et al., 2011; Lin et al., 2010a; Pacilé et al., 2008; Shi et al., 2010; Song et al., 2010). The number of layers can be calculated by looking at the step height typically with respect to the surface (Fig. 2D and E). Most monolayer h-BN sheets from mechanical exfoliation and chemical vapor deposition (CVD) have a height of ~0.4 nm (Oliveira et al., 2012). For the chemical exfoliation sheets, the height obtained can vary as much as 1 nm because of the solvent that can be trapped between the h-BN flakes and the substrate (Lin et al., 2011a). Hence, HRTEM is usually used to report the number of layers and when reporting the height through AFM, more characterization techniques should be used to confirm the findings.

Another simple and convenient technique using an optical microscope was developed to observe the number of layers of graphene sheets and can be employed for h-BN as well (Blake and Hill, 2007). Usually, a standard silicon wafer coated with ~300 nm of an SiO_2 layer (Fig. 2C) is used as a reference substrate and the number of layers is predicted based on the optical contrast that was observed using the optical microscope. In the case of h-BN, as it does not absorb in visible region, the substrate oxide layer had to be changed to 80 nm in order to optically identify the mono to few-layered materials. As shown in Fig. 2F, the white light contrast reaches a low of 2.5% for monolayer h-BN, which is four times less than that of graphene, and the contrast increases with the increase in the number of layers (Hod, 2012). Another way to identify the number of layers is by using a different wavelength light, which can be observed in Fig. 2C. By using a light as shown in Fig. 2G with ~590 nm wavelength, the contrast between the number of layers was clearly evident.

This layer identification study can further be extended to Raman microscopy (Gorbachev et al., 2011). Boron nitride has a vibrational mode at 1364–1371 cm^{-1}, which is dependent on the number of layers. Fig. 2H and I shows the comparative study made by Gorbachev et al. for different layers of BNNS. It can be observed in Fig. 2H that for monolayer, the Raman peak broadens and shifts to a higher wavenumber of 1370 cm^{-1} when compared to the bulk, which is at 1366 eV. This study clearly shows the red shift for BNNS formation, which could later be used as a standard to confirm the presence of BNNS.

2.2 Thermal Stability

BN allotropes, especially h-BN and the cubic phase of boron nitride (c-BN), exhibit high thermal and chemical stability. h-BN is stable without decomposing at temperatures of over 1000°C in air, 1400°C in vacuum, and up to 2850°C in an inert atmosphere (Paine and Narula, 1990; Sinclair and Simmons, 1987). The theoretical thermal conductivity values for h-BN can be close to graphene (~1700–2000 W/m K) (Jiang et al., 2009; Lan et al., 2009), one of the best thermally conductive materials available to date (Moore et al., 1992). Depending on the type of BNNR structure, the zigzag-edged BNNRs are 20% larger than the arm chair-edged nanoribbons at room temperature (Ouyang et al., 2010). In-plane thermal conductivity has been reported as high as 390 W/m K at room temperature, 280 times higher than SiO_2, making h-BN an attractive dielectric material for heat-generating electronic devices (Ouyang et al., 2010). Similar to graphite, h-BN has a strongly anisotropic coefficient of thermal expansion (CTE) due to its anisotropic bond strength. The CTE in the *a*-direction (in-plane) is $-2.90\times10^{-6}\ K^{-1}$ at room temperature, while the CTE in the *c*-direction is over 10 times larger and $4.05\times10^{-5}\ K^{-1}$ at room temperature (Yates et al., 1975). The large positive thermal expansion in the *c*-direction is due to weak van der Waals bonding between planes.

The most commonly accepted phase diagram of boron nitride was calculated from thermodynamic properties of boron nitride phases by Solozhenko et al. (1999) after refining their original work from 1988. It was found that the c-BN, rather than h-BN, is thermodynamically stable at ambient conditions (Solozhenko et al., 1999). This is in contrast to the carbon phase diagram, where the hexagonal phase (graphite) is the stable phase at ambient conditions (Bundy, 1963). The phase diagram of boron nitride is shown in Fig. 3A and B, where the dashed lines indicate the original calculations by Solozhenko et al. in 1988, and the solid lines indicate the refined diagram. The h-BN/c-BN/liquid triple point occurs at 3480 ± 10 K and 5.9 ± 0.1 GPa, while the h-BN/liquid/vapor triple point occurs at 3400 ± 20 K and 400 ± 20 Pa (Solozhenko et al., 1999).

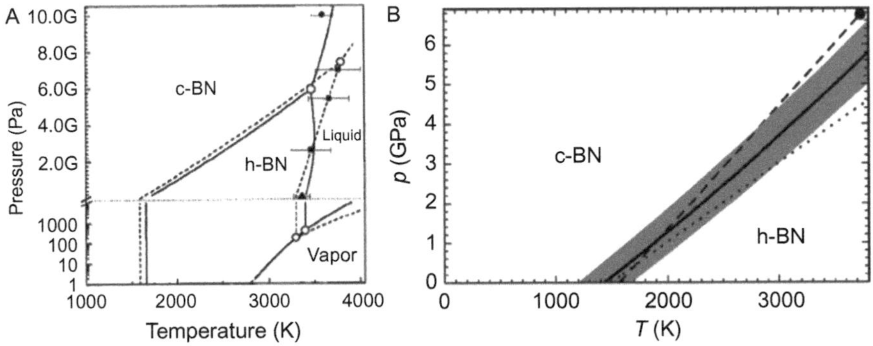

Fig. 3 (A) Phase diagram of h-BN showing the stability of h-BN vs c-BN at various temperatures and pressures. (B) Close up image of the phase transition and calculations from Solozenko (*dotted line*) (Solozhenko et al., 1999).

From the above phase diagram, it is evident that c-BN is more favorable than the h-BN at temperatures below 1600 K. Although this is true, the CVD growth of h-BN, which will be discussed in Section 3.6, was reported on transition metals at much lower temperatures. The typical growth temperatures for CVD are around 700–1100°C, which are at least 300°C cooler than that observed in the general phase diagram. This unusual behavior can be because of the Gibbs free energy of the system. The transition temperature between h-BN and c-BN shifts as a function of Gibbs free energy (Shift ±10 meV/atom). This change in the free energy can vary the transition temperature between 1200 and 1800 K, where the lowest value is typically used for the growth of h-BN. These small variations are generally caused by grain size, defects, contaminants, or sometimes interactions with the substrate transition metal itself. Further details about the growth of h-BN is discussed in Section 3.6.

3. SYNTHESIS OF BNNS

BNNS can be obtained via top-down (typical exfoliation-type approaches) or bottom-up approaches (usually CVD or other deposition techniques). The most common methods that have been used are mechanical exfoliation (Chen et al., 2012; Ghosh et al., 2008; Hua Li et al., 2012; Lee et al., 2015; Li et al., 2011a), chemical exfoliation (Anota et al., 2015; Bhimanapati et al., 2014; Cui et al., 2014; Lee et al., 2015; Li et al., 2013; Lin et al., 2010a; Paton et al., 2014; Roy et al., 2014a; Sainsbury et al., 2012, 2014; Wang et al., 2011a; Warner et al., 2010), CVD (Bresnehan et al., 2014; Kim et al., 2012a,b, 2013; Lee et al., 2012a,b; Shi et al., 2010; Song et al., 2010), and pulsed laser deposition (PLD) (Glavin et al., 2014). Most of these methods have their own advantages and disadvantages, as they are used for specific targeted applications. BNNS isolated from exfoliation techniques are usually crystalline, although obtaining a high percentage of exfoliated sheets is very difficult. Using these exfoliation techniques, obtaining a higher density of monolayer flakes is really difficult and time consuming. Also, because of the strong lip–lip interactions between the layers, there is high probability of layer stacking on top of each other. In contrast, CVD techniques use the breakdown of precursors and results in a highly controllable process of making BNNS. Typically, the crystallinity obtained via CVD is less than that when compared with exfoliation techniques.

3.1 Mechanical Exfoliation

Mechanical exfoliation, also known as the "scotch tape method," or micromechanical cleavage method, uses scotch tape as a very simple tool for exfoliation (Novoselov et al., 2004). Using this technique, isolation of layers can be controlled down to a monolayer and the flake size depends on the biggest flake available for exfoliation. The final BNNS obtained via this technique retains the original lateral size and low defect density as the parent h-BN crystal used. As a variety of nanosheets can be easily obtained with this technique, fundamental studies in physics and electronics have been done mostly using the materials obtained via exfoliation (Alem et al., 2009; Gorbachev et al., 2011; Jin et al., 2009; Meyer et al., 2009; Novoselov et al., 2005; Pacilé et al., 2008). Compared to other exfoliation methods, the mechanical exfoliation method is the least scalable technique, especially for large-scale applications. Other reports using mechanical cleavage methods beyond simple mechanical exfoliation have been explored to produce large quantities of BNNS. As these materials are bonded by weak van der Waals forces, techniques with shear forces like a ball mill technique were used (Hua Li et al., 2012; Li et al., 2011a). Traditional ball mill techniques are very aggressive, nonselective, and yielded significant number of defects in the exfoliated material (Ghosh et al., 2008; Lee et al., 2015; Li et al., 2011b; Lin et al., 2010b). Although this technique yields a better output than mechanical exfoliation, the defect density in the final exfoliated films is higher. Further improvements to the ball milling technique were also used by using a solvent along with the ball milling process (Fig. 4A). Li et al. reported that using a mild wet ball milling process (Li et al., 2011b), few-layered crystalline BNNS with only a slight

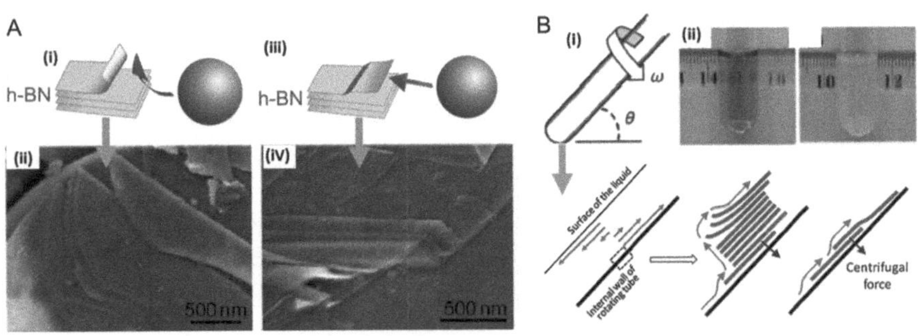

Fig. 4 (A) (i and iii) Wet ball mill process used for the mechanical exfoliation process (Li et al., 2011b), (ii and iv) SEM images showing the exfoliation process mechanism. (B) (i) Mechanism of vortex fluidic process (Chen et al., 2012) which was operated at 8000 rpm. This process works on centrifugal force for exfoliation, which can be observed in the figure and (ii) image showing the uniform dispersion before and after the process.

lateral size reduction was obtained. Various solvents have been tested in wet ball milling. Benzyl benzoate was an effective lubricant because of its weight, which provided better exfoliation results than water, ethanol, or dodecane (Deepika et al., 2014). As selection of the solvent is important for exfoliation, which can be obtained by looking at the weight of the solvent and the surface tension properties of the solvent and h-BN. Further, use of planetary mill instead of the more aggressive high-energy mills provide uniform rolling motions of the balls applying predominantly shear force to the h-BN flakes (Hua Li et al., 2012; Yao et al., 2012). Further, use of smaller diameter balls instead of the heavier balls added additional lubrication for the interaction, which reduced the loss of the BNNS sheet size.

Another setup that was used for this type of exfoliation is the vortex fluidic device (Chen et al., 2012). As it can be observed in Fig. 4B, a small amount of h-BN suspension in *N*-methyl-2-pyrrolidone (NMP) was rotated at 8000 rpm in a glass tube with a fixed angle of the fork which resulted in significant exfoliation of the h-BN crystals. The presence of NMP formed a thin layer on the inner walls of the centrifuge tube and the rotating shaft, which at an angle, provides enough shear force to partially exfoliate the layers in the liquid mixture. Further, a more scalable approach was proposed based on this technique (Fig. 5A). Here, a high-pressure microfluidization process (Yurdakul et al., 2012) was used in a continuous

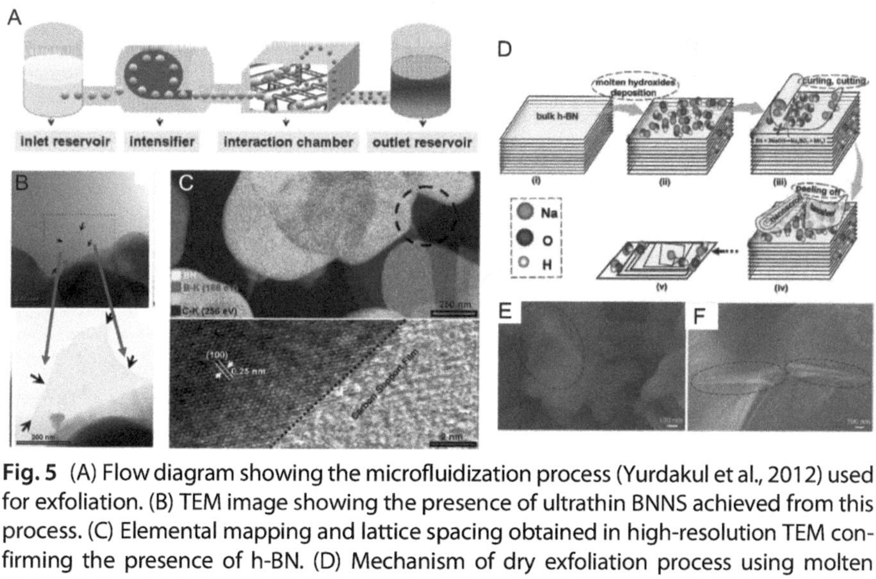

Fig. 5 (A) Flow diagram showing the microfluidization process (Yurdakul et al., 2012) used for exfoliation. (B) TEM image showing the presence of ultrathin BNNS achieved from this process. (C) Elemental mapping and lattice spacing obtained in high-resolution TEM confirming the presence of h-BN. (D) Mechanism of dry exfoliation process using molten hydroxides (Li et al., 2013) (NaOH, KOH). (E and F) SEM images showing the curling of h-BN sheets and formation of nanoscrolls of h-BN.

flow setup by using h-BN flakes and a polar organic solvent which was pumped through a microfluidic processor at a high pressure of 207 MPa, circulated continuously for multiple times. At these high pressures, the liquid generates a large shear force, which results in exfoliation of h-BN particles with a reported yield efficiency of 45%. High-resolution TEM mapping clearly showed that the final exfoliated material was h-BN without any functional groups and defects.

Dry mechanical process can also be used for the exfoliation process (Li et al., 2013), which is achieved by using molten hydroxides. The most common chemicals used in this case are sodium hydroxide (NaOH) and potassium hydroxide (KOH). This process involves that the hydroxide and the h-BN powder was ground together then transferred into a polytetrafluoroethylene-lined stainless steel autoclave where the reaction takes place at 180°C for 2 h. This process is used for making nanoscrolls of BNNS. The exfoliation process follows a sequence of steps that are shown in Fig. 5D. The first step involves the self-curling of the h-BN sheets at the edges after the molten hydroxide is attached on the surface layer. Subsequently, the anions and the cations enter the interlayer space and the adsorbed anion (OH^-) radicals start curling process in the adsorbed layer and finally, the peeling away of the parent material is observed as a result of this curling process. As this process is easy and involves low-cost method, it can be scalable. The final product obtained can be transferred to any solvents or to any substrates.

3.2 Solvent-Assisted Ultrasonication

Sonication-assisted exfoliation has been a common approach and has been used widely across various 2D material systems. The general idea in this process is to disperse the boron nitride in a solvent and with the aid of sonication power, the sample tends to exfoliate because of the energy that is generated by the sonication process. Surfactants can also be added to improve the dispersion of the nanoparticles.

In order to understand the dispersion ability for 2D materials, Coleman et al. (2011) and Shen et al. (2015) attempted to understand the direct solvent dispersion of 2D materials by using Hansen solubility parameter theory. They postulated that in order to select a good solvent or a mixture of solvents for direct dispersion enthalpy of mixing must be minimized, and hence the energy of the exfoliation is minimized. This means that the solvent should possess a similar surface energy as that of the nanosheets (Fig. 6A). In the case

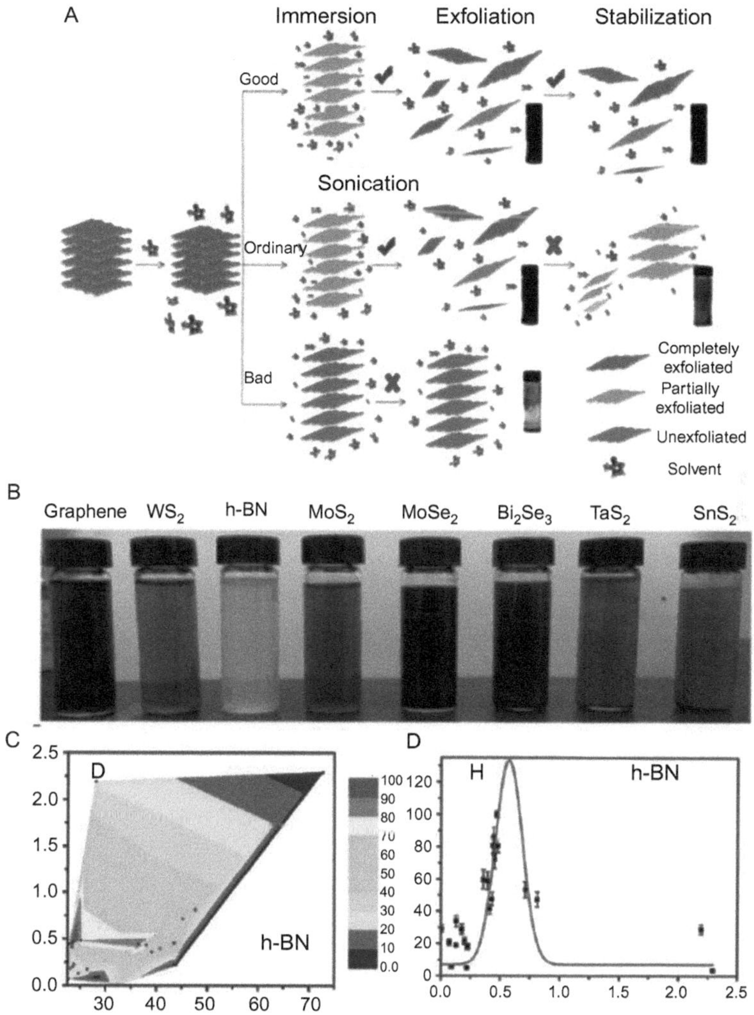

Fig. 6 (A) Proposed mechanism for liquid-phase exfoliation by matching the solvent surface tension with the 2D material (Shen et al., 2015). (B) Image showing the dispersion of various solvents. (C and D) Dispersion of h-BN as a function of total surface tension and polar dispersive ratio (panel D is fitted with a Lorentzian peak function).

of boron nitride, a surface energy of around 65 mJ/m^2 was proposed to be ideal and because of this, BNNS usually were formed in solvents such as isopropylalcohol (IPA) or NMP. This theory is true for most of the 2D materials (Fig. 6B) where the surface energy ~70 mJ/m^2 or a surface tension of ~40 mJ/m^2 would provide the best possible results (Coleman et al., 2011). The dispersion ratio for h-BN with varying concentrations is listed in

Fig. 6C and D. Certain mixture of solvents based on ethanol and water were also used which showed improved exfoliation and dispersion than when used individually.

Although this model holds for most solvents, certain solvents tend to deviate from this behavior because they tend to interact with the nanomaterials much more strongly and can even chemically react with the material. One good example is water. Although water does not wet h-BN at room temperature and was proposed to be hydrophobic, it was found to be counter intuitive when reports claimed to show high dispersion of BNNS in water (Li et al., 2013; Lin et al., 2011a). Also, it was shown that BNNS dispersion contains a very high fraction of exfoliated sheets (~20%) than that of the other solvents. This unusual behavior can be accounted to the reactivity of h-BN, which could yield boron oxide and ammonia when it was treated in hydrolysis conditions. The mechanism of exfoliation can be observed in Fig. 7A. Hence, after the exfoliation process, a small amount of

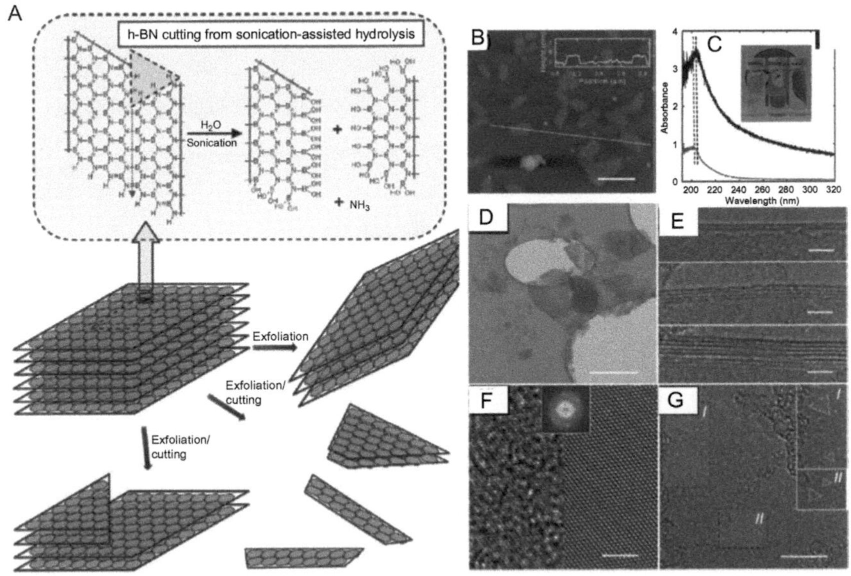

Fig. 7 (A) Image showing the sonicated assisted hydrolysis of h-BN to form BNNS (Lin et al., 2011a). Using a high polar solvent, exfoliation and cutting of the nanosheets are possible. (B) Image showing the BNNS flakes obtained after the exfoliation process with a thickness of 1 nm. (C) Optical absorption of BNNS suspended in solvents. (D and E) TEM images showing the BNNS flakes obtained and by looking at the edges, different thickness BNNS are obtained. (F and G) HRTEM image showing the monolayer BNNS structure and the defects in the material.

boron oxide and dissolved ammonia was found in the solvent and OH functionality was present on the surface of the h-BN. This presence of OH functionality on the surface further yielded the additional stability and dispersibility in water (Fig. 7B–G).

3.3 Acid Exfoliation

Boron nitride is typically neutral when reacting to acids. General acids do not show a high degree of exfoliation, although using a high protic acid such as (Wang et al., 2011b), methyl sulfonic acid (MSA) exfoliation was much more successful. Concentrations of up to 0.3 mg/mL were obtained when bulk BN was dispersed and sonicated in MSA, which was comparable to water or other solvents. Similar to the graphite exfoliation by hummer's process, BNNSs are also protonated at the edges and surface during this process, which lead to repulsive forces between the layers. This was further supported by looking at the color change that occurred after the reaction. The BNNS dispersed in MSA turned orange which could be because of the charge transfer between acid molecules and nanosheets. This color was later changed to milky white when redispersed back into other organic solvents. Further characterization using FTIR and X-ray photoelectron spectroscopy (XPS) showed that there was no presence of additional bonding which indicated that the BNNS was not functionalized (Wang et al., 2011b). Another approach using modified hummer's process was made by Bhimanapati et al. (2014) for exfoliation of h-BN. Hummer's process did not affect the boron nitride although, tampering the temperature and time of reaction in the hummers process provided enough time for the permanganate ions and the acid solution to intercalate between the h-BN sheets thereby forming BNNS. The BNNS could be later dispersed in various solvents still retaining the functional groups.

3.4 Chemical Functionalization of h-BN

Utilizing chemical functionalization for exfoliation has proved effective for the case of graphite allotropes, especially CNTs (Cui et al., 2010; Kuila et al., 2012). Similar strategies have been used in the case of BN to achieve exfoliation via functionalization of the surface. Typically, BN is resistant to wet and dry oxidation, although 2D and 1D BN are ionically active to functionalization due to the local polarization of B–N bonding. A simple way to functionalize BNNS includes heating h-BN in air to a high temperature (~800°C) (Cui et al., 2014). Heating h-BN in air leads to the incorporation of small amounts of oxygen into the lattice of the h-BN (Cui et al., 2014), which is followed by stirring in DI water to incorporate a B–O functionality. The stirring mechanism leads to the formation of the functional groups that result in separation of the h-BN layers into BNNSs. As this process involves heating and dispersion in water, it could be termed as a similar approach to sonication in DI water process. The yield obtained through this process is small (~2–4%). In order to completely functionalize the BNNS, the functionalization approaches can be divided into three major categories: noncovalent type functionalization, ionic functionalization or Lewis acid–base functionalization, and covalent functionalization.

3.4.1 Noncovalent Functionalization

Noncovalent functionalization is typically obtained by using a polymer or a copolymer. These polymers typically tend to functionalize via the π–π interactions between the polymer and the h-BN sheets. The polarity of the functional molecule plays a major role in its optimal orientation on the h-BN sheets (Gao et al., 2014; Grimme, 2011; Grimme et al., 2007). The lowest energy molecular interactions will be governed by the electrostatic interactions between the functional polymer and the boron nitride sheets. Several polymers such as poly(*m*-phenylenevinylene), 2,5-dioctoxy-*p*-phenylenevinylene, tetracyanoquinodimethane, and tetrathiafulvalene were used to study the functionalization mechanism. Recently, BNNS were noncovalently functionalized using polyvinyl alcohol (PVA) and later attached to a paper with natural nacre (Zeng et al., 2015) (Fig. 8A). This noncovalent functionalized approach drastically improved the mechanical and thermal performance of the natural nacre, which were also proposed to be used as substrates for electronics. Also, several transition metal arene complexes were also predicted to noncovalently functionalize h-BN. It was predicted from theory that the stable configuration for the transition metal arenes to functionalize is by having a metal atom at the center of the borazine rings. Most of the time after functionalization, band gap reduction and spin polarization are observed in h-BN, indicating the success of functionalization. Further ab initio studies showed that h-BN can also be attached with DNA and RNA molecules (Ding et al., 2013; Lin et al., 2011b) (Fig. 8B), where the N and O atoms of the nucleobases are located above the B atoms. Further, the nucleobases prefer to attach facially rather than perpendicularly because the noncovalent interaction is higher than the π–π interactions. This could lead to a better biomedical applications using h-BN.

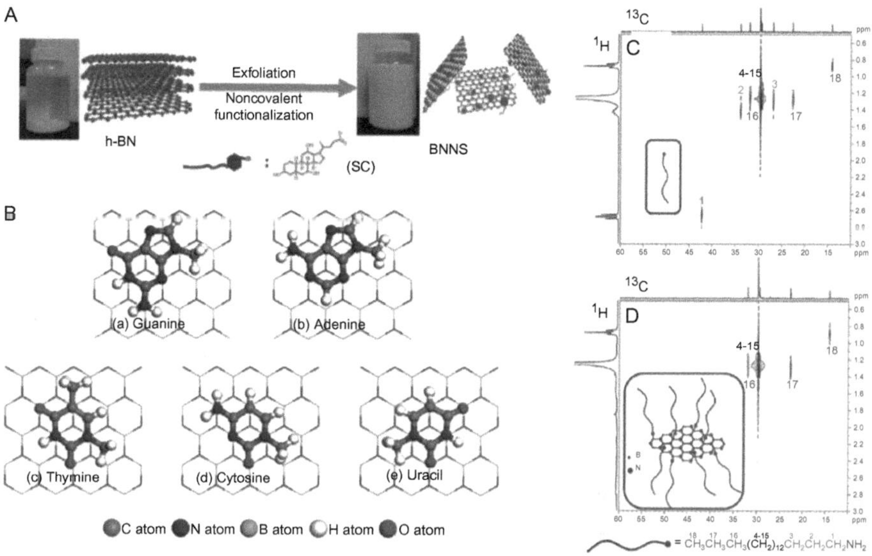

Fig. 8 (A) Noncovalent functionalization of BNNS using polyvinyl alcohol (PVA) which improves the dispersion of the BNNS in solvents, used in thermal and mechanical composites (Zeng et al., 2015). (B) Ab initio calculations for optimized structure configurations for different nucleobases adsorbed on BNNS for bioapplications (Lin et al., 2011b). (C) Amine functional group addition on the BNNS and NMR spectra confirming the presence of ODA-BN functionalization (Lin et al., 2010a). (D) Dibromocarbene (DBC) addition on BNNS via aqueous heating in sodium hydroxide (Sainsbury et al., 2014).

3.4.2 Ionic Functionalization (or Lewis Acid–Base Interactions)

The boron atom in the h-BN lattice is electron deficient and possesses Lewis acid characteristics, making it susceptible to attack by a Lewis base molecule such as amines or phosphines to form stable Lewis acid–base complexes. Although there are reports on the boron atom being ionically functionalized (Lin et al., 2010a; Xie et al., 2005; Yu et al., 2012), there was no experimental reports on nitrogen atom although it is electrophilic in nature. Because of the B–N bonding being locally polarized, in theory it should be a convenient target for both the atoms to get ionically functionalized.

Several attempts have been made to demonstrate the Lewis acid–base interactions on h-BN using amino groups of octadecylamine (ODA), amine-terminated oligomeric polyethylene glycol, trioctylamine, and trioctylphosphine (Lin et al., 2010a,b). The lipophilic and hydrophilic chains of these molecules intercalated into the h-BN layers and functionalized the sheets which were later extracted by dispersing in water or organic solvents. The final dispersed material was still transparent for lower concentrations and milky white for higher concentrations. Lateral sizes of ~1 μm and thickness of about 1–7 nm were obtained via this process. Trace amounts of monolayer materials were also obtained, but the lateral sizes were <100 nm. To further improve the efficiency of functionalization, defects were created on h-BN sheets by ball milling the h-BN powder (Lin et al., 2010b). These defects were confirmed by X-ray diffraction and Raman as the (002) peak and the in-plane mode (E_{2g} peak) broaden when defects are introduced. After defects were introduced, there was better solubility after the functionalization. This was further confirmed using NMR spectroscopy which is observed in Fig. 8C and D (Lin et al., 2010a; Yu et al., 2012). The alkyl radicals on the amino-terminated region were observed for its C–H signals, which were reduced from the parent material, indicating that this end was attached to the h-BN. Further studies on the defect h-BN starting material showed a shifting position of the same C–H signals of the ODA functional group. This indicated that the amino end of the ODA molecule that was attached to the B atoms and the boron sites contained the defects.

3.4.3 Covalent Functionalization

Direct covalent functionalization is not possible, as BN is very chemically stable material. Several functional groups have been added covalently onto h-BN sheets and typically adhere to the edges and basal planes of the material. In order to attach the functionalization and exfoliation via covalent functionalization, several treatments have to be made. Recent reports showed that Hummer's process with modifications in the temperature worked effectively for creating oxygen bonding on the boron sites. The confirmation of the bonding was observed through XPS and FTIR where B–O bond was present. Other approaches were also made by attaching hydroxyl groups, a nitrene (methylphenylcarbonate) via a thermal oxidation and treatment with strong oxidants. These were mostly used in the applications for polymer composites which improved the performance of the composite. For specific targeted applications, silver nanoparticles were further decorated after hydroxyl functionalization on the BN sheets (Fig. 9A and B) (Roy et al., 2014a). Coleman et al. also functionalized BNNS using dibromocarbene (DBC). In their work, they used BNNS as a 2D phase transfer catalyst for the carbine migration across the organic–aqueous phase boundary. The BNNS formed B–CBr_2 bonding in order to stabilize the carbenes which also act as the reactive substrate for the functionalization. The functionalization reaction is shown in Fig. 9D. This functionalization resulted in the formation of B–C and B–N bonds on the BNNS lattice by forming dibromo-bridged bicycle BCN linkage. This covalent chemistry

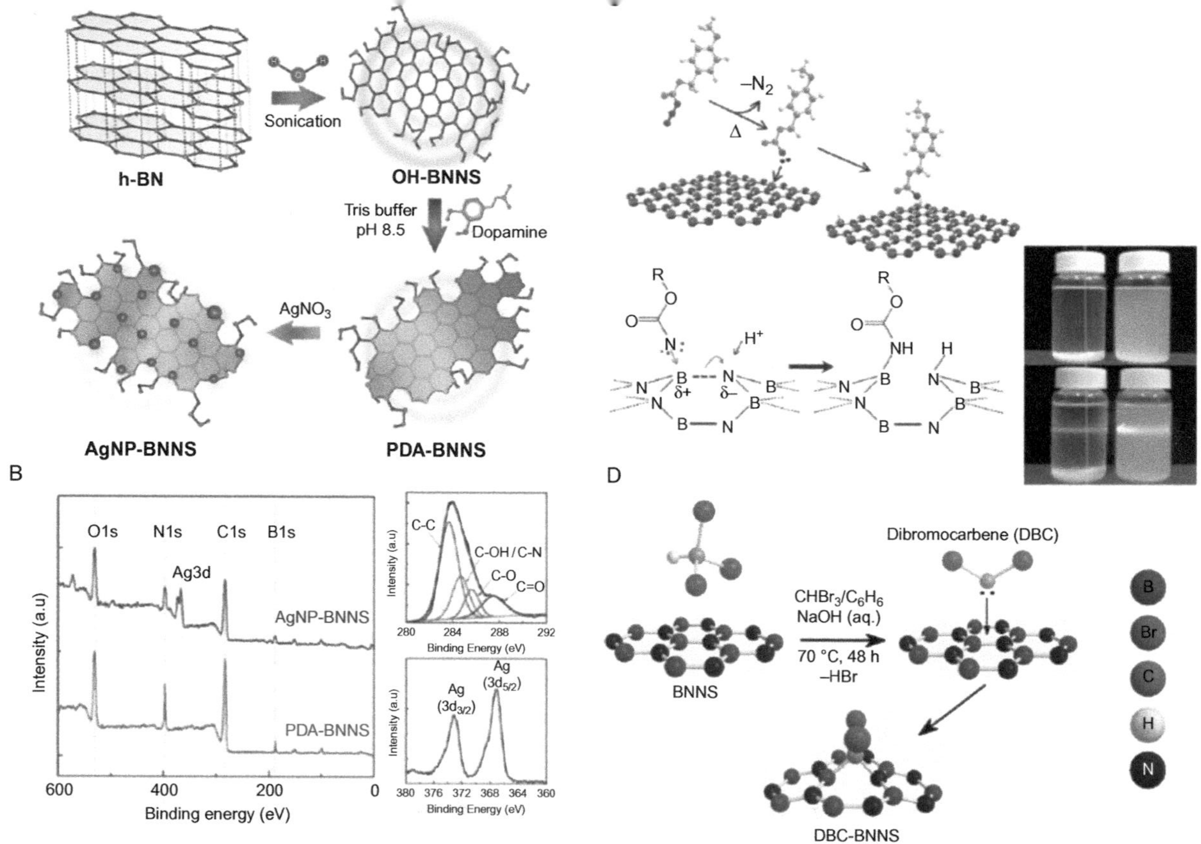

Fig. 9 Covalent functionalization of h-BN nanosheets using different functional groups. (A) Hydroxyl functionalization of the BNNS sheets and later decorating it with silver nanoparticles (Roy et al., 2014a). (B) XPS spectra confirming the presence of the functionalization and the decoration of Ag on BNNS. (C) Nitrene addition on BNNS, covalently bonded on to the phase and edge sites (Sainsbury et al., 2012) (inset image showing the dispersion and absorption change in the BNNS before and after functionalization). (D) Dibromocarbene (DBC) addition on BNNS via aqueous heating in sodium hydroxide (Sainsbury et al., 2014).

was confirmed using multiple different characterization techniques. This CBr_2 linkage on the BNNS was later used to interface with solvents, which can later be integrated in polymer composites easily (Sainsbury et al., 2014). Further, similar work was extended to functionalize the BNNS by nitrene addition (Sainsbury et al., 2012). 4-Methoxybenzyloxycarbonylazide was used as the source to functionalize it. This type of functionalization mostly attacks the B atoms in the h-BN lattice, resulting in breaking of B–N bonds in the structure. This reaction changes the milky white color of BNNS to dark brown generating N_2 bubbles and creating reactive nitrene radicals. This later yields the formation of methoxylphenyl carbomate bound to BNNS via the B–N bond as shown in Fig. 9C. Most of the functionalization approaches result in highly dispersed BNNS in solvents which can later be integrated into many different applications. Typically, most of these processes were used to make polymer composites to improve thermal and mechanical strength of the polymers.

3.5 Unzipping of BNNTs

Another method to synthesize h-BN sheets is by unzipping BNNTs. Various approaches have been utilized to unzip BNNTs. The process of obtaining BNNS through unzipping of BNNTs is typically used to study the properties of the edges that are obtained through this process. One interesting property to note is that BNNS obtained from BNNTs are only ~15 nm wide and semiconducting because of the small dimensions.

The first reported approach described first dispersing the BNNTs in polymers, followed by etching the polymer away (Zeng et al., 2010). In this case, the BNNTs need to be well dispersed in solution and dried on a substrate—typically Si or SiO_2. This unzipping process can be observed in Fig. 10A. A polymer solution (PMMA) is then spin coated on the silicon substrate until the polymer matrix is formed on the surface. It is subsequently peeled off from the substrate leaving a BNNT encapsulated polymer film. Finally, the BNNT is subjected to etching via an inert gas or plasma treatment. The etching process takes place until a part of the BNNT is cut leaving us with layers of BNNS (Fig. 10A). This process can be treated as a controllable way to synthesize h-BN nanosheets as most of the samples are protected by the polymer matrix and only a selected few area is etched away. The etched BNNTs are then dispersed in solvents and sonicated to obtain individual sheets of h-BN. Further confirmation of the unzipping process was observed using TEM (Fig. 10B and C).

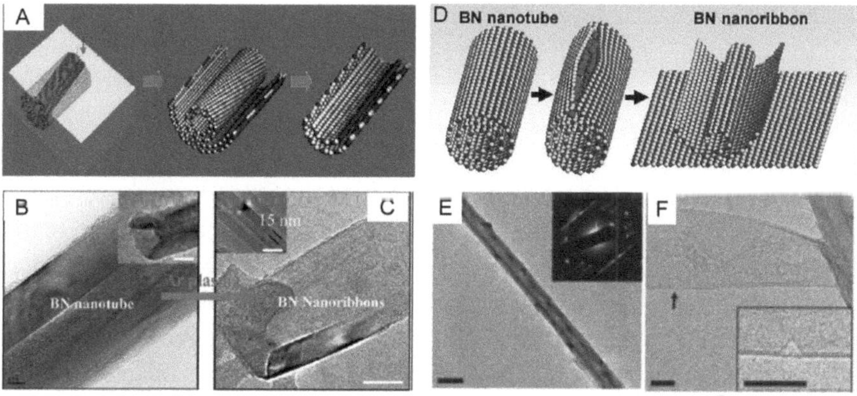

Fig. 10 (A) Unfolding mechanism of BNNTs encapsulated in a polymer matrix (Zeng et al., 2010). (B) TEM image showing the BNNT before exfoliation and (C) TEM showing the BNNR obtained after the controlled plasma etching of BNNTs. (D) Alkali salt intercalation mechanism of BNNT where the alkali salt is attached on the lateral side of the BNNT and the exfoliation along the longitudinal side (Erickson et al., 2011). (E and F) TEM images showing the exfoliation of BNNT to BNNR after the alkali intercalation and washing the salt in water.

Another approach to unzipping BNNTs is by using an alkali metal intercalant (Erickson et al., 2011), as shown in Fig. 10D. This process is more difficult than that of graphite, as h-BN is less vulnerable to intercalants. Erickson et al. (2011) vaporized potassium metal into the walls of the nanotubes for intercalation as shown in Fig. 10E. This process was performed at 300°C for almost 72 h in vacuum and is efficient for inducing longitudinal splitting of BNNTs. Although this process used an external deposition of metal to break the sheets, sheets over 1-μm long with widths up to 50 nm were obtained (Fig. 10F). The interplanar spacing for BNNS intercalated with potassium was found to be 5.8 Å (Okada and Otani, 2010). This significant increase in the wall spacing for BNNT (from ~3.33 Å) will result in circumferential strain on the bonds. This strain later breaks the bonds at the weakest points, resulting in exfoliation. This technique suffers from low yield (~1%) and this is a not a bulk process. The intercalated potassium atoms will form islands on the surface which continuously grew inducing circumferential strain around the nanotube surface which led to bond breakage at the weakest points. As this is a slow process, most of the etched nanosheets were still attached to the parent nanotubes, which were then removed later by sonication in solvents such as IPA. Using this process, the nanoribbons obtained will inherit the same structural properties as that of the parent nanotubes (Kosynkin et al., 2011). This process could potentially be scalable approach to make nanoribbons of h-BN by looking into the cleaning process after the reaction.

3.6 Single and Few-Layer h-BN via CVD

Nucleation and growth methodologies for high-quality mono- and few-layer h-BN films, as well as many other 2D material counterparts, have been of significant interest within the past decade. The necessity for direct growth, in contrast to the previously discussed exfoliation techniques, is driven by the desire for large area film coverage with strict control of layered thickness as well as eliminating the difficulties associated with the lift-off and transfer processes seen in many exfoliation variations (Bresnehan et al., 2013; Geim and Grigorieva, 2013). To date, the progress in growth of 2D insulators including h-BN has significantly lagged behind comparable conductors and semiconductors (eg, graphene, transition metal dichalcogenides, and other 2D material systems), as the difficulty in materials processing presents many challenges for device usage (Osada and Sasaki, 2012). These difficulties include lack of large area growth required for electronic devices, compatibilities with substrate materials, and required high-deposition temperatures (typically >1000°C), each that will be discussed in following sections.

Even as the overall interest in 2D materials has increased substantially since the isolation and initial studies at the University of Manchester (Novoselov et al., 2004), monolayer and few-layer h-BN has been directly grown in small domains since the mid-1990s. One of these first isolations of monolayer h-BN was made possible by Nagashima (Nagashima et al., 1996; Rokuta et al., 1997) and was performed by thermal decomposition of borazene ($B_3N_3H_6$) at 700–800°C on metallic substrates including nickel (Ni(111)), palladium (Pd(111)), and platinum (Pt(111)), which continue to be common substrates for h-BN growth. Initial studies using EELS and XPS led to key observations including the realization of very low interaction energy between h-BN and the substrate, as well as the extremely slow growth rate for the second layer of h-BN from a lack of reactivity on the surface of the monolayer islands. These studies were performed with limited interest at the time, but would later influence the 2D materials community substantially nearly a decade later.

Once the exciting discovery of graphene ignited the research in the topic of 2D materials in the mid-2000s, CVD become the most prominent growth technique used for single and few-layer 2D materials, including h-BN. Demonstration of large area growth of few-layer h-BN by atmospheric pressure CVD (APCVD) in early 2010 yielded highly uniform films on copper foil with thicknesses of a few atomic layers (Shi et al., 2010; Song et al., 2010). The precursor reactants in these initial studies included ammonia borane (NH_3-BH_3), a solid powder requiring sublimation at temperatures of 120–130°C, as well as a gaseous precursor borazene, also used in initial studies by Nagashima (Nagashima et al., 1996). Both studies were performed at substrate temperatures of 600–700°C for 20–30 min of growth in an Ar/H_2 gas mixture, followed by a postannealing step at 1000°C to further crystallize the multilayer film. The similarities in film growth and structure for initial efforts in CVD h-BN coatings are represented in Fig. 11, where Fig. 11A–C shows an optical image of the few-layer h-BN film, high-resolution "top-down" TEM view of the atomic structure, and cross-sectional TEM image of the few-layer h-BN film, respectively, from

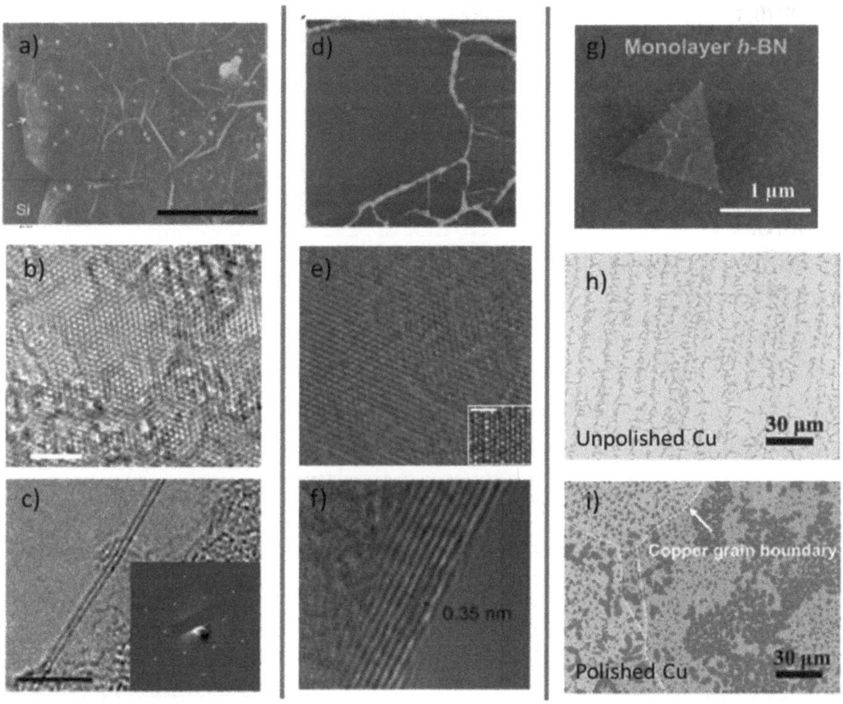

Fig. 11 (A) An optical top-down view of the as-deposited h-BN on Silicon, (B) a close-up TEM micrograph of the hexagonal patterning in h-BN, and (C) a cross-sectional view of few-layer h-BN growth (Song et al., 2010). Similar images from Shi et al. (2010) in (D)–(F). The last column in Fig. 1 is from Kim et al. (2012a) with (G) an SEM image of monolayer triangle formation of h-BN, (H) optical image of h-BN growth on unpolished, and (I) highly polished copper foil.

Song et al. (2010). Fig. 11D–F displays complementary images using the same techniques from Shi et al. (2010). The first growth procedures for the ultrathin h-BN coatings identified many critical issues, common in current growth methods as well, including the necessity to reduce wrinkles, cranks, and pinholes, as the need for uniform, large area coverage is required for many applications.

It was not until 2012, when Kim et al. (2012a) were able to isolate repeatable, triangular domains of single layer h-BN by performing the growth at lower pressures (technique known as LPCVD) using borazene, a gaseous species slightly more environmentally stable than borazene when decomposed. Advantages of the LPCVD synthesis include increased cleanliness in a vacuum environment, reduced activation energy for h-BN formation at lower pressures (Vlassiouk et al., 2013), as well as the reduced effect of kinetics and dynamics of the inlet gas flow for growth (Bhaviripudi et al., 2010). Fig. 1E is an optical image of one of the triangular domains, with grain sizes on the order of 1 μm. At a single layer, the reactivity of the h-BN is very low and can lead to incomplete film coverage in this instance of monolayer triangles. Additionally, the surface conditions of the substrate play a critical role in the nucleation density as well as the growth dynamics (Lee et al., 2012b), evidenced by the optical images of the distribution of domains of h-BN seen in Fig. 11H and I of the same growth conditions on unpolished and polished copper foil, respectively (Kim et al., 2012a).

At the time, it was a surprise that the first domains of h-BN were triangular, in contrast to the traditional hexagonal pattern seen in graphene synthesis on copper foil (Cu) (Tang et al., 2013a). Origins of these triangle domains are attributed to low reactivity of the nitrogen-terminated edges (Fig. 12A) (Liu et al., 2011) that restrict the lateral growth of the triangular domains. As graphene is a monatomic structure, the lack of contrasting edge site nucleation forms a hexagonal pattern similar to the internal structure of the graphene rings. It was observed by Stehle et al. (2015) that by altering the location of the substrate in the tube (ie, further or closer to reactant source), and thus altering the nitrogen-to-boron active species concentrations in the gas phase at the substrate location, the shape of these few micron-sized domains can be altered from triangle to semi-hexagons, as seen in Fig. 12B. Further manipulation of gas flow dynamics in the CVD tube (more substantial impact at atmospheric pressure growth) has also proven to control the nucleation shape of the h-BN domains from controllable triangles to more advanced structures (Yin et al., 2015).

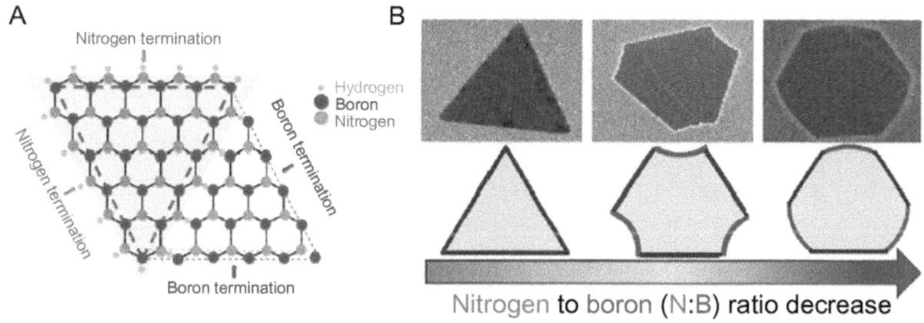

Fig. 12 (A) Representative schematic of the triangle formation of h-BN, where the nitrogen termination is the more energetically favorable termination and (B) the change in h-BN initial grain formation shape as a function of nitrogen and boron chemical concentrations in the CVD furnace. *Panel (A) from Kim, K.K., Hsu, A., Jia, X., Kim, S.M., Shi, Y., Hofmann, M., et al., 2012. Synthesis of monolayer hexagonal boron nitride on Cu foil using chemical vapor deposition. Nano Lett. 12, 161–6. doi:10.1021/nl203249a. Panel (B) from Stehle, Y., Meyer, H.M., Unocic, R.R., Kidder, M., Polizos, G., Datskos, P.G., et al., 2015. Synthesis of hexagonal boron nitride monolayer: control of nucleation and crystal morphology. Chem. Mater. 27, 8041–8047. doi:10.1021/acs.chemmater.5b03607, acs.chemmater.5b03607.*

Beyond h-BN growth on copper foils, 2D h-BN nucleation and growth have continued to expand to other metals including various crystal orientations of Ni (Cho et al., 2015; Preobrajenski et al., 2005; Yang et al., 2015; Zhang et al., 2014), Pt (Kim et al., 2013), Pd (Morscher et al., 2006), and alloys of these materials (Lu et al., 2015). Nickel, in particular the (111) crystal orientation, has shown to facilitate high-quality h-BN growth due to the small lattice mismatch (0.4%). This strain induced by the difference in lattice mismatch is compensated by a slight corrugation of the ad layer, and the periodicity of this depends on the degree of lattice mismatch between the substrates (Morscher et al., 2006). Grain size of the 2D h-BN can be directly influenced by the surface structure, as exemplified by record high triangular grain size of 7500 μm^2 on Cu/Al alloys at growth temperatures of 1050°C (Lu et al., 2015). The authors attribute the large grain size to the small addition of Al in the alloy, where the nucleation density can be decreased to 60 sites per mm^2, leading to unimpeded, large crystal growth on the surface.

Other interesting 2D structures can be formed by epitaxial lattice interactions between the substrate and h-BN. h-BN nanomesh, a porous material consisting of 12 substrate unit cells with aperture structures 2 nm in width, can be formed on ruthenium (Ru(0001)) and rhodium (Rh(111)) by LPCVD techniques used for monolayer growth on other foils (Corso et al., 2004; Goriachko et al., 2007). These structures form upon the decomposition of borazine vapor at temperatures at 1100°C and can have a particular benefit for use in patterning and assembly of nanoparticles such as gold nanoclusters. The change in nanostructure simply by altering the surface lattice structure and texturing conditions highlights the fact that h-BN in both the monolayer and few-layer form is extremely surface sensitive during the nucleation and growth process.

The majority of CVD synthesis routes have been focused on metallic substrates, as the foils can withstand high-temperature synthesis (800–1000°C) and can have epitaxial-like growth effects. Ideally, the 2D material can be grown directly on substrates of greater interest for electronics and other applications. If this would be possible, the requirement for the time and personnel-intensive lift-off and transfer steps would be eliminated, streamlining the device construction process. Prospects for direct growth of h-BN on alumina (Al_2O_3) and silica (SiO_2), both common substrates for nanoelectronic applications, are discussed by Bresnehan et al. (2013, 2014). Key aspects of the h-BN growth such as stoichiometry as a function of growth temperature, crystal size, and thermal transport properties were evaluated to elucidate how the thin h-BN films would behave in a graphene device construct. In addition, the enhancement of graphene hole concentration and mobility was evaluated with respect to the quality of the as-deposited films on Al_2O_3 and SiO_2. Direct growth of the turbostatic, 2D h-BN films processed at 400°C was shown to possess a higher boron content, however, show the largest increase in mobility of the graphene layered on top of the film, with a 1.5 × and 2.5 × increase compared to Al_2O_3 and SiO_2, respectively.

In parallel to the initial studies involving CVD nucleation and growth, physical vapor deposition techniques have also become of great interest for growth of few-layer h-BN. Physical vapor deposition techniques including PLD, magnetron sputtering, and molecular beam epitaxy (MBE) have all proven to facilitate h-BN growth at various conditions. In these cases, the use of a high-energy plasma (in the case of PLD and sputtering) or molecular beam is initiated in an ultrahigh vacuum chamber (typically residual base pressures of 10^{-7}–10^{-8} Pa). The plasma or molecular beam then comes in contact with the substrate and condenses to form thin films of h-BN.

With the use of high-energy plasmas, several benefits over conventional CVD approaches including reduced required temperature for h-BN formation, large area coverage, and versatility in device fabrication. The plasma-based techniques including magnetron sputtering as well as PLD have been used to coat very large wafers with uniform, cohesive thin films on the order of 4″ in diameter and larger. Due to the high-energy plasma species, however, special considerations need to be addressed as to how to control the kinetically driven processes to resist damage to the substrate and reduce the nucleation site density to allow for large area, unimpeded grain formation. Sutter et al. (2013) using reactive magnetron sputtering from a boron source in an Ar/N_2 gas mixture (see Fig. 13A) was able to generate a very slow flux of energetic ions and neutral species to the surface to condense and form uniform, highly crystalline multilayer h-BN on a Ru substrate, seen in Fig. 13B. With a three-step process combination of initial CVD to seed the growth to less than a monolayer, following by sputtering at high and low temperatures for subsequent multilayer growth, multiple layers can be formed with a varying degree of thicknesses and stoichiometry, as measured by XPS in Fig. 13C and D. Glavin et al. (2014) described that in PLD from an amorphous boron nitride target, a highly stoichiometric B–N plasma can be formed, and a recipe can be constructed where the stoichiometry of the as-deposited BN film can be uniform regardless of substrate. By appropriately choosing the working distance from target to substrate and adjusting the background pressure to facilitate shock-wave-like plasma dynamics, the appropriate precursors for stoichiometric h-BN formation can be realized, regardless of surface conditions. The structure, however, is very dependent upon the surface, as a highly dense amorphous boron nitride film is formed on most substrates, and a nanocrystalline h-BN is grown on nearly lattice-matched substrates including HOPG at temperatures of 700°C. The growth procedure is fairly quick, at ~0.4 nm/laser pulse, where the laser pulses are at 1 Hz repetition rate. Because of the high deposition rate, the nucleation density is also very high, resulting in grain size formation of 3 nm. The film, however, is cohesive and uniform over the substrates of interest, yielded similar electronic tunneling properties at all measured locations on the substrate. Other PLD studies using very high-energy CO_2 lasers have shown to produce high-quality h-BN nanosheets that overlap and form on the substrate surface at an even lower temperature of 400°C (Sajjad et al., 2013).

Studies of MBE of h-BN growth on nickel foils by Nakhaie et al. (2015) were able to use energy from the molecular ion beam, namely an elemental B and N source, to form large area, cohesive thin films of h-BN. The mixture of the high-energy molecular flux, coupled with the high surface temperatures, can lead to interesting metastable structures that are not necessarily bound to triangular film formation seen in CVD. Fig. 13A and a close-up in Fig. 13B provide images of MBE growth of h-BN that can result in

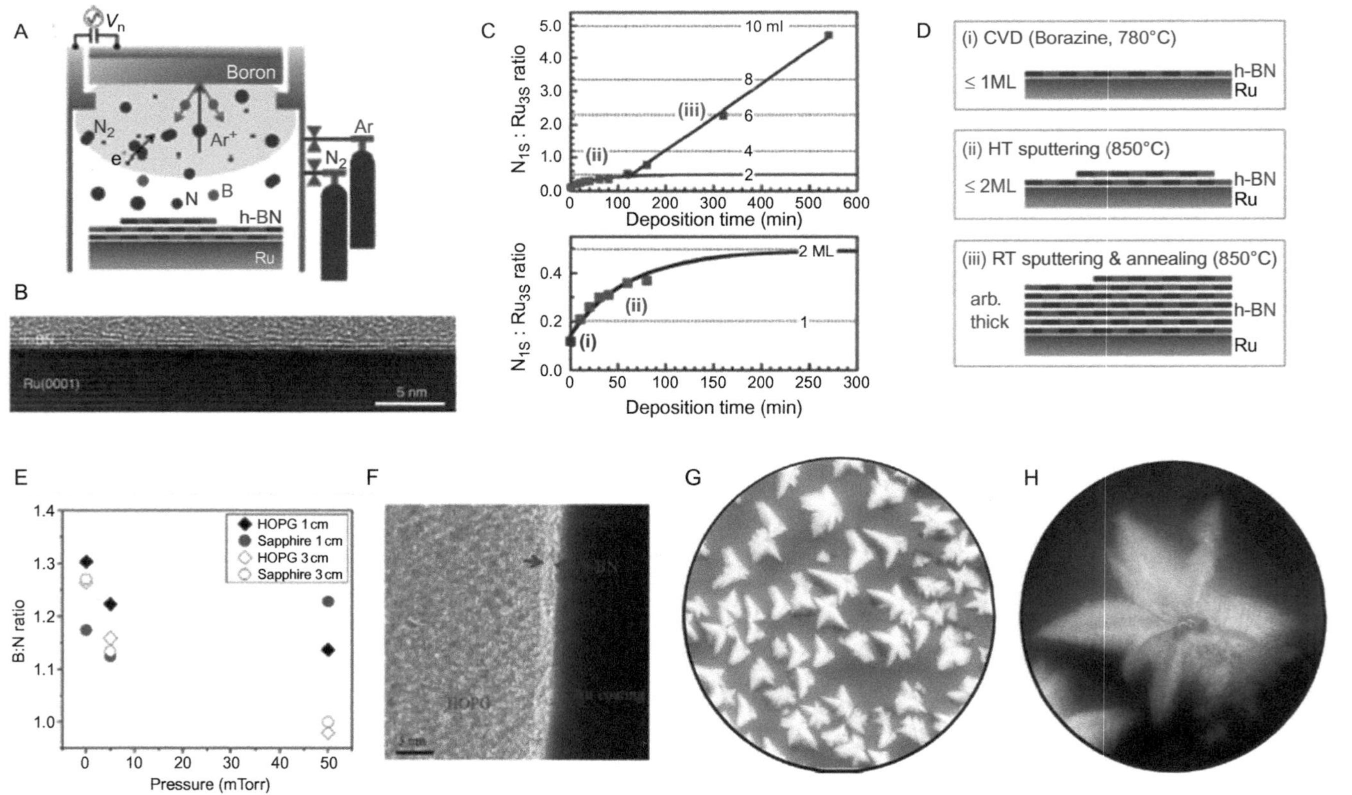

Fig. 13 (A) Setup of reactive magnetron sputtering for h-BN films, (B) cross-sectional TEM image of few-layer growth on Ru (0001), (C) XPS analysis of thickness for the different growth conditions, and (D) outlining the three-step process for arbitrarily thick h-BN coatings. The next two images portray h-BN growth with PLD processing of (E) the B:N ratio measured by XPS as a function of background gas nitrogen pressure and working distance in the chamber and (F) a cross-sectional image of few-layer h-BN on HOPG. The two final images are of MBE of h-BN with (G) initial h-BN grain formation and (H) a close-up of a similar image, window sizes of 30 and 9 μm, respectively. *Panels (A–D) from Sutter, P., Lahiri, J., Zahl, P., Wang, B., Sutter, E., 2013. Scalable synthesis of uniform few-layer hexagonal boron nitride dielectric films. Nano Lett. 13, 276–281. doi:10.1021/nl304080y. Panel (H) from Nakhaie S., Wofford J.M., Schumann T., Jahn U., Ramsteiner M., Hanke M., et al., 2015. Synthesis of atomically thin hexagonal boron nitride films on nickel foils by molecular beam epitaxy. Appl. Phys. Lett. 106, 213108. doi:10.1063/1.4921921.*

flower-like configurations, which eventually nucleate and grow in ways that are not traditional in CVD reactor chambers. These can be attributed to the additional kinetic energy (Barth et al., 2005) supplied by the incoming beam, where the addition of this energy can lead to a reduction in required growth temperature, evidence by growth of these initial flower-like growths at 730°C.

2D h-BN is one of the more interest 2D materials to date, and direct growth methods to understand how to enhance and control the film formation are critical to the future of nanoelectronic devices, anticorrosive coatings, and other uses. Advances in CVD processing have resulted in extremely large grain size, stoichiometric films, and alternative processes including PLD and plasma-based processing techniques could allow for realization of direct-growth device constructs.

3.7 Defect Manipulation in h-BN Using Scanning Tunneling Microscopy

h-BN is insulating, hence characterization of defects using scanning tunneling microscopy (STM) requires the use of a graphene capping layer because of the lack of conduction drain path for the electrical current (Wong et al., 2015). In order to observe the different defects in boron nitride, a graphene/boron nitride heterostructure was made. This allowed the visualization of the defect phenomenon in the underlying boron nitride.

Electronic configuration of the BN defects from the STM dI/dV signal was measured and quantified from the graphene capping layer. For the ring defects, the gate (V_g) and the bias (V_b) voltages were analyzed which were dependent on the ring radius. This was attributed to the charging of the adsorbate or the defect. Depending on the response time when the STM tip was present, the defect can be narrowed to the top most layer which is strongly coupled with the graphene electronic structure. Further, the defect manipulation was performed using STM. Fig. 14A shows the presence of numerous charged defects in the h-BN sheet. In order to manipulate the defects, the STM tip was placed 1 nm over the center of the defect area and a bias voltage was applied. After the voltage pulse, the dI/dV map was acquired again for the same region at low bias as observed in Fig. 14B. From Fig. 14B and C, it can be observed that the defect configuration was significantly altered by applying voltage pulses. The defects would switch reversibly between charged state and neutral state by applying the voltage across the sheet. These defects also switched between states having opposite charges. This can be observed in Fig. 14B and C with the red, blue, and green arrows. This tip-induced manipulation can be explained by electric field

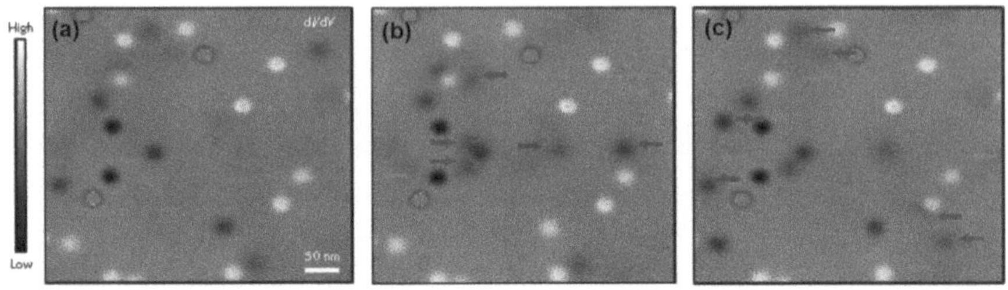

Fig. 14 Defect manipulation in h-BN using an STM tip. A tip pulse of $V_s=5$ V and $t=10$ s was used to alter the charge states of the defects. (A) d*I*/d*V* map of graphene/h-BN reveals various dot and ring defects in the h-BN sheet, (B) d*I*/d*V* map of the same region after one-tip pulse applied at the center, and (C) d*I*/d*V* map of the same region after two-tip pulses on the h-BN. The *red* (*gray* in the print version) *arrow* corresponds to the disappearance of the dot defects relative to previous image, *blue* (*dark gray* in the print version) *arrow* shows the appearance of the same defects, and the *green* (*light gray* in the print version) *arrow* shows the dot defects that have changed sign of their charge (Wong et al., 2015).

induction emission of the charge carriers from the defect states of h-BN. The STM tip causes charge carriers to tunnel through the ionization barrier between different defects, which allows partial charging and partial neutralization of the other points. This would explain the appearance and disappearance of the dot defects, which are not as a part of defect migration in BN lattice. Hence, this allowed the imaging and manipulation of individual defects in the insulating BN substrate with the help of a graphene capping layer.

4. APPLICATIONS FOR 2D h-BN ATOMIC LAYERS

Two-dimensional h-BN has been considered for use in many different applications including gate and substrate dielectrics for nanoelectronic devices, anticorrosive coatings, frictional layered coatings, high-temperature resistive layers, and other advanced applications for 2D material systems. The large bandgap (~6 eV), high temperature and chemical resistivity, and other properties discussed earlier in the chapter make both the exfoliated h-BN and direct-grown h-BN material ideal for such applications.

4.1 Dielectrics in Next-Generation Nanoelectronic Devices

Graphene is known for having one of the highest electron and hole mobility of any material to date, making the 2D material a natural candidate for next generation high-speed electronic devices (Bolotin et al., 2008). Initial graphene devices transferred onto traditional SiO_2 substrates were observed to have a reduction in intrinsic transport properties due to adverse interactions with the substrate (>120,000 cm^2/Vs compared to device performance on SiO_2 of ~2000 cm^2/Vs) (Liao et al., 2010; Lin et al., 2010c). The carrier mobility in the devices was reduced due to the scattering from charged surface states and impurities in SiO_2, the relatively high surface roughness of the substrate, and the low-frequency surface optical phonon modes in SiO_2 (Chen et al., 2008b; Martin et al., 2008; Morozov et al., 2008). Dean et al. (2010) was the first to investigate the use of bulk h-BN as a substrate material to counteract these effects seen in SiO_2 devices. The strong in-plane bonding of the h-BN substrate was hypothesized to allow for an atomically smooth, inert surface free of dangling bonds, or surface charge traps. This was found to be true, as the electronic properties of graphene including enhanced mobility, reduction in carrier inhomogeneity, and reduced intrinsic doping from the substrate were substantially improved. Wang et al. (2011c) demonstrated the first exfoliated GFET with h-BN integrated in both the substrate and gate dielectric, effectively creating a h-BN/graphene/h-BN sandwiched device. When compared to a traditional GFET device with SiO_2 and Al_2O_3 as dielectrics, an increase of 70% in the peak transconductance (g_m) was observed with a higher frequency of operation, as well as a greater than 5× increase in graphene mobility from 1200 to 6500 cm^2/Vs. Additionally, Dean et al. (2012) demonstrated a similar device construct using exfoliated materials, where the flakes can be seen in Fig. 15A and a hall device setup in Fig. 15B for electron mobility measurements. The devices display a mean electron mobility of ~35,000 cm^2/Vs, with measurements as high as ~70,000 cm^2/Vs. These mobility values are of a substantial increase of nearly an order of magnitude higher than similar graphene devices on SiO_2. Also, the charge neutrality point of the graphene is shifted to near zero gate voltage, as seen in Fig. 15C on a series of devices. This increase in mobility is also observed in polycrystalline CVD graphene on h-BN, where an increase of greater than three times is observed in Fig. 15D (Gannett et al., 2011), an effect seen in other device constructs as well (Nayfeh et al., 2013).

h-BN is also of interest as a key component as a gate dielectric in graphene and other 2D material-based electronic devices. With a dielectric constant of 6, breakdown voltage of 8–10 MV/cm, and bandgap nearing 6 eV (Lee et al., 2011; Watanabe et al., 2004), advantageous over conventional SiO_2 dielectrics can be realized. In the first demonstration of a 2D material component device, exfoliated h-BN was the choice of gate

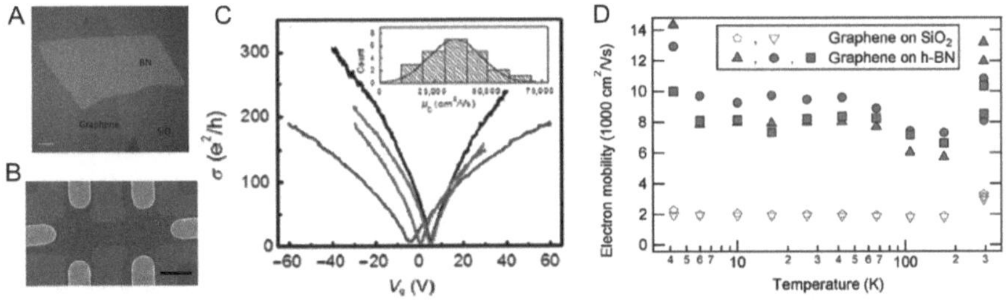

Fig. 15 (A) Optical image of graphene and h-BN flakes (scale bar 10 μm), (B) optical image of a hall bar measurements device for graphene electron mobility (scale bar 1 μm), (C) electron mobility as a function of gate voltage, with a histogram of mobility seen in inset, and (D) electron mobility enhancement in graphene on h-BN at various temperature as compared to graphene on SiO_2. *Panels (A–C) Dean, C., Young, A.F., Wang, L., Meric, I., Lee, G.H., Watanabe, K., et al., 2012. Graphene based heterostructures. Solid State Commun. 152, 1275–1282. doi:10.1016/j.ssc.2012.04.021. Panel (D) from Gannett, W., Regan, W., Watanabe, K., Taniguchi, T., Crommie, M.F., Zettl, A., 2011. Boron nitride substrates for high mobility chemical vapor deposited graphene. Appl. Phys. Lett. 98, 242105. doi:10.1063/1.3599708.*

dielectric with MoS_2 as the working material and graphene as the electrical contacts, providing for a unique device architecture based on all van der Waals solids (Roy et al., 2014b). The thickness of h-BN in this instance was very thick in this case, around 55 nm. While this device exemplifies rectifying behavior with high-quality interfaces between the materials, all of the 2D components were carefully exfoliated and placed in their given location. Thus, the meticulous device construction does not elude to further scale-up required for real world applications. Ideally, devices in the future can incorporate a directly grown synthesis method, such as CVD growth of h-BN, described earlier in this chapter. In addition, the prospect toward flexible, transparent devices has also been explored (Lee et al., 2013). Kim et al. (2012b) describe device-scale dielectric properties of the CVD grown h-BN, where at 15–19 nm, the film measures a dielectric constant of 2–4 with a breakdown of 2.0 ± 0.5 MV/cm. The decrease in breakdown and dielectric constant is attributed to the polycrystallinity of the h-BN films, potentially allowing for leakage through the grain boundaries. In order to incorporate h-BN as a gate dielectric material in next generation 2D devices, it appears that leakage through these boundaries will be critical with regard to dielectric and device performance.

4.2 Vertical Tunneling Device and Behavior

In addition to BN as a substrate material for graphene electronics, few-layer boron nitride has been investigated as a high-quality, low-dielectric constant barrier material for the 2D vertical electron tunneling devices where electrodes of different compositions are separated by the layer thickness of the dielectric. These devices rely on the tunneling through the ultrathin crystalline films layers and allow for the smallest allowable gate lengths down to a single atom, allowing for extremely fast transport (Tang et al., 2013b).

Tunneling measurements through few-layer, exfoliated h-BN for device constructs have been performed on a conductive substrate using conductive atomic force microscopy (C-AFM). With effective tip areas typically on the order of 10^3 nm^2, intrinsic property values of the dielectric can be evaluated that are not dependent on typical macro- or microscopic dielectric failure mechanisms including dust particles, cross-plane defects, and grain edges. From these measurements, *I–V* curves were generated on pristine crystalline BN flakes of various thicknesses from Lee et al. (2011). At mono-, bi-, and trilayer h-BN films, direct tunneling was observed as the dominant transport mechanism, seen in Fig. 16A. Beyond this thickness, the resulting *I–V* curves obeyed a more traditional breakdown behavior. This study indicates that four atomic layers of h-BN is the required thickness to restrict direct tunneling behavior in 2D h-BN. Similar studies were performed using a top and bottom graphene electrode setup, seen in Fig. 16C from Britnell et al. (2012) Using a C-AFM setup, as depicted in Fig. 16D, can help identify the tunneling through the BN flake on a device-scale contact setup. With this technique, the thickness and corresponding resistance values can be mapped out as portrayed in Fig. 16E and F.

Fig. 16B clearly shows that there are two regimes associated with tunneling in ultrathin h-BN materials of thicknesses greater than three layers (not direct tunneling). At low biases, the *I–V* curves will start out linear and then eventually become exponentially dependent at higher biases (typically >0.5 V for monolayer h-BN). At the lower voltages, the tunneling barrier has not been significantly deformed by the applied electric field. Thus, the tunneling current is linearly dependent upon the applied voltage and exponentially dependent upon the thickness of the layered material, as expected for direct tunneling (Bumm et al., 1999). The tunneling current at this regime can be modeled using the following relationship:

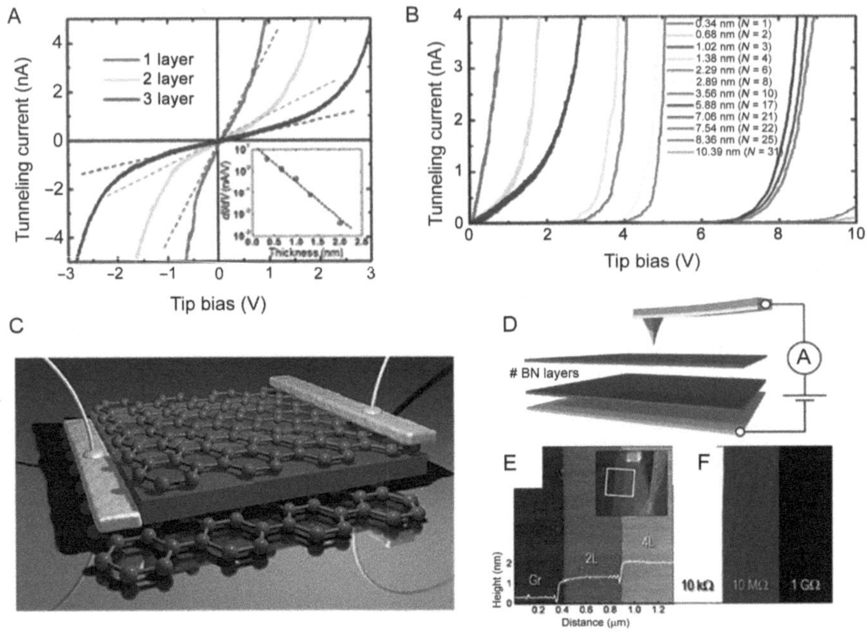

Fig. 16 (A) *I–V* curves generated from C-AFM studies on mono-, bi-, and trilayer films, as well as (B) thicker films up to 31 layers. In addition, device studies from Britnell et al. (2012) using a setup described in (C) and (D), and resulting electrical resistivity behavior in multiple layers of h-BN in (E) and (F). *Panel (A) from Lee, G.-H., Yu, Y.-J., Lee, C., Dean, C., Shepard, K.L., Kim, P., et al., 2011. Electron tunneling through atomically flat and ultrathin hexagonal boron nitride. Appl. Phys. Lett. 99, 243114. doi:10.1063/1.3662043.*

$$I(V) = \frac{A_{\text{eff}}\sqrt{m\phi_B}q^2 V}{h^2 d} * \exp\left[\frac{-4\pi\sqrt{m\phi_B}d}{h}\right]$$

where q, m, d, and h are the charge of an electron, free electron mass, film thickness, and Plank's constant, respectively. The effective area of contact A_{eff} is typically the area of the contacted tip, and ϕ_B is the barrier height. Direct tunneling was only able to be completely resolved at thicknesses typically less than about 2 nm, as direct tunneling is not measurable in thicker films because of the low probability of electrons tunneling through the barrier.

At higher applied biases, the tunneling is dominated by field-emission tunneling across the barrier, and the voltage dependence becomes quadratic. In this regime, modeling using the Fowler–Nordheim tunneling (Britnell et al., 2012) theory is performed with the following I–V equation:

$$I(V) = \frac{A_{\text{eff}} q^3 m V^2}{8\pi h \phi_B d^2 m^*} * \exp\left[\frac{-8\pi\sqrt{2m^*}\phi_B^{2/3} d}{3hqV}\right],$$

where effective mass $(m^*)/m$ is 0.26 for h-BN. The voltage is high enough at this point to alter the barrier characteristics and a significant reduction in resistivity is observed, leading to a breakdown in the dielectric characteristics within the material.

Determination of the barrier height is critical in understanding electronic behavior of a semiconductor or insulator and the interface of a metal. Depending upon the barrier height and the interface qualities, the Fermi level can potentially be pinned, resulting in rectifying characteristics within the Schottky barrier. Using the above equations, the barrier height can be determined if the affective contact area is known. Lee et al. (2011) measured the barrier height to be 3.07 ± 0.3 eV in h-BN, which is very close to that of traditional SiO_2 dielectric (3.25 eV). Another critical piece of information that can be obtained from the C-AFM experiments on tunneling through crystalline materials is the dielectric breakdown strength. In Fig. 16B, the breakdown strength is shown to be linearly dependent upon film thickness up to relatively thick films. Using the constant current method (Boucart and Ionescu, 2008), pristine h-BN exfoliated flakes appear to have measured dielectric breakdown strength of 8–10 MV/cm (Britnell et al., 2012; Lee et al., 2011), which, like to the barrier height, happens to be a very similar value to that of comparable SiO_2.

4.3 h-BN in Protective Coatings

While research on h-BN has focused on nanoelectronic applications, other technologies utilizing the ultrathin, layered materials have also been impactful. Thin h-BN for corrosion and antioxidation coatings, in particular, is a technology that is enabled by the chemical inertness and high temperature stability of h-BN (Jacobson et al., 1999), to effectively seal the underlying material from exposure to a variety of elements (Liu et al., 2013). Initial studies indicate that grown of few-layer h-BN on Ni can withstand oxidation up to temperatures as high as 1100°C, even in oxygen-rich atmospheres, as displayed in Fig. 17, with an SEM before (A) and after (B) image (Liu et al., 2013). In addition, the weight gain as a function of temperature and several different coating thicknesses can be resolved in Fig. 17C. The coating technique can also be used for other 2D materials, as graphene was shown in the same study to withstand oxidation up to temperatures of 1000°C. The main

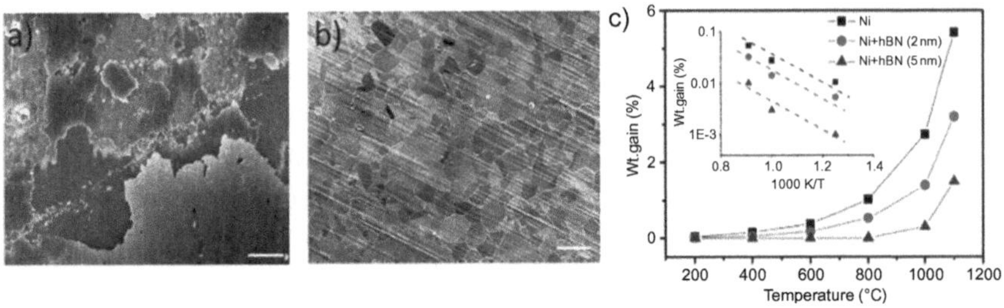

Fig. 17 (A) SEM images of nickel oxidation at 1100°C without h-BN coating and (B) with h-BN coating (no oxidation), and (C) weight gain percentage at various temperature, showing the change in oxygenation at these conditions (Liu et al., 2013).

reaction associated with BN and oxygen involves the decomposition of the BN itself into boric acid under the following reaction (Jacobson et al., 1999):

$$2BN + 3/2O_2 = B_2O_3 + N_2$$

The large activation energy required to break this BN molecule (~13 eV) lends itself to the explanation as a great antioxidation coating, where other oxidation reactions of the underlying substrates can be much lower energies of formation (Liu et al., 2013).

The antioxidative properties, as well as the ultrathin nature of 2D h-BN, can also be utilized in sensing platforms to effectively seal the active material from the element, but not impede in the sensing capabilities. Take for instance, the study performed by Liu et al. (2015) where an MoS_2 transistor setup was used to measure various gases including acetonitrile, ethanol, methanol, and various other gases. When the exfoliated h-BN cap protecting the MoS_2, the sensing ability was not degraded, and the lifetime of the device was extended to 7 days when capped, compared to 1 day uncapped.

Due to the inert nature and lack of dangling bonds, as discussed earlier, h-BN can be an excellent antifriction coating, similar to many other 2D materials that share the same characteristics (Lee et al., 2010). When reducing friction, a monolayer of coverage may actually be detrimental, and a thickness of at least a few layers may be required to reduce the frictional characteristics. The mechanism describing this phenomenon involves the fact that single layers tend to induce "puckering," where the AFM tip creates out of plane deformation of the 2D material. This effect is mitigated as the thicknesses increases, and the stiffness of the resulting layers restricts this. Still, flakes of h-BN and other materials pose to have a great impact in antifrictional coatings in the near future.

4.4 h-BN in Gas Sensing

More recently, h-BN has been used as a substrate to detect gases such as ammonia and ethanol (Liu et al., 2015; Postole et al., 2005; Xu et al., 2013). Gas sensors were made using atomic layers of h-BN and a very low concentration (100 ppm) of gases was passed through the device (Lin et al., 2016). The mechanism of the detection is shown in Fig. 18A, where the conductance change is measured which is caused by the chemical absorption and

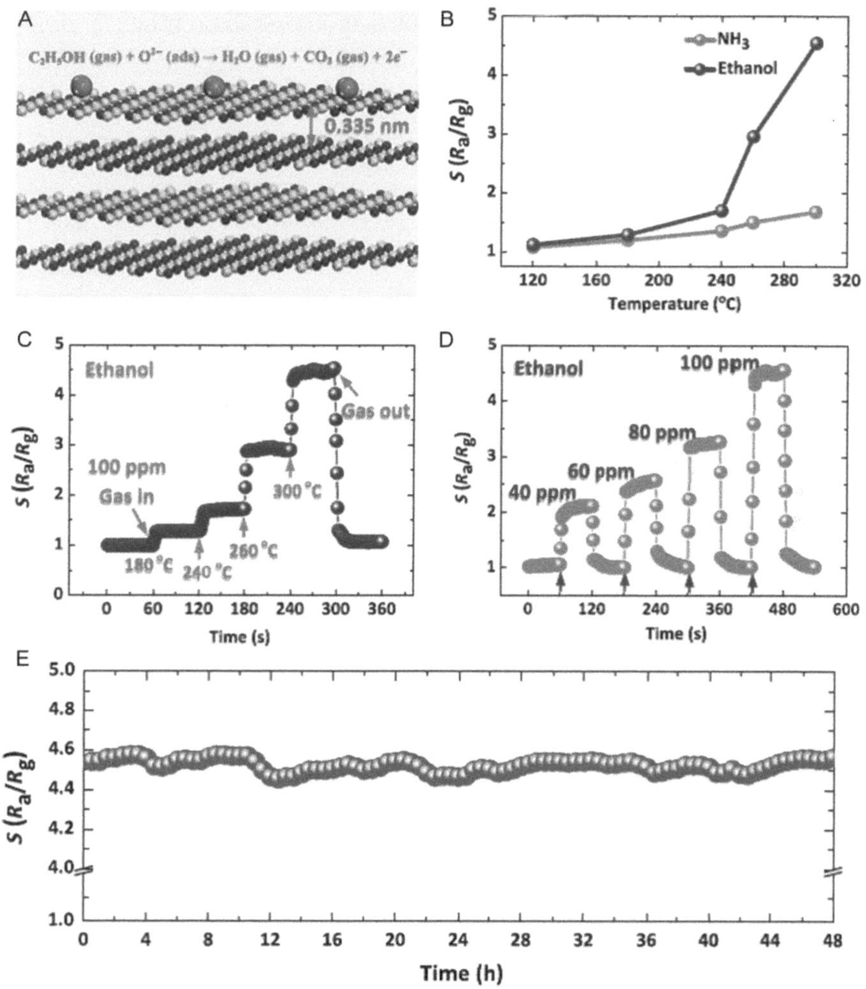

Fig. 18 (A) Schematic showing the gas detection mechanism on the surface of the h-BN nanosheets. (B and C) Gas sensing for ammonia and ethanol at various temperatures. (D) Ethanol detection at various concentrations of gas input. (E) Change in the resistance over time (Lin et al., 2016).

desorption process of the oxygen molecules on the sensor surface. When the sensor is exposed to air, the oxygen molecules gets absorbed and are ionized by the available free electrons from the conduction band of the h-BN, resulting in an increase of the resistance. Similarly, when ethanol is introduced, it causes a reaction on the surface, which releases electrons back to the sensor, reducing the resistance. Hence, this mechanism is used to test the efficiency at various temperatures and concentrations of ethanol as observed in Fig. 18B–D. Although the sensitivity of this sensor is not as low as the traditional zinc oxide or tinoxide sensors, the response time is very short and is reproducible over a period of time (Fig. 18E), which still makes it a good gas sensor.

4.5 h-BN as Filler for Binder-Free Anode for Lithium Ion Batteries

BNNSs were also used to make composite material with graphene oxide for binder-free anode for lithium ion battery application (Li et al., 2015a). The h-BN/rGO composite was first made via vacuum filtration (Fig. 19A),

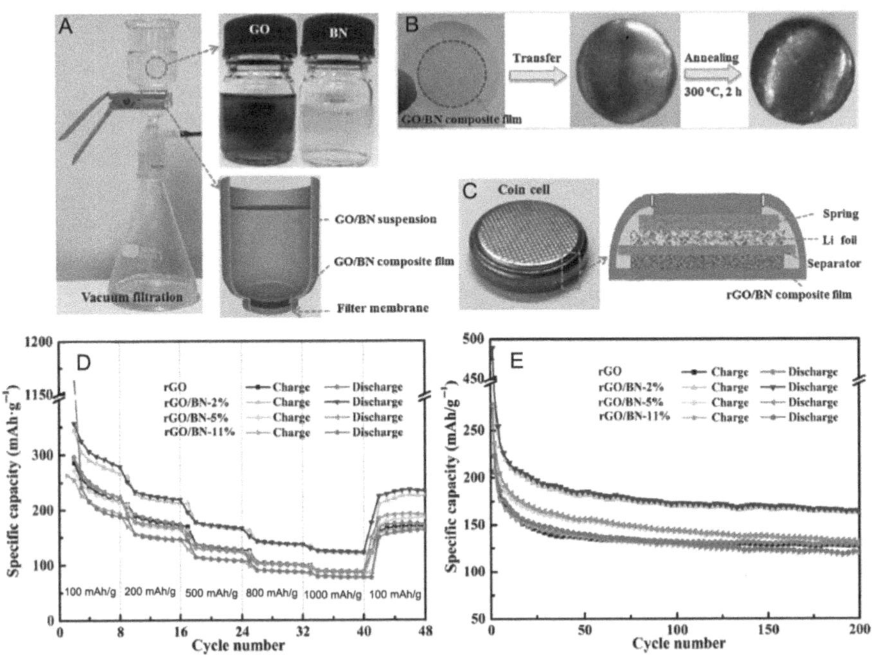

Fig. 19 (A) Vacuum filtration process used to create the heterostructure of h-BN/GO. (B) Dried film obtained after the vacuum filtration which is transferred onto copper electrode. (C) Coin cell assembly structure with the h-BN/GO composite as anode and Li foil as cathode. (D) Rate performance for different current densities. (E) Cyclic performance of the composite showed a constant cycling capacity of 200 mAh/g (Li et al., 2015a).

which was then dried to form a film as observed in Fig. 19B, transferred onto a copper metal, and then annealed at 300°C for 24 h (Fig. 19C), which formed the anode material. This is placed against a lithium foil as cathode and a coin cell was made to test the performance of this heterostructure.

Varying concentrations of h-BN was tested to understand the effect of BN on the electrochemical performances of the coin cell as observed in Fig. 19D and E. This composite exhibited a high reversible cyclic capacity of 278 mAh/g at a high current density of 100 mA/g. It also exhibited a high rate capability and excellent cyclic stability and showed no obvious signs of decay even after 200 cycles. This is attributed to the synergistic effect between the rGO and the h-BN nanosheets. This addition of h-BN showed a way to improve on the fabrication of other graphene-based composite films for high-performance energy storage devices without the use of binders or additives.

4.6 Hyperbolicity in h-BN

Previously, graphene was shown to be an ideal material for tunable plasmonics in the mid-infrared and terahertz range because of the electrostatic doping and its ability to produce higher confinement and low loss when compared to metals. Similar to graphene, h-BN possesses extremely high confinement and even lower low loss (Caldwell et al., 2014; Cortes et al., 2014; Dai et al., 2015a). Also, h-BN is a natural hyperbolic material which can be used for its photonic properties such as spontaneous emission enhancement, negative refraction, and thermal radiation enhancement (Caldwell et al., 2015, 2016; Dai et al., 2015b; Gilburd et al., 2016; Ju et al., 2014; Kumar et al., 2015; Li et al., 2015b; Woessner et al., 2015). Hence, making a heterostructure with graphene–h-BN will allow the use of the electrical tenability and high quality through the hybrid plasmon–phonon polaritons. h-BN being a hyperbolic material, has one of the relative permittivity tensor's opposite to the other two components. This property leads to indefinite dispersion for electromagnetic waves to propagate inside the material, resulting in photonic properties. Hence, h-BN would act as an atomic scale hyperbolic metamaterials. It also has both type I and II hyperbolic responses in different spectra bands, which allows a direct comparison between them. Typically, this optical property determination is performed using a near-field optical microscopy combined with an infrared nanospectroscopy, as represented in Fig. 20A. The broadband nano-infrared spectra of the normalized scattering amplitude plotted as a

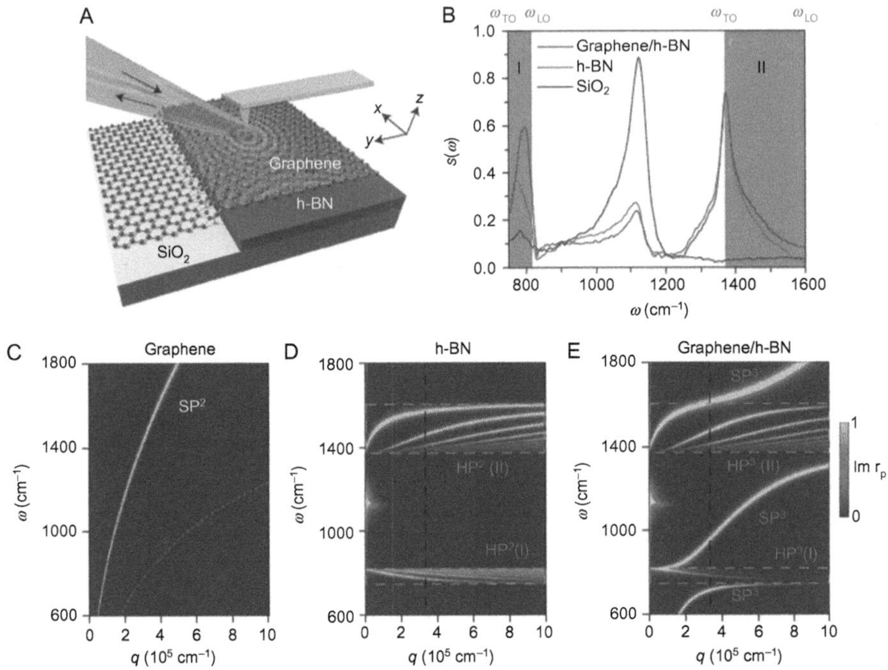

Fig. 20 (A) Schematic showing the infrared beams from a monochromatic broadband laser which are incident on an AFM tip and backscattered light collected for extraction of near-field signal. (B) Broadband nano-infrared spectra of the metastructure with hBN thickness of 58 nm. (C) Calculated dispersion of the SPP (SP^2) in freestanding graphene with Fermi energies of EF = 0.37, 0.15, and 0.08 eV. (D) Calculated dispersion of the hyperbolic phonon polaritons (HP^2) in h-BN of thickness 58 nm. (E) Similar to (D) but for a graphene/h-BN structure with EF = 0.37 eV. The false-color map reveals the dispersion of the hyperbolic plasmon–phonon polaritons (HP^3) and the surface plasmon–phonon polaritons (SP^3). Weak resonances around $\omega = 1130\ \text{cm}^{-1}$ in (D) and (E) originate from the SiO_2 substrate (Dai et al., 2015b).

function of frequency is represented in Fig. 20B for h-BN, SiO_2 substrate, and the graphene/h-BN metastructure. The hyperbolic regions of h-BN are highlighted in Fig. 20B where the type I has a frequency range of 746–819 cm^{-1} and the type II has a range at 1370–1610 cm^{-1}. Both the types of the resonances are modified for the metastructure as it can be observed in Fig. 20B. The presence of graphene significantly enhanced and blue shifted by ~25 cm^{-1} compared to that of h-BN alone. The surface plasmon polaritons (SPP) as measured are plotted in Fig. 20C and D for graphene and h-BN alone on SiO_2. As it can be observed that h-BN has multiple distinct branches which correspond to the quantized

HP^2 waveguide modes with a scalar potential oscillating across the h-BN with a different number of nodes. Similarly, we can observe new dispersion modes, ie, hyperbolic plasmon–phonon polaritons (HP^3) which arise from the mixing of HP^2 and SPP from graphene/h-BN structure. The graphene influence is clearly observed in the blue shift of the HP^3 frequencies with respect to the HP2, with a negative shift for type II band and positive shift in type I. This arises from the polariton dispersion being negative and positive in type I and type II regions. Hence, we can say that van der Waal polaritonic heterostructures with locally tunable properties will fulfil the essential prerequisites for the transformation of 2D plasmonics. The hybridization and graphene-induced tunability can be generic for most other electromagnetic materials and van der Waal's heterostructures.

Where only a few sets of applications were touched upon in this section, there are many potential exciting opportunities for ultrathin and 2D h-BN thin films. Coupled with the exciting material properties, the emergence of a greater understanding of nucleation and growth procedures, and continued push for 2D implantation into commercial devices, h-BN has a very promising future to directly impact many aspects of our daily lives for years to come.

REFERENCES

Alem, N., Erni, R., Kisielowski, C., Rossell, M., Gannett, W., Zettl, A., 2009. Atomically thin hexagonal boron nitride probed by ultrahigh-resolution transmission electron microscopy. Phys. Rev. B 80, 155425. http://dx.doi.org/10.1103/PhysRevB.80.155425.

N. No, 1958. Manufacture of boron compounds, United States Patent Office, 2–4.

Anota, E.C., Tlapale, Y., Villanueva, M.S., Márquez, J.A.R., 2015. Non-covalent functionalization of hexagonal boron nitride nanosheets with guanine. J. Mol. Model. 21, 215. http://dx.doi.org/10.1007/s00894-015-2768-0.

Balmain, W.H., 1842. Bemerkungen uber die Bildung von Verbindungen des Bors und Siliciums mit Stickstoff und gewissen Metallen. Philos. Mag. J. Sci. 62, 422–430. http://dx.doi.org/10.1002/prac.18420270164.

Barth, J.V., Costantini, G., Kern, K., 2005. Engineering atomic and molecular nanostructures at surfaces. Nature 437, 671–679. http://dx.doi.org/10.1038/nature04166.

Bhaviripudi, S., Jia, X., Dresselhaus, M.S., Kong, J., 2010. Role of kinetic factors in chemical vapor deposition synthesis of uniform large area graphene using copper catalyst. Nano Lett. 10, 4128–4133. http://dx.doi.org/10.1021/nl102355e.

Bhimanapati, G.R., Kozuch, D., Robinson, J., 2014. Large-scale synthesis and functionalization of hexagonal boron nitride nanosheets. Nanoscale 6, 11671–11675. http://dx.doi.org/10.1039/C4NR01816H.

Bhimanapati, G.R., Lin, Z., Meunier, V., Jung, Y., Cha, J.J., Das, S., et al., 2015. Recent advances in two-dimensional materials beyond graphene. ACS Nano 9, 11509–11539. http://dx.doi.org/10.1021/acsnano.5b05556.

Blake, P., Hill, E.W., 2007. Making graphene visible. Appl. Phys. Lett. 91, 063124. http://dx.doi.org/10.1063/1.2768624.
Bolotin, K.I., Sikes, K.J., Hone, J., Stormer, H.L., Kim, P., 2008. Temperature-dependent transport in suspended graphene. Phys. Rev. Lett. 101, 096802. http://dx.doi.org/10.1103/PhysRevLett.101.096802.
Boucart, K., Ionescu, A.M., 2008. A new definition of threshold voltage in Tunnel FETs. Solid State Electron. 52, 1318–1323. http://dx.doi.org/10.1016/j.sse.2008.04.003.
Bresnehan, M., Hollander, M.J., Wetherington, M.T., Wang, K., Miyagi, T., Pastor, G., et al., 2013. Prospects of direct growth boron nitride films as substrates for graphene electronics. J. Mater. Res. 29, 1–13. http://dx.doi.org/10.1557/jmr.2013.323.
Bresnehan, M.S., Bhimanapati, G.R., Wang, K., Snyder, D.W., Robinson, J.A., 2014. Impact of copper overpressure on the synthesis of hexagonal boron nitride atomic layers. ACS Appl. Mater. Interfaces 6, 16755–16762. http://dx.doi.org/10.1021/am503844u.
Britnell, L., Gorbachev, R.V., Jalil, R., Belle, B.D., Schedin, F., Katsnelson, M.I., et al., 2012. Electron tunneling through ultrathin boron nitride crystalline barriers. Nano Lett. 12, 1707–1710. http://dx.doi.org/10.1021/nl3002205.
Bumm, L.A., Arnold, J.J., Dunbar, T.D., Allara, D.L., Weiss, P.S., 1999. Electron transfer through organic molecules. J. Phys. Chem. B 103, 8122–8127. http://dx.doi.org/10.1021/jp9921699.
Bundy, F.P., 1963. Direct conversion of graphite to diamond in static pressure apparatus. J. Chem. Phys. 38, 631. http://dx.doi.org/10.1063/1.1733716.
Caldwell, J.D., Kretinin, A., Chen, Y., Giannini, V., Fogler, M.M., Francescato, Y., et al., 2014. Sub-diffraction, volume-confined polaritons in the natural hyperbolic material, hexagonal boron nitride. Nat. Commun. 5, 1–9. http://dx.doi.org/10.1038/ncomms6221.
Caldwell, J.D., Vurgaftman, I., Tischler, J.G., 2015. Probing hyperbolic polaritons. Nat. Publ. Gr. 9, 638–640. http://dx.doi.org/10.1038/nphoton.2015.183.
Caldwell, J.D., Vurgaftman, I., Tischler, J.G., Glembocki, O.J., Owrutsky, J.C., Reinecke, T.L., 2016. Atomic-scale photonic hybrids for mid-infrared and terahertz nanophotonics. Nat. Publ. Gr. 11, 9–15. http://dx.doi.org/10.1038/nnano.2015.305.
Chen, H., Zhang, H., Fu, L., Chen, Y., Williams, J.S., Yu, C., et al., 2008a. Nano Au-decorated boron nitride nanotubes: conductance modification and field-emission enhancement. Appl. Phys. Lett. 92, 243105. http://dx.doi.org/10.1063/1.2943653.
Chen, J.-H., Jang, C., Xiao, S., Ishigami, M., Fuhrer, M.S., 2008b. Intrinsic and extrinsic performance limits of graphene devices on SiO_2. Nat. Nanotechnol. 3, 206–209. http://dx.doi.org/10.1038/nnano.2008.58.
Chen, X., Dobson, J.F., Raston, C.L., 2012. Vortex fluidic exfoliation of graphite and boron nitride. Chem. Commun. 48, 3703. http://dx.doi.org/10.1039/c2cc17611d.
Cho, H., Park, S., Won, D.-I., Kang, S.O., Pyo, S.-S., Kim, D.-I., et al., 2015. Growth kinetics of white graphene (h-BN) on a planarised Ni foil surface. Sci. Rep. 5, 11985. http://dx.doi.org/10.1038/srep11985.
Coleman, J.N., Lotya, M., O'Neill, A., Bergin, S.D., King, P.J., Khan, U., et al., 2011. Two-dimensional nanosheets produced by liquid exfoliation of layered materials. Science 331, 568–571. http://dx.doi.org/10.1126/science.1194975.
Constantinescu, G., Kuc, A., Heine, T., 2013. Stacking in bulk and bilayer hexagonal boron nitride. Phys. Rev. Lett. 111, 036104. http://dx.doi.org/10.1103/PhysRevLett.111.036104.
Corso, M., Auwärter, W., Muntwiler, M., Tamai, A., Greber, T., Osterwalder, J., 2004. Boron nitride nanomesh. Science 303, 217–220. http://dx.doi.org/10.1126/science.1091979.
Cortes, C.L., Newman, W., Molesky, S., Jacob, Z., 2014. Corrigendum: quantum nanophotonics using hyperbolic metamaterials (2012 J. Opt. 14063001). J. Opt. 16, 129501. http://dx.doi.org/10.1088/2040-8978/16/12/129501.

Cui, Y., Kim, S.N., Jones, S.E., Wissler, L.L., Naik, R.R., McAlpine, M.C., 2010. Chemical functionalization of graphene enabled by phage displayed peptides. Nano Lett. 10, 4559–4565. http://dx.doi.org/10.1021/nl102564d.
Cui, Z., Oyer, A.J., Glover, A.J., Schniepp, H.C., Adamson, D.H., 2014. Large scale thermal exfoliation and functionalization of boron nitride. Small 10, 2352–2355. http://dx.doi.org/10.1002/smll.201303236.
Dai, S., Ma, Q., Andersen, T., Mcleod, A.S., Fei, Z., Liu, M.K., et al., 2015a. Sub-diffractional focusing and guiding of polaritonic rays in a natural hyperbolic material. Nat. Commun. 6, 6963. http://dx.doi.org/10.1038/ncomms7963.
Dai, S., Ma, Q., Liu, M.K., Andersen, T., Fei, Z., Gold, M.D., et al., 2015b. Graphene on hexagonal boron nitride as a tunable hyperbolic metamaterial. Nat. Nanotechnol. 10, 682–686. http://dx.doi.org/10.1038/nnano.2015.131.
Dean, C.R., Young, A.F., Meric, I., Lee, C., Wang, L., Sorgenfrei, S., et al., 2010. Boron nitride substrates for high-quality graphene electronics. Nat. Nanotechnol. 5, 722–726. http://dx.doi.org/10.1038/nnano.2010.172.
Dean, C., Young, A.F., Wang, L., Meric, I., Lee, G.H., Watanabe, K., et al., 2012. Graphene based heterostructures. Solid State Commun. 152, 1275–1282. http://dx.doi.org/10.1016/j.ssc.2012.04.021.
Deepika, Li, L.H., Glushenkov, A.M., Hait, S.K., Hodgson, P., Chen, Y., 2014. High-efficient production of boron nitride nanosheets via an optimized ball milling process for lubrication in oil. Sci. Rep. 4, 7288. http://dx.doi.org/10.1038/srep07288.
Ding, N., Chen, X., Wu, C.-M.L., Li, H., 2013. Adsorption of nucleobase pairs on hexagonal boron nitride sheet: hydrogen bonding versus stacking. Phys. Chem. Chem. Phys. 15, 10767–10776. http://dx.doi.org/10.1039/c3cp50912e.
Erickson, K.J., Gibb, A.L., Sinitskii, A., Rousseas, M., Alem, N., Tour, J.M., et al., 2011. Longitudinal splitting of boron nitride nanotubes for the facile synthesis of high quality boron nitride nanoribbons. Nano Lett. 11, 3221–3226.
Gannett, W., Regan, W., Watanabe, K., Taniguchi, T., Crommie, M.F., Zettl, A., 2011. Boron nitride substrates for high mobility chemical vapor deposited graphene. Appl. Phys. Lett. 98, 242105. http://dx.doi.org/10.1063/1.3599708.
Gao, Z., Zhi, C., Golberg, D., Bando, Y., Serizawa, T., 2014. Noncovalent functionalization of boron nitride nanotubes in aqueous media. Nanobiomedicine 1, 1–14. http://dx.doi.org/10.5772/60000.
Geim, A.K., Grigorieva, I.V., 2013. Van der Waals heterostructures. Nature 499, 419–425. http://dx.doi.org/10.1038/nature12385.
Ghosh, J., Mazumdar, S., Das, M., Ghatak, S., Basu, A.K., 2008. Microstructural characterization of amorphous and nanocrystalline boron nitride prepared by high-energy ball milling. Mater. Res. Bull. 43, 1023–1031. http://dx.doi.org/10.1016/j.materresbull.2007.04.022.
Gilburd, L., Xu, X.G., Bando, Y., Golberg, D., Walker, G.C., 2016. Near-field infrared pump–probe imaging of surface phonon coupling in boron nitride nanotubes. J. Phys. Chem. Lett. 7 (2), 289–294. http://dx.doi.org/10.1021/acs.jpclett.5b02438.
Glavin, N.R., Jespersen, M.L., Check, M.H., Hu, J., Hilton, A.M., Fisher, T.S., et al., 2014. Synthesis of few-layer, large area hexagonal-boron nitride by pulsed laser deposition. Thin Solid Films 572, 245–250. http://dx.doi.org/10.1016/j.tsf.2014.07.059.
Golberg, D., Bando, Y., Huang, Y., Terao, T., Mitome, M., Tang, C., et al., 2010. Boron nitride nanotubes and nanosheets. ACS Nano 4, 2979–2993. http://dx.doi.org/10.1021/nn1006495.
Gorbachev, R.V., Riaz, I., Nair, R.R., Jalil, R., Britnell, L., Belle, B.D., et al., 2011. Hunting for monolayer boron nitride: optical and Raman signatures. Small 7, 465–468. http://dx.doi.org/10.1002/smll.201001628.
Goriachko, A., He, Y., Knapp, M., Over, H., Corso, M., Brugger, T., et al., 2007. Self-assembly of a hexagonal boron nitride nanomesh on Ru(0001). Langmuir 23, 2928–2931. http://dx.doi.org/10.1021/la062990t.

Grimme, S., 2011. Density functional theory with London dispersion corrections. WIREs Comput. Mol. Sci. 1, 211–228. http://dx.doi.org/10.1002/wcms.30.

Grimme, S., Muck-Lichtenfeld, C., Antony, J., 2007. Noncovalent interactions between graphene sheets and in multishell (hyper)fullerenes. J. Phys. Chem. C 111, 11199–11207. http://dx.doi.org/10.1021/jp0720791.

Han, W.-Q., Wu, L., Zhu, Y., Watanabe, K., Taniguchi, T., 2008. Structure of chemically derived mono- and few-atomic-layer boron nitride sheets. Appl. Phys. Lett. 93, 223103. http://dx.doi.org/10.1063/1.3041639.

Hod, O., 2012. Graphite and hexagonal boron-nitride have the same interlayer distance. Why? J. Chem. Theory Comput. 8, 1360–1369. http://dx.doi.org/10.1021/ct200880m.

Hua Li, L., Chen, Y., Cheng, B.M., Lin, M.Y., Chou, S.L., Peng, Y.C., 2012. Photoluminescence of boron nitride nanosheets exfoliated by ball milling. Appl. Phys. Lett. 100, 261108. http://dx.doi.org/10.1063/1.4731203.

Jacobson, N., Farmer, S., Moore, A., Sayir, H., 1999. High-temperature oxidation of boron nitride: I. Monolithic boron nitride. J. Am. Ceram. Soc. 82, 393–398. http://dx.doi.org/10.1111/j.1151-2916.1999.tb01944.x.

Jiang, J.W., Wang, J.S., Li, B., 2009. Thermal conductance of graphene and dimerite. Phys. Rev. B 79, 205418. http://dx.doi.org/10.1103/PhysRevB.79.205418.

Jin, C., Lin, F., Suenaga, K., Iijima, S., 2009. Fabrication of a freestanding boron nitride single layer and its defect assignments. Phys. Rev. Lett. 102, 3–6. http://dx.doi.org/10.1103/PhysRevLett.102.195505.

Ju, L., Jr, J.V., Huang, E., Kahn, S., Nosiglia, C., Tsai, H., et al., 2014. Photoinduced doping in heterostructures of graphene and boron nitride. Nat. Nanotechnol. 9, 348–352. http://dx.doi.org/10.1038/nnano.2014.60.

Kim, K.K., Hsu, A., Jia, X., Kim, S.M., Shi, Y., Hofmann, M., et al., 2012a. Synthesis of monolayer hexagonal boron nitride on Cu foil using chemical vapor deposition. Nano Lett. 12, 161–166. http://dx.doi.org/10.1021/nl203249a.

Kim, K.K., Hsu, A., Jia, X., Kim, S.M., Shi, Y., Dresselhaus, M., et al., 2012b. Synthesis and characterization of hexagonal boron nitride film as a dielectric layer for graphene devices. ACS Nano 6, 8583–8590. http://dx.doi.org/10.1021/nn301675f.

Kim, G., Jang, A.-R., Jeong, H.Y., Lee, Z., Kang, D.J., Shin, H.S., 2013. Growth of high-crystalline, single-layer hexagonal boron nitride on recyclable platinum foil. Nano Lett. 13, 1834–1839. http://dx.doi.org/10.1021/nl400559s.

Kosynkin, D.V., Lu, W., Sinitskii, A., Pera, G., Sun, Z., Tour, J.M., 2011. Highly conductive graphene nanoribbons by longitudinal splitting of carbon nanotubes using potassium vapor. ACS Nano 5, 968–974. http://dx.doi.org/10.1021/nn102326c.

Kuila, T., Bose, S., Mishra, A.K., Khanra, P., Kim, N.H., Lee, J.H., 2012. Chemical functionalization of graphene and its applications. Prog. Mater. Sci. 57, 1061–1105. http://dx.doi.org/10.1016/j.pmatsci.2012.03.002.

Kumar, A., Low, T., Fung, K.H., Avouris, P., Fang, N.X., 2015. Tunable light–matter interaction and the role of hyperbolicity in graphene–hBN system. Nano Lett. 15, 3172–3180. http://dx.doi.org/10.1021/acs.nanolett.5b01191.

Lan, J., Wang, J.-S., Gan, C.K., Chin, S.K., 2009. Edge effects on quantum thermal transport in graphene nanoribbons: tight-binding calculations. Phys. Rev. B 79, 115401. http://dx.doi.org/10.1103/PhysRevB.79.115401.

Lee, C., Li, Q., Kalb, W., Liu, X.-Z., Berger, H., Carpick, R.W., et al., 2010. Frictional characteristics of atomically thin sheets. Science 328, 76–80. http://dx.doi.org/10.1126/science.1184167.

Lee, G.-H., Yu, Y.-J., Lee, C., Dean, C., Shepard, K.L., Kim, P., et al., 2011. Electron tunneling through atomically flat and ultrathin hexagonal boron nitride. Appl. Phys. Lett. 99, 243114. http://dx.doi.org/10.1063/1.3662043.

Lee, Y.-H., Liu, K.-K., Lu, A.-Y., Wu, C.-Y., Lin, C.-T., Zhang, W., et al., 2012a. Growth selectivity of hexagonal-boron nitride layers on Ni with various crystal orientations. RSC Adv. 2, 111. http://dx.doi.org/10.1039/c1ra00703c.

Lee, K.H., Shin, H.J., Lee, J., Lee, I.Y., Kim, G.H., Choi, J.Y., et al., 2012b. Large-scale synthesis of high-quality hexagonal boron nitride nanosheets for large-area graphene electronics. Nano Lett. 12, 714–718. http://dx.doi.org/10.1021/nl203635v.

Lee, G.H., Yu, Y.J., Cui, X., Petrone, N., Lee, C.H., Choi, M.S., et al., 2013. Flexible and transparent MoS2 field-effect transistors on hexagonal boron nitride-graphene heterostructures. ACS Nano 7, 7931–7936. http://dx.doi.org/10.1021/nn402954e.

Lee, D., Lee, B., Park, K.H., Ryu, H.J., Jeon, S., Hong, S.H., 2015. Scalable exfoliation process for highly soluble boron nitride nanoplatelets by hydroxide-assisted ball milling. Nano Lett. 15, 1238–1244. http://dx.doi.org/10.1021/nl504397h.

Li, L.H., Chen, Y., Behan, G., Zhang, H., Petravic, M., Glushenkov, A.M., 2011. Large-scale mechanical peeling of boron nitride nanosheets by low-energy ball milling. J. Mater. Chem. 21, 11862. http://dx.doi.org/10.1039/c1jm11192b.

Li, X., Hao, X., Zhao, M., Wu, Y., Yang, J., Tian, Y., et al., 2013. Exfoliation of hexagonal boron nitride by molten hydroxides. Adv. Mater. 25, 2200–2204. http://dx.doi.org/10.1002/adma.201204031.

Li, H., Tay, R.Y., Tsang, S.H., Liu, W., Teo, E.H.T., 2015a. Reduced graphene oxide/boron nitride composite film as a novel binder-free anode for lithium ion batteries with enhanced performances. Electrochim. Acta 166, 197–205. http://dx.doi.org/10.1016/j.electacta.2015.03.109.

Li, P., Lewin, M., Kretinin, A.V., Caldwell, J.D., Novoselov, K.S., Taniguchi, T., et al., 2015b. Hyperbolic phonon-polaritons in boron nitride for near-field optical imaging and focusing. Nat. Commun. 6, 1–9. http://dx.doi.org/10.1038/ncomms8507.

Liao, L., Lin, Y.-C., Bao, M., Cheng, R., Bai, J., Liu, Y., et al., 2010. High-speed graphene transistors with a self-aligned nanowire gate. Nature 467, 305–308. http://dx.doi.org/10.1038/nature09405.

Lin, Y., Connell, J.W., 2012. Advances in 2D boron nitride nanostructures: nanosheets, nanoribbons, nanomeshes, and hybrids with graphene. Nanoscale 4, 6908. http://dx.doi.org/10.1039/c2nr32201c.

Lin, Y., Williams, T.V., Connell, J.W., 2010a. Soluble, exfoliated hexagonal boron nitride nanosheets. J. Phys. Chem. Lett. 1, 277–283. http://dx.doi.org/10.1021/jz9002108.

Lin, Y., Williams, T.V., Cao, W., Elsayed-Ali, H.E., Connell, J.W., 2010b. Defect functionalization of hexagonal boron nitride nanosheets. J. Phys. Chem. C 114, 17434–17439. http://dx.doi.org/10.1021/jp105454w.

Lin, Y.-M., Dimitrakopoulos, C., Jenkins, K.A., Farmer, D.B., Chiu, H.-Y., Grill, A., et al., 2010c. 100-GHz transistors from wafer-scale epitaxial graphene. Science 327, 662. http://dx.doi.org/10.1126/science.1184289.

Lin, Y., Williams, T.V., Xu, T.-B., Cao, W., Elsayed-Ali, H.E., Connell, J.W., 2011a. Aqueous dispersions of few-layered and monolayered hexagonal boron nitride nanosheets from sonication-assisted hydrolysis: critical role of water. J. Phys. Chem. C 115, 2679–2685. http://dx.doi.org/10.1021/jp110985w.

Lin, Q., Zou, X., Zhou, G., Liu, R., Wu, J., Li, J., et al., 2011b. Adsorption of DNA/RNA nucleobases on hexagonal boron nitride sheet: an ab initio study. Phys. Chem. Chem. Phys. 13, 12225–12230. http://dx.doi.org/10.1039/c1cp20783k.

Lin, L., Liu, T., Zhang, Y., Sun, R., Zeng, W., Wang, Z., 2016. Synthesis of boron nitride nanosheets with a few atomic layers and their gas-sensing performance. Ceram. Int. 42, 971–975. http://dx.doi.org/10.1016/j.ceramint.2015.08.109.

Liu, Y., Bhowmick, S., Yakobson, B.I., 2011. BN white graphene with "colorful" edges: the energies and morphology. Nano Lett. 11, 3113–3116. http://dx.doi.org/10.1021/nl2011142.

Liu, Z., Gong, Y., Zhou, W., Ma, L., Yu, J., Idrobo, J.C., et al., 2013. Ultrathin high-temperature oxidation-resistant coatings of hexagonal boron nitride. Nat. Commun. 4, 1–8. http://dx.doi.org/10.1038/ncomms3541.

Liu, G., Rumyantsev, S.L., Jiang, C., Shur, M.S., Balandin, A.A., 2015. Gas sensing with h-BN capped MoS2 heterostructure thin film transistors. IEEE Electron Device Lett. 36, 1202–1204.

Lu, G., Wu, T., Yuan, Q., Wang, H., Wang, H., Ding, F., et al., 2015. Synthesis of large single-crystal hexagonal boron nitride grains on Cu–Ni alloy. Nat. Commun. 6, 6160. http://dx.doi.org/10.1038/ncomms7160.

Marom, N., Bernstein, J., Garel, J., Tkatchenko, A., Joselevich, E., Kronik, L., et al., 2010. Stacking and registry effects in layered materials: the case of hexagonal boron nitride. Phys. Rev. Lett. 105, 046801. http://dx.doi.org/10.1103/PhysRevLett.105.046801.

Martin, J., Akerman, N., Ulbricht, G., Lohmann, T., Smet, J.H., von Klitzing, K., et al., 2008. Observation of electron–hole puddles in graphene using a scanning single-electron transistor. Nat. Phys. 4, 144–148. http://dx.doi.org/10.1038/nphys781.

Meyer, J.C., Chuvilin, A., Algara-Siller, G., Biskupek, J., Kaiser, U., 2009. Selective sputtering and atomic resolution imaging of atomically thin boron nitride membranes. Nano Lett. 9, 2683–2689. http://dx.doi.org/10.1021/nl9011497.

Moore, A.W., Duclaux, L., Nysten, B., Issi, J.-P., 1992. Structure and low temperature thermal conductivity of pyrolytic boron nitride. Phys. Rev. B 46, 3362.

Morozov, S.V., Novoselov, K.S., Katsnelson, M.I., Schedin, F., Elias, D.C., Jaszczak, J.A., et al., 2008. Giant intrinsic carrier mobilities in graphene and its bilayer. Phys. Rev. Lett. 100, 11–14. http://dx.doi.org/10.1103/PhysRevLett.100.016602.

Morscher, M., Corso, M., Greber, T., Osterwalder, J., 2006. Formation of single layer h-BN on Pd(1 1 1). Surf. Sci. 600, 3280–3284. http://dx.doi.org/10.1016/j.susc.2006.06.016.

Nagashima, A., Tejima, N., Gamou, Y., Kawai, T., Oshima, C., 1996. Electronic states of monolayer hexagonal boron nitride formed on the metal surfaces. Surf. Sci. 357–358, 307–311. http://dx.doi.org/10.1016/0039-6028(96)00134-3.

Nakhaie, S., Wofford, J.M., Schumann, T., Jahn, U., Ramsteiner, M., Hanke, M., et al., 2015. Synthesis of atomically thin hexagonal boron nitride films on nickel foils by molecular beam epitaxy. Appl. Phys. Lett. 106, 213108. http://dx.doi.org/10.1063/1.4921921.

Nayfeh, O.M., Glen Birdwell, A., Tan, C., Dubey, M., Gullapalli, H., Liu, Z., et al., 2013. Increased mobility for layer-by-layer transferred chemical vapor deposited graphene/boron-nitride thin films. Appl. Phys. Lett. 102, 103115. http://dx.doi.org/10.1063/1.4794533.

Novoselov, K.S., Geim, A.K., Morozov, S.V., Jiang, D., Zhang, Y., Dubonos, S.V., et al., 2004. Electric field effect in atomically thin carbon films. Science 306, 666–669. http://dx.doi.org/10.1126/science.1102896.

Novoselov, K.S., Jiang, D., Schedin, F., Booth, T.J., Khotkevich, V.V., Morozov, S.V., et al., 2005. Two-dimensional atomic crystals. Proc. Natl. Acad. Sci. U.S.A. 102, 10451–10453. http://dx.doi.org/10.1073/pnas.0502848102.

Okada, S., Otani, M., 2010. Stability and electronic structure of potassium-intercalated hexagonal boron nitride from density functional calculations. Phys. Rev. B 81, 3–6. http://dx.doi.org/10.1103/PhysRevB.81.233401.

Oliveira, C.K., Matos, M.J.S., Mazzoni, M.S.C., Chacham, H., Neves, B.R.A., 2012. Anomalous response of supported few-layer hexagonal boron nitride to DC electric fields: a confined water effect? Nanotechnology 23, 175703. http://dx.doi.org/10.1088/0957-4484/23/17/175703.

Osada, M., Sasaki, T., 2012. Two-dimensional dielectric nanosheets: novel nanoelectronics from nanocrystal building blocks. Adv. Mater. 24, 210–228. http://dx.doi.org/10.1002/adma.201103241.

Ouyang, T., Chen, Y., Xie, Y., Yang, K., Bao, Z., Zhong, J., 2010. Thermal transport in hexagonal boron nitride nanoribbons. Nanotechnology 21, 245701. http://dx.doi.org/10.1088/0957-4484/21/24/245701.

Pacilé, D., Meyer, J.C., Girit, C.O., Zettl, A., 2008. The two-dimensional phase of boron nitride: few-atomic-layer sheets and suspended membranes. Appl. Phys. Lett. 92, 133107. http://dx.doi.org/10.1063/1.2903702.

Paine, R., Narula, C., 1990. Synthetic routes to boron nitride. Chem. Rev. 90, 73–91. http://pubs.acs.org/doi/abs/10.1021/cr00099a004 (accessed February 19, 2014).

Pakdel, A., Bando, Y., Golberg, D., 2014. Nano boron nitride flatland. Chem. Soc. Rev. 43, 934–959. http://dx.doi.org/10.1039/c3cs60260e.

Paton, K.R., Varrla, E., Backes, C., Smith, R.J., Khan, U., O'Neill, A., et al., 2014. Scalable production of large quantities of defect-free few-layer graphene by shear exfoliation in liquids. Nat. Mater. 13, 624–630. http://dx.doi.org/10.1038/nmat3944.

Postole, G., Caldararu, M., Ionescu, N.I., Bonnetot, B., Auroux, A., Guimon, C., 2005. Boron nitride: a high potential support for combustion catalysts. Thermochim. Acta 434, 150–157. http://dx.doi.org/10.1016/j.tca.2005.01.007.

Preobrajenski, A.B., Vinogradov, A.S., Mårtensson, N., 2005. Monolayer of h-BN chemisorbed on Cu(111) and Ni(111): the role of the transition metal 3d states. Surf. Sci. 582, 21–30. http://dx.doi.org/10.1016/j.susc.2005.02.047.

Rafiee, M.A., Narayanan, T.N., Hashim, D.P., Sakhavand, N., Shahsavari, R., Vajtai, R., et al., 2013. Hexagonal boron nitride and graphite oxide reinforced multifunctional porous cement composites. Adv. Funct. Mater. 23, 5624–5630. http://dx.doi.org/10.1002/adfm.201203866.

Rokuta, E., Hasegawa, Y., Suzuki, K., Gamou, Y., Oshima, C., Nagashima, A., 1997. Phonon dispersion of an epitaxial monolayer film of hexagonal boron nitride on Ni(111). Phys. Rev. Lett. 79, 4609–4612. http://dx.doi.org/10.1103/PhysRevLett.79.4609.

Roy, A.K., Park, B., Lee, K.S., Park, S.Y., In, I., 2014a. Boron nitride nanosheets decorated with silver nanoparticles through mussel-inspired chemistry of dopamine. Nanotechnology 25, 445603. http://dx.doi.org/10.1088/0957-4484/25/44/445603.

Roy, T., Tosun, M., Kang, J.S., Sachid, A.B., Desai, S.B., Hettick, M., et al., 2014b. Field-effect transistors built from all two-dimensional material components. ACS Nano 8, 6259–6264. http://dx.doi.org/10.1021/nn501723y.

Sainsbury, T., Satti, A., May, P., O'Neill, A., Nicolosi, V., Gun'Ko, Y.K., et al., 2012. Covalently functionalized hexagonal boron nitride nanosheets by nitrene addition. Chem. Eur. J. 18, 10808–10812. http://dx.doi.org/10.1002/chem.201201734.

Sainsbury, T., O'Neill, A., Passarelli, M.K., Seraffon, M., Gohil, D., Gnaniah, S., et al., 2014. Dibromocarbene functionalization of boron nitride nanosheets: toward band gap manipulation and nanocomposite applications. Chem. Mater. 26, 7039–7050. http://dx.doi.org/10.1021/cm503475t. 141208064817004.

Sajjad, M., Morell, G., Feng, P., 2013. Advance in novel boron nitride nanosheets to nanoelectronic device applications. ACS Appl. Mater. Interfaces 5, 8–13.

Shen, J., He, Y., Wu, J., Gao, C., Keyshar, K., Zhang, X., et al., 2015. Liquid phase exfoliation of two-dimensional materials by directly probing and matching surface tension components. Nano Lett. 15, 5449–5454. http://dx.doi.org/10.1021/acs.nanolett.5b01842. 150724081634000.

Shi, Y., Hamsen, C., Jia, X., Kim, K.K., Reina, A., Hofmann, M., et al., 2010. Synthesis of few-layer hexagonal boron nitride thin film by chemical vapor deposition. Nano Lett. 10, 4134–4139. http://dx.doi.org/10.1021/nl1023707.

Sinclair, W., Simmons, H., 1987. Microstructure and thermal shock behaviour of BN composites. J. Mater. Sci. Lett. 6, 627–629.

Solozhenko, V.L., Turkevich, V.Z., Holzapfel, W.B., 1999. Refined phase diagram of boron nitride. J. Phys. Chem. B 103, 2903–2905. http://dx.doi.org/10.1021/jp984682c.

Song, L., Ci, L., Lu, H., Sorokin, P.B., Jin, C., Ni, J., et al., 2010. Large scale growth and characterization of atomic hexagonal boron nitride layers. Nano Lett. 10, 3209–3215. http://dx.doi.org/10.1021/nl1022139.

Song, W.-L., Wang, P., Cao, L., Anderson, A., Meziani, M.J., Farr, A.J., et al., 2012. Polymer/boron nitride nanocomposite materials for superior thermal transport performance. Angew. Chem. Int. Ed. Engl. 51, 6498–6501. http://dx.doi.org/10.1002/anie.201201689.

Stehle, Y., Meyer, H.M., Unocic, R.R., Kidder, M., Polizos, G., Datskos, P.G., et al., 2015. Synthesis of hexagonal boron nitride monolayer: control of nucleation and crystal morphology. Chem. Mater. 27, 8041–8047. http://dx.doi.org/10.1021/acs.chemmater.5b03607.

Sutter, P., Lahiri, J., Zahl, P., Wang, B., Sutter, E., 2013. Scalable synthesis of uniform few-layer hexagonal boron nitride dielectric films. Nano Lett. 13, 276–281. http://dx.doi.org/10.1021/nl304080y.

Tang, S., Wang, H., Zhang, Y., Li, A., Xie, H., Liu, X., et al., 2013a. Precisely aligned graphene grown on hexagonal boron nitride by catalyst free chemical vapor deposition. Sci. Rep. 3, 1–7. http://dx.doi.org/10.1038/srep02666.. http://www.nature.com/srep/2013/130916/srep02666/abs/srep02666.html#supplementary-information.

Tang, S., Wang, H., Zhang, Y., Li, A., Xie, H., Liu, X., et al., 2013b. Precisely aligned graphene grown on hexagonal boron nitride by catalyst free chemical vapor deposition. Sci. Rep. 3, 1–7. http://dx.doi.org/10.1038/srep02666.

Vlassiouk, I., Smirnov, S., Regmi, M., Surwade, S.P., Srivastava, N., Feenstra, R., et al., 2013. Graphene nucleation density on copper: fundamental role of background pressure. J. Phys. Chem. C 117, 18919–18926. http://dx.doi.org/10.1021/jp4047648.

Wang, Y., Shi, Z., Yin, J., 2011a. Boron nitride nanosheets: large-scale exfoliation in methanesulfonic acid and their composites with polybenzimidazole. J. Mater. Chem. 21, 11371–11377. http://dx.doi.org/10.1039/c1jm10342c.

Wang, H., Taychatanapat, T., Hsu, A., Watanabe, K., Taniguchi, T., Jarillo-herrero, P., et al., 2011b. BN/graphene/BN transistors for RF applications. Nano 32, 1209–1211.

Warner, J.H., Rümmeli, M.H., Bachmatiuk, A., Büchner, B., 2010. Atomic resolution imaging and topography of boron nitride sheets produced by chemical exfoliation. ACS Nano 4, 1299–1304. http://dx.doi.org/10.1021/nn901648q.

Watanabe, K., Taniguchi, T., Kanda, H., 2004. Direct-bandgap properties and evidence for ultraviolet lasing of hexagonal boron nitride single crystal. Nat. Mater. 3, 404–409. http://dx.doi.org/10.1038/nmat1134.

Woessner, A., Lundeberg, M.B., Gao, Y., Principi, A., Alonso-gonzález, P., Carrega, M., et al., 2015. Highly confined low-loss plasmons in graphene–boron nitride heterostructures. Nat. Mater. 14, 421–425. http://dx.doi.org/10.1038/NMAT4169.

Wong, D., Jr., J.V., Ju, L., Lee, J., Kahn, S., Tsai, H., et al., 2015. Characterization and manipulation of individual defects in insulating hexagonal boron nitride using scanning tunnelling microscopy. Nat. Nanotechnol. 10, 949–953. http://dx.doi.org/10.1038/nnano.2015.188.

Xie, S.-Y., Wang, W., Fernando, K.A.S., Wang, X., Lin, Y., Sun, Y.-P., 2005. Solubilization of boron nitride nanotubes. Chem. Commun. (Camb.), 3670–3672. http://dx.doi.org/10.1039/b505330g.

Xu, Chao, Qiu, Xiaoqing, Huang, Caijin, Chen, Cheng, Ye, Xinxin, Ye, Weiqing, Hu, Jinli, 2013. Stable colloidal boron nitride nanosheet dispersion and its potential application in catalysis. J. Mater. Chem. A 1, 12192–12197. http://dx.doi.org/10.1039/c3ta12231j.

Yang, Y., Fu, Q., Li, H., Wei, M., Xiao, J., Wei, W., et al., 2015. Creating a nanospace under an h-BN cover for adlayer growth on Nickel(111). ACS Nano 9, 11589–11598.

Yao, Y., Lin, Z., Li, Z., Song, X., Moon, K.-S., Wong, C., 2012. Large-scale production of two-dimensional nanosheets. J. Mater. Chem. 22, 13494. http://dx.doi.org/10.1039/c2jm30587a.

Yates, B., Overy, M.J., Pirgon, O., 1975. The anisotropic thermal expansion of boron nitride I. Experimental results and their analysis. Philos. Mag. 32, 847–857.

Yin, J., Yu, J., Li, X., Li, J., Zhou, J., Zhang, Z., et al., 2015. Large single-crystal hexagonal boron nitride monolayer domains with controlled morphology and straight merging boundaries. Small 11, 1497–4502. http://dx.doi.org/10.1002/smll.201500210.

Yu, J., Huang, X., Wu, C., Wu, X., Wang, G., Jiang, P., 2012. Interfacial modification of boron nitride nanoplatelets for epoxy composites with improved thermal properties. Polymer 53, 471–480. http://dx.doi.org/10.1016/j.polymer.2011.12.040.

Yurdakul, H., Göncü, Y., Durukan, O., Akay, A., Seyhan, A.T., Ay, N., et al., 2012. Nanoscopic characterization of two-dimensional (2D) boron nitride nanosheets (BNNSs) produced by microfluidization. Ceram. Int. 38, 2187–2193. http://dx.doi.org/10.1016/j.ceramint.2011.10.064.

Zeng, H., Zhi, C., Zhang, Z., Wei, X., Wang, X., Guo, W., et al., 2010. "White graphenes": boron nitride nanoribbons via boron nitride nanotube unwrapping. Nano Lett. 10, 5049–5055. http://dx.doi.org/10.1021/nl103251m.

Zeng, X., Ye, L., Yu, S., Li, H., Sun, R., Xu, J., et al., 2015. Artificial nacre-like papers based on noncovalent functionalized boron nitride nanosheets with excellent mechanical and thermally conductive properties. Nanoscale 7, 6774–6781. http://dx.doi.org/10.1039/C5NR00228A.

Zhang, Y., Chang, T.-R., Zhou, B., Cui, Y.-T., Yan, H., Liu, Z., et al., 2014. Direct observation of the transition from indirect to direct bandgap in atomically thin epitaxial MoSe2. Nat. Nanotechnol. 9, 111–115. http://dx.doi.org/10.1038/nnano.2013.277.

Zhi, C., Bando, Y., Tang, C., Golberg, D., Xie, R., Sekigushi, T., 2005. Phonon characteristics and cathodolumininescence of boron nitride nanotubes. Appl. Phys. Lett. 86, 213110. http://dx.doi.org/10.1063/1.1938002.

Zhi, C., Bando, Y., Tang, C., Kuwahara, H., Golberg, D., 2009. Large-scale fabrication of boron nitride nanosheets and their utilization in polymeric composites with improved thermal and mechanical properties. Adv. Mater. 21, 2889–2893. http://dx.doi.org/10.1002/adma.200900323.

CHAPTER FOUR

Elemental Group IV Two-Dimensional Materials Beyond Graphene

M.E. Dávila*, L.C. Lew Yan Voon†, J. Zhao‡, G. Le Lay§,1
*Instituto de Ciencia de Materiales de Madrid-ICMM-CSIC, Madrid, Spain
†School of Science and Mathematics, The Citadel, Charleston, SC, United States
‡Laboratory of Materials Modification by Laser, Ion and Electron Beams (Dalian University of Technology), Ministry of Education, Dalian, China
§Aix Marseille Université, CNRS, PIIM UMR 7345, Marseille, France
[1]Corresponding author: e-mail address: guy.lelay@univ-amu.fr

Contents

1. INTRODUCTION

The concept of a thermodynamically stable two-dimensional (2D) material came as a surprise when graphene was finally fabricated and identified in 2004 (Geim and Novoselov, 2007). It has been proposed that either ripples or the fabrication route from graphite stabilizes the free-standing sheet. Other sheets, such as BN or MoS_2, were quickly proposed and fabricated experimentally soon after that.

Semiconductors and Semimetals, Volume 95
ISSN 0080-8784
http://dx.doi.org/10.1016/bs.semsem.2016.04.003

It was little known, through, that other group-IV sheets had been proposed as possibly stable from total energy calculations using ab initio calculations already in 1994 (Takeda and Shiraishi, 1994), and experimentally though functionalized (Nakano et al., 2006). Coming before graphene's fame, the early work on silicene and germanene, or corrugated silicon and germanium as they were then called in the theoretical work and silicon monolayer in the experimental work, did not focus on the band structure and the existence of Dirac cones was missed. The reason for the paucity of work on silicene and germanene is due to the known scarcity of layered structures made out of silicon (Si) and germanium (Ge), as opposed to carbon (C) with graphite, and to the misconception that silicene and germanene would need to have sp^2 bonded atoms like graphene does, with every self-respecting chemist knowing that Si and Ge do not like sp^2 bonding (Hoffmann, 2013). Hence, silicene and germanene remained in obscurity even though silicene was theoretically rediscovered twice in 2005 (Durgun et al., 2005; Yang and Ni, 2005). The final prediction which catalyzed the experimental search for silicene was the demonstration in 2007 that free-standing silicene, in spite of having a buckled structure and sp^3-like bonding, would have a Dirac cone just like graphene (Guzmán-Verri and Lew Yan Voon, 2007).

Six years after this prediction, where the name "silicene" was introduced for the first time, it was noticed that the number of times that silicene was mentioned in abstract titles for the American Physical Society's March meetings had shot up in 2013 (Brumfiel, 2013). The reason is easy to trace back to the publication in Spring 2012 of the seminal paper on the first compelling experimental evidence of the synthesis by dry epitaxy of silicene on a Silver (111) substrate (Vogt et al., 2012), soon followed by another article showing its growth by segregation on a ZrB_2 thin film in epitaxy on a Si(111) template (Fleurence et al., 2012). Later, Thomson Reuters has released its citation-based study, covering years 2011–14, of the 100 "hottest research fronts" of 2014, pointing to 10 research fronts within 10 broad disciplines, such as clinical medicine, geoscience, and physics, as discussed by Day (2015). Roughly speaking, a top-10 physics front in 2014 ended up being one whose core papers, published no earlier than 2011, had already generated about 2000 citations. Remarkably, "The growth and properties of silicene" is ranked fourth, preceded by "Observation of Higgs boson," "Global neutrino data analysis," and "Nonlinear massive gravity" and immediately followed by "MoS_2 and transistors." This confirms the trend already noticed in 2013: silicene and its group-IV brothers, germanene and stanene, constitute a novel emblematic research field on its own.

In this chapter, we will focus, on the one hand, on the theoretical aspects of standalone and of epitaxial silicene and germanene, and on the other hand, on the experimental realization of epitaxial silicene and germanene sheets, either bare or functionalized, and, finally, of the first transistors based on a silicene channel.

2. BARE SILICENE AND GERMANENE

2.1 Free-Standing Properties: Theory

As of today, no free-standing silicene and germanene has reportedly been made. The closest would be functionalized monolayers made in solution (Brumfiel, 2013; Nakano et al., 2006). Silicene and germanene have been deposited on substrates in ultrahigh vacuum and their properties will be treated in appropriate sections. We, therefore, discuss exclusively predictions of the properties of free-standing silicene and germanene.

2.1.1 Structural Properties

The first prediction of the structure of silicene and germanene was by Takeda and Shiraishi (1994) using ab initio calculations. It was assumed that the lowest energy structure would have a hexagonal structure with two atoms per unit cell (Fig. 1), very much like graphene but with the internal atom possibly out-of-plane to produce a sp^3-like bonding. Numerous calculations since and, in particular, a detailed study of Cahangirov et al. (2009) involving using a (2×2) supercell, phonon spectra, and ab initio finite temperature molecular dynamics (MD) appear to confirm the earlier results.

The structural parameters for the structure in Fig. 1 are given in Table 1.

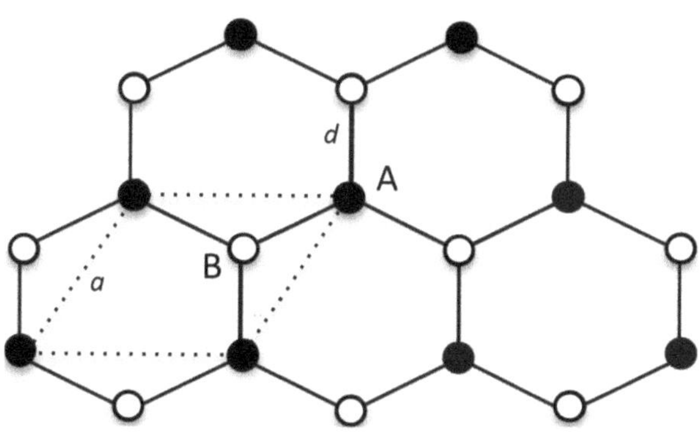

Fig. 1 Hexagonal structure of silicene and germanene.

All of the earlier structural optimizations started with the honeycomb structure with or without a supercell. It is well known that current ab initio codes provide structures for local energy minima with respect to the geometrical constraints imposed and a global minimum is not guaranteed. This point is well illustrated for these materials as a calculation with different initial geometries resulted in new local energy minima (Kaltsas and Tsetseris, 2013). What is, of course, important here is the fact that these new local minima are lower in energy than the previously found ones. The starting geometries used were those of the top atomic layer of reconstructed silicon. Thus, the most favorable structure found for both silicene and germanene is the so-called $\left(\sqrt{3}\times\sqrt{3}\right)$ structure (Table 2 and Fig. 2).

The lowest energy structure found, the $\left(\sqrt{3}\times\sqrt{3}\right)$ structure, consists of silicon atoms in a planar graphene-like structure with one of the two types of atoms having adatoms above and below. For silicene, the lattice constant is 6.52 Å, and the distance between an adatom and its nearest neighbor is 2.40 Å while the other Si–Si bond lengths are 2.28 Å. The corresponding lattice constant for germanene is 6.89 Å (Kaltsas and Tsetseris, 2013). Stability of the structures was confirmed by studying the phonon spectrum and looking for the absence of negative frequencies.

Table 1 Structural Parameters of Silicene and Germanene (Δ, a, d Being the Buckling Height, Lattice Constant, and Bond Length in Å), and with the Corresponding Cohesive Energy E_c (in eV/Atom) (Cahangirov et al., 2009)

	Δ	a	d	E_c
Silicene	0.44	3.83	2.25	5.16
Germanene	0.64	3.97	2.38	4.15

Table 2 Energy (in eV/Atom) of Silicene and Germanene Relative to the Most Stable Structure Obtained (the $\left(\sqrt{3}\times\sqrt{3}\right)$ Structure)

Structure	Silicene	Germanene
Hexagonal	0.048	0.031
$\left(\sqrt{3}\times\sqrt{3}\right)$	0.000	0.000
(5×5)	0.042	0.009
(7×7)	0.031	0.002

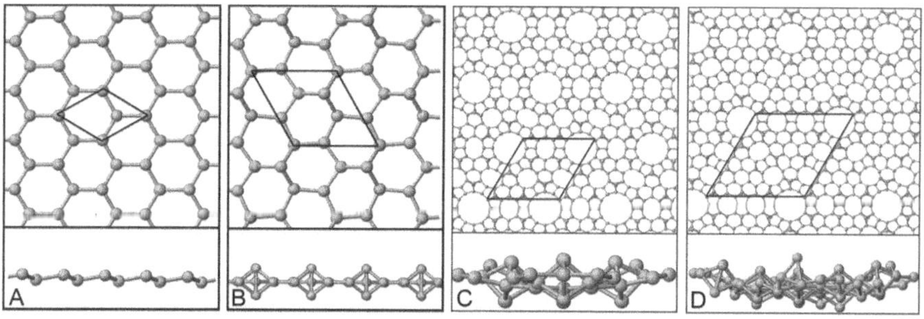

Fig. 2 New structures of silicene and germanene. From *left* to *right*, hexagonal (same as Fig. 1), $\left(\sqrt{3}\times\sqrt{3}\right)$, (5×5), and (7×7). *Reproduced from Kaltsas, D., Tsetseris, L., 2013. Stability and electronic properties of ultrathin films of silicon and germanium. Phys. Chem. Chem. Phys. 15, 9710–9715 with permission from the PCCP Owner Societies.*

Another approach that has been used is to simulate a superstructure of silicene and germanene on a silver substrate, then removing the substrate, and finally reoptimizing the structure (Yang et al., 2014b). It was found that the systems did not relax to the ideal free-standing structures, showing again the dependence of the optimized structure to the initial conditions.

2.1.2 Electronic Properties

The reason why silicene is as equally interesting as graphene is because of the same reason: the prediction of a Dirac cone (Guzmán-Verri and Lew Yan Voon, 2007), linear dispersions in the electronic band structure. Because of the buckled and lower symmetry of silicene, this result was perceived as "somehow unexpected" (Cahangirov et al., 2009) and even went unnoticed (Takeda and Shiraishi, 1994) for a while. Nevertheless, more recent studies have shown Dirac cones to be more prevalent in materials than originally thought.

Thus, ideal silicene and germanene have both been found to be semimetallic, with linear bands crossing at the Fermi energy and *K* point in the first Brillouin zone when spin–orbit effects are neglected; inclusion of the latter leads to small gaps of 1.55 and 23.9 meV, respectively (Liu et al., 2011). The density of states is very small at the Fermi energy (Fig. 3). The slope of the linear bands is known as the Fermi velocity and is found to be decreasing from graphene to silicene to germanene and between 1 and 10×10^5 m/s (Lew Yan Voon et al., 2010). The corresponding band structures for the $\left(\sqrt{3}\times\sqrt{3}\right)$ structures have also been computed; the latter was found to be metallic due to the strong band crossing (Fig. 4). One qualitative

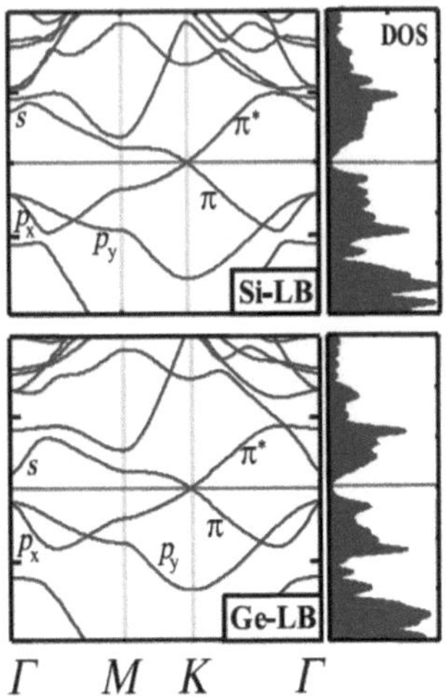

Fig. 3 Band structure and density of states (DOS) for silicene and germanene. *Reprinted figure with permission from Cahangirov, S., Topsakal, M., et al., 2009. Two- and one-dimensional honeycomb structures of silicon and germanium. Phys. Rev. Lett. 102, 236804. Copyright 2009 by the American Physical Society.*

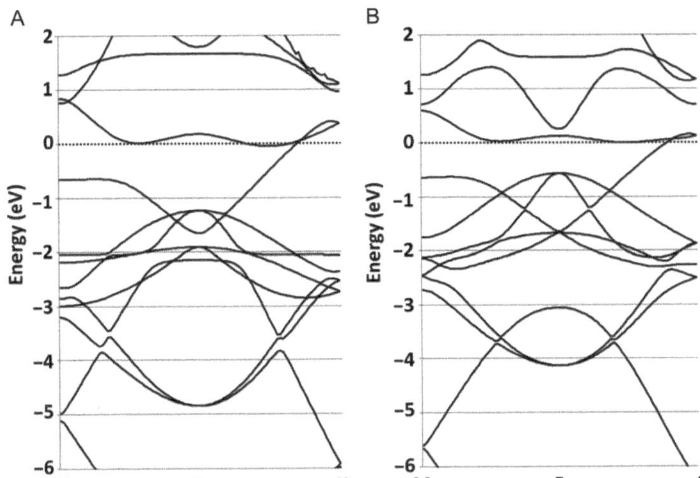

Fig. 4 Band structures of $\left(\sqrt{3} \times \sqrt{3}\right)$ silicene and germanene. *Reproduced from Kaltsas, D., Tsetseris, L., 2013. Stability and electronic properties of ultrathin films of silicon and germanium. Phys. Chem. Chem. Phys. 15, 9710–9715 with permission from the PCCP Owner Societies.*

difference between the Dirac bands above and those of graphene is that the graphene bands are pure p_z bands whereas they are mixed s–p_z in character for the other materials due to the lack of mirror symmetry in the latter case. Quantitative predictions of the band structure have been achieved using conventional density functional theory (DFT) calculations (Takeda and Shiraishi, 1994) whereas semianalytical results can be obtained using the tight-binding method (Guzmán-Verri and Lew Yan Voon, 2007) throughout the Brillouin zone and using k.p near the K point (Lew Yan Voon et al., 2015; Winkler and Zülicke, 2010).

The electronic properties can be modified by external perturbations such as strain, electric field, and temperature. To date, all of these studies (as well as the properties to be discussed in the following subsections) have all been carried out only for the ideal (1×1) structures.

Changes to the band structure under biaxial and uniaxial strains have been widely studied for silicene (Durajski et al., 2014; Kaloni et al., 2013; Lew Yan Voon et al., 2015; Lin and Ni, 2015; Liu et al., 2012; Manjanath et al., 2014; Mohan et al., 2014; Qin et al., 2012, 2014; Wang and Ding, 2013; Wang et al., 2014a; Yang et al., 2014a; Zhao, 2012) and much less for germanene (Kaloni and Schwingenschlögl, 2013; Manjanath et al., 2014; Wang and Ding, 2013). Under a biaxial strain, it is widely expected that no gap would open since the symmetry is not changed; this has, indeed, been found to be the case. What happens instead is similar to bulk semiconductors: the band edges shift in energy due to the deformation. Thus, for silicene under a biaxial tensile strain, the conduction band at the Γ point in the Brillouin zone decreases in energy while the Dirac point at the K point increases, leading to a semimetal–metal transition when the strain is larger than 7% (Liu et al., 2012; Qin et al., 2012) and to the phenomenon of p-type self-doping (Kaloni et al., 2013). A biaxial compressive strain leads to the opposite effect. A biaxial strain has also been found to lead to the possibility of superconductivity in silicene (Durajski et al., 2014). Uniaxial strains would, instead, be expected to lead to a band-gap opening due to a lowering of the crystal symmetry. Indeed, some DFT calculations appear to confirm this intuitive result (Mohan et al., 2014; Zhao, 2012). However, other DFT calculations (Qin et al., 2014; Yang et al., 2014a) did not obtain a band gap and the previous result was attributed to computing the band structure at insufficient number of k points. The absence of a band-gap opening has since been confirmed using an analytical deformation–potential theory (Lew Yan Voon et al., 2015).

For germanene, a biaxial tensile strain similarly leads to a lowering of the conduction band at the Γ point and an increase of the Dirac point at the K point (Kaloni and Schwingenschlögl, 2013; Wang and Ding, 2013). For example, at 5% strain, the Γ point conduction band shifts down by 0.6 eV, the Dirac point shifts up by 0.24 eV, and the gap decreases only slightly to 23 meV (Fig. 5). Thus, hole doping occurs. Wang and Ding (2013) also pointed out that self-doping occurred with a uniaxial strain (instead of a band-gap opening).

Very early on, it was realized that silicene and germanene would behave differently from graphene under the influence of an external vertical electric field, i.e., perpendicular to the layer. DFT calculations indeed revealed that a band gap can be induced with an external electric field (Drummond et al.,

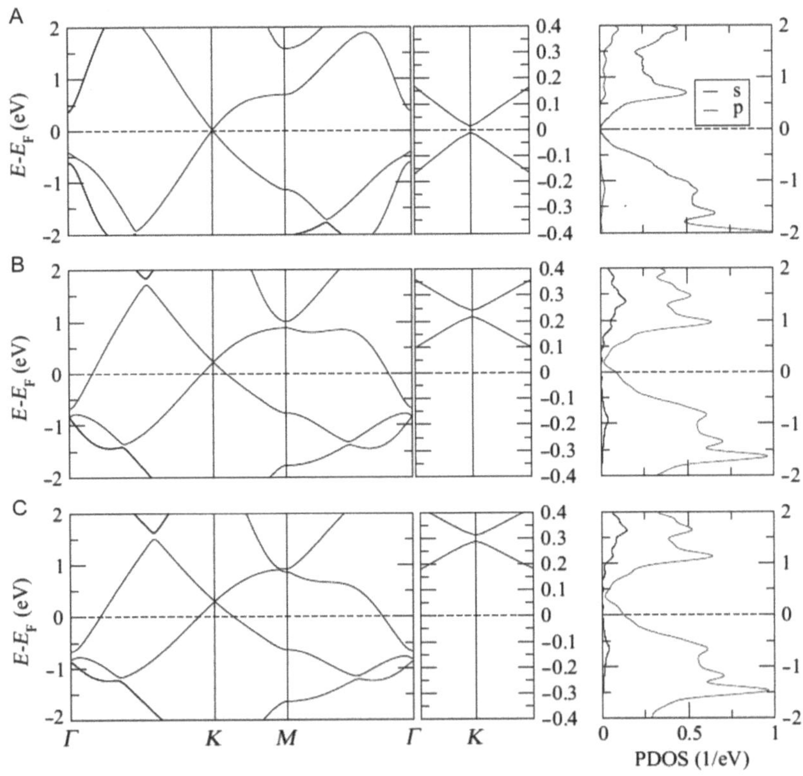

Fig. 5 Band structure and partial density of states (PDOS) of germanene under (A) no strain, (B) 5%, and (C) 16% biaxial tensile strain. *Reprinted figure with permission from Kaloni, T.P., Schwingenschlögl, U., 2013. Stability of germanene under tensile strain. Chem. Phys. Lett. 583, 137–140. Copyright 2013 by Elsevier.*

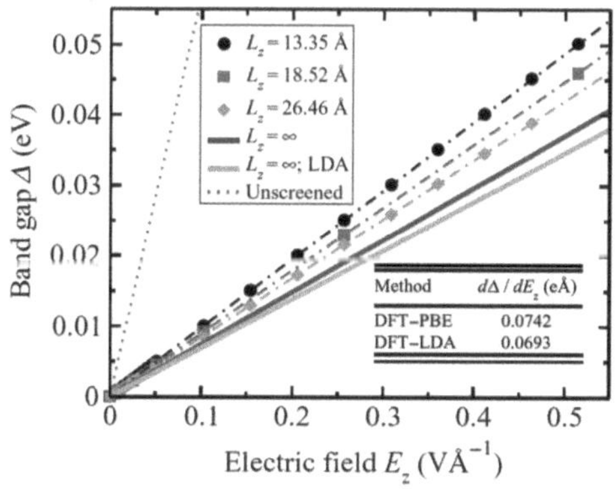

Fig. 6 Silicene band gap as a function of the external vertical electric field. *Reprinted figure with permission from Drummond, N.D., Zólyomi, V., et al., 2012. Electrically tunable band gap in silicene. Phys. Rev. B 85, 075423. Copyright 2012 by the American Physical Society.*

2012; Ni et al., 2012). This is due to the fact that the two basis atoms are at different vertical heights and, therefore, would feel a different electric potential. The linear rate of gap opening has been calculated to be 0.0742–0.157 eÅ for silicene (Drummond et al., 2012; Ni et al., 2012) (Fig. 6) up to about 0.5 V/Å when the gap starts closing again (Drummond et al., 2012) due to the overlap of the Dirac valence bands with the conduction band at the Γ point; the rate for germanene was found to be slightly less (Ni et al., 2012). Thus, any band gap is not expected to be larger than 30–40 meV. The linear slope has been confirmed using k.p theory (Lew Yan Voon et al., 2015) though higher-order variations have been shown to exist as well.

2.1.3 Phonons

Phonons in elemental 2D hexagonal crystals consist of six modes since there are two atoms per unit cell and modes transverse to the plane exist. Of these, three modes are acoustic and three are optical. For graphene, two of the modes are purely out-of-plane and have been labeled as out-of-plane or transverse. For silicene and germanene, the lack of mirror symmetry with respect to the layer plane means that the in-plane (x,y) and out of-plane (z) modes are coupled.

Phonons in silicene and germanene (Fig. 7) have often been studied mainly to determine the stability of the structure obtained (Cahangirov et al., 2009; Roome and Carey, 2014). Thus, a structure is stable if the phonon frequencies are all positive. Nevertheless, phonons have also been

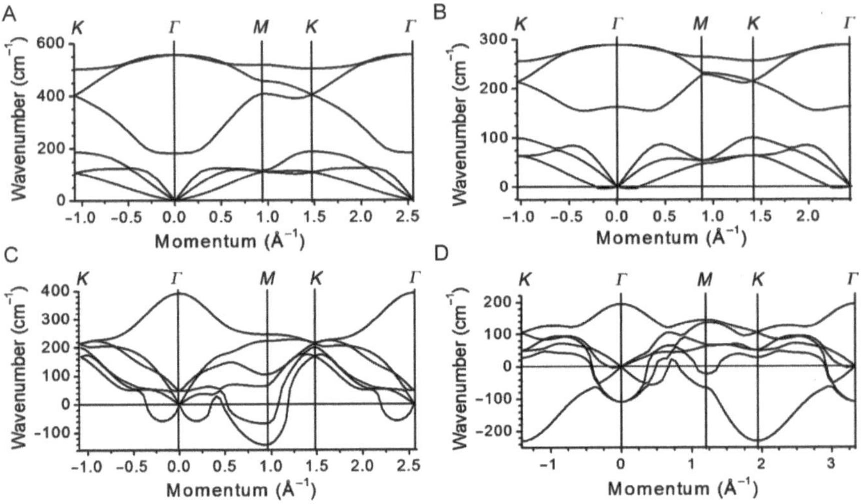

Fig. 7 Phonon spectra for (A) low-buckled silicene, (B) low-buckled germanene, (C) high-buckled silicene, and (D) high-buckled germanene. *Reprinted with permission from Roome, N.J., Carey, J.D., 2014. Beyond graphene: stable elemental monolayers of silicene and germanene. ACS Appl. Mater. Interfaces 6, 7743–7750. Copyright 2014 American Chemical Society.*

Table 3 Phonon Modes of Silicene and Germanene Compared to Graphene (Kaloni et al., 2013; Kaloni and Schwingenschlögl, 2013)

Phonon (cm^{-1})	Silicene	Germanene	Graphene
G	550	427	1580
D	545	366	1360

computed and compared to graphene (Table 3) (Kaloni and Schwingenschlögl, 2013; Kaloni et al., 2013). The method of density functional perturbation theory (Baroni et al., 2001) has been exclusively used to compute the phonons in these materials.

The most studied phonon modes are the G and D modes, the highest optical modes at the Γ (E_{2g}) and K (A'_1) point, respectively. These modes have been well studied using Raman spectroscopy in graphene (Ferrari et al., 2006), though the D mode requires defects to be Raman activated; graphene does have a very strong second-order D' peak at $\sim$2700 cm^{-1}. As one moves down the periodic table column, the weaker bonds lead to smaller force constants and, hence, optical frequencies.

Kaloni and coworkers (Kaloni and Schwingenschlögl, 2013; Kaloni et al., 2013) computed the change in the phonon frequencies of silicene and germanene. The phonon frequencies decrease with increasing biaxial tensile strain due to the softening of the bonds as they are stretched.

2.1.4 Mechanical Properties

It can be expected that a 2D material would be more "flexible" than a bulk solid. Thus, it might be easier to stretch and to bend. This has indeed been the case for graphene, which also has great mechanical strength. In the isotropic and harmonic approximation, the linear elastic properties of silicene and germanene can be characterized using two independent parameters: an in-plane stiffness C and the Poisson ratio ν defined as

$$\nu = \frac{\varepsilon_{trans}}{\varepsilon_{axial}},$$

$$C = \frac{1}{A_0}\frac{\partial^2 E_s}{\partial \varepsilon^2},$$

where A_0 is the equilibrium cross-sectional area, ε is the strain, and E_s is the strain energy (the difference in total energy with and without strain). One often introduces additionally the elastic constants C_{11} and C_{12}. Thus, if the strain energy is written as Şahin et al. (2009):

$$E_s(\varepsilon_{xx}, \varepsilon_{yy}) = a_1 {\varepsilon_{xx}}^2 + a_2 {\varepsilon_{yy}}^2 + a_3 \varepsilon_{xx}\varepsilon_{yy},$$

the elastic parameters are given by

$$C_{11} = \frac{2}{hA_0} a_1,$$

$$C_{12} = \frac{1}{hA_0} a_3,$$

$$\nu = \frac{C_{12}}{C_{11}},$$

$$C = hC_{11}\left[1 - \left(\frac{C_{11}}{C_{12}}\right)^2\right],$$

where h is the effective thickness of the supercell in a DFT calculation. The latter is always done by assuming 3D periodicity so an artificial super-periodicity is assumed perpendicular to the material plane. In the elastic regime, calculations are often done using a biaxial strain on a (1×1)

Table 4 Elastic Properties of Silicene and Germanene Compared to Graphene (Şahin et al., 2009)

	ν	C (J/m^2)
Silicene	0.30	62
Germanene	0.33	48
Graphene	0.16	335

hexagonal or (1×2) rectangular unit cell (Qin et al., 2012; Yang et al., 2014a). Computed values of the elastic parameters within the harmonic approximation are given in Table 4.

The softening of the mechanical properties down the periodic table is clear from the in-plane stiffness. The closer similarity of silicene and germanene is a reflection of the closer similarity in their structures (buckled) compared to the planar structure of graphene.

Studies beyond the linear elastic regime have also been carried out (Kaloni and Schwingenschlögl, 2013; Kaloni et al., 2013; Qin et al., 2012; Yang et al., 2014a). In Kaloni et al. (2013) and Yang et al. (2014a), stability was investigated on a (1×1) structure both by applying a biaxial tensile strain (and looking for a maximum in the stress–strain curve) and by computing the phonon spectrum (and looking for negative frequencies). More detailed studies on larger unit cells have been carried out to display the structural instability (Qin et al., 2012; Yang et al., 2014a). Thus, a (1×2) cell has been used with uniaxial tension (Yang et al., 2014a) and a cell as large as (4×4) has been used with biaxial tension (Qin et al., 2012), both for silicene. Indeed, the latter identified two regimes beyond the linear elastic one. For tensile strain beyond 4%, the strain energy continued to increase with strain but $dE_s/d\varepsilon$ no longer increases linearly with strain; this is the anharmonic regime. A plastic regime is entered when $dE_s/d\varepsilon$ reaches a maximum, i.e., beyond which less tension is required to expand the structure. Finally, the yield point is reached when the stress reaches a maximum; for silicene (germanene), this was found to occur for a biaxial tensile strain of 17% (Qin et al., 2012) (16%; Yang et al., 2014a). Calculations with uniaxial strains have also been carried out for silicene (Yang et al., 2014a) and it was found that there is a small anisotropy between the behaviors for armchair (AC) and zigzag (ZZ) strains, and they also both differ slightly from the biaxial results (Fig. 8).

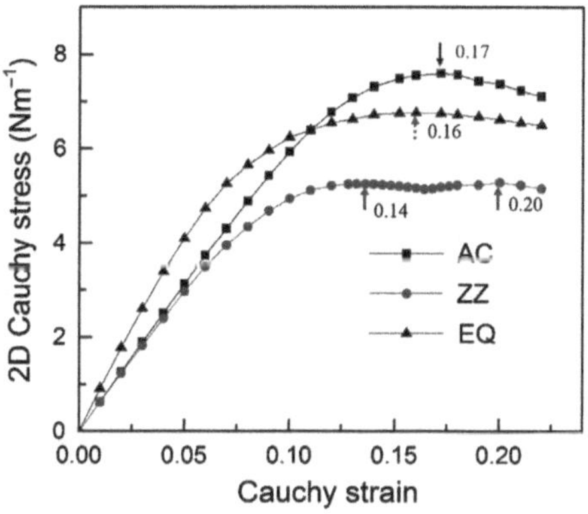

Fig. 8 Stress vs strain for silicene under armchair (AC), zigzag (ZZ), and biaxial (EQ) strain. *Reprinted figure with permission from Yang, C.-H., Yu, Z.-Y., et al., 2014a. Phonon instability and ideal strength of silicene under tension. Comput. Mater. Sci. 95, 420. Copyright 2014 by Elsevier.*

2.1.5 Thermal Properties

Most of the work on the thermal properties of silicene and germanene have focused on the lattice thermal conductivity and on the thermoelectric effect.

Thermal conductivity calculations have focused on the lattice contribution since this is expected to be the dominant contribution for a semimetal with a vanishing electron density of states at the Fermi energy. These calculations are based on the simulation of the heat transport in a finite sample and the lattice contribution has been done using classical models. Most calculations can then be divided into those using an equilibrium molecular dynamics (EMD) approach and a nonequilibrium molecular dynamics (NEMD) version, both with an empirical potential. At room temperature, values of the thermal conductivity obtained for silicene have ranged 3–65 W/mK (Hu et al., 2013; Kamatagi and Sankeshwar, 2015; Li and Zhang, 2012; Liu et al., 2014a; Ng et al., 2013; Pei et al., 2013; Wang and Sun, 2012; Xie et al., 2014; Yeo et al., 2014; Zhang et al., 2014), compared to values of 3000–5000 W/mK for suspended graphene (Balandin et al., 2008). In the EMD method, the atoms in the simulation region are given an initial velocity distribution and allowed to relax. At each step of the MD simulation, the heat current is calculated from the displacements, velocities, and forces on the atoms; the Kubo–Green formalism is then used to calculate the thermal conductivity from the heat current autocorrelation function (Kamatagi and Sankeshwar, 2015; Li and Zhang, 2012):

$$\kappa = \frac{1}{V\kappa_B T^2}\int_0^\infty \langle J(0)\cdot J(t)\rangle dt$$

where V is the volume and J is the heat current. Simulations were done with up ~600 atoms. Other methods have been used, such as the NEMD (Hu et al., 2013; Liu et al., 2014a; Ng et al., 2013; Pei et al., 2013; Wang and Sun, 2012). Thus, the EMD calculation of Li and Zhang (2012) obtained 20 W/mK, the NEMD calculations of Hu et al. (2013) and Pei et al. (2013) obtained ~40 W/mK while Ng et al. (2013) obtained 65 W/mK, Calculations based upon the Tersoff and Stillinger–Weber potentials for bulk silicon are questionable since they have not reproduced the correct silicene structure, including predicting a flat structure. Hence, Zhang et al. (2014) obtained a new parameterization of the Stillinger–Weber potential that reproduces the structure and phonons of silicene better and they obtained $k \sim 5-10\,\mathrm{W/mK}$. Finally, in an attempt to resolve the earlier discrepancies, a first-principles calculation of k was carried out, by calculating phonon scattering rates and using in the Boltzmann transport equation (Xie et al., 2014). They obtained 9.4 W/mK and explained the small value to be the result of enhanced scattering of the flexural modes in silicene as compared to graphene because the purely out-of-plane character of these modes in graphene reduces scattering. The strain dependence of thermal conductivity has been studied and shown to be different for silicene and graphene (Hu et al., 2013; Pei et al., 2013). For graphene, all the phonon modes soften with tensile strain and κ decreases. For silicene, some of the phonon modes (particularly the flexural ones) stiffen and, therefore, the behavior is more complicated. Kamatagi and Sankeshwar (2015) and Xie et al. (2014) have computed the temperature dependence of κ and the latter found κ to vary from 11 to 6 in the range 200–600 K.

Another important thermal property of silicene studied is the thermoelectric figure of merit,

$$ZT = \frac{\sigma S^2 T}{\kappa},$$

where σ is the electrical conductivity and S is the Seebeck coefficient. Good thermoelectric materials would need large σ and small κ and are defined to have $ZT \sim 1$. Graphene, with its high thermal conductivity, turns out not to be a good thermoelectric. Thus, silicene and germanene are better candidates. The electronic figure of merit (an upper limit to the total figure of merit) for free-standing silicene and germanene is given in Fig. 9. It is seen

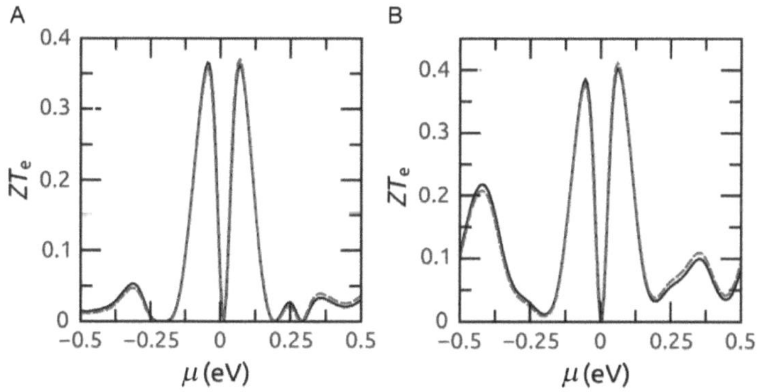

Fig. 9 Electronic figure of merit for free-standing (A) silicene and (B) germanene, as a function of chemical potential μ. *Reprinted figure with permission from Yang, K., Cahangirov, S., et al., 2014b. Thermoelectric properties of atomically thin silicene and germanene nanostructures. Phys. Rev. B 89, 125403. Copyright 2014 by the American Physical Society.*

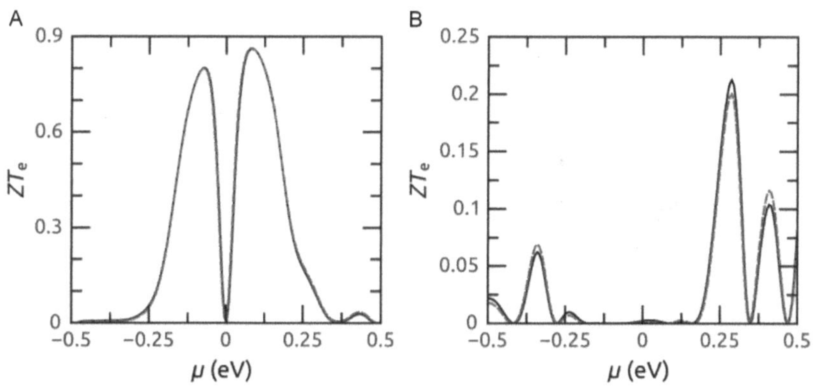

Fig. 10 Electronic figure of merit for distorted (A) silicene and (B) germanene, as a function of chemical potential μ. *Reprinted figure with permission from Yang, K., Cahangirov, S., et al., 2014b. Thermoelectric properties of atomically thin silicene and germanene nanostructures. Phys. Rev. B 89, 125403. Copyright 2014 by the American Physical Society.*

that they can reach close to 0.4 for both (Yang et al., 2014b). Yang and coworkers (2014b) also considered "distorted" structures (resulting from creating a superstructure on a Ag(111) substrate, "peeling" off the Ag and relaxing the geometry); those are found to have $ZT_e \sim 1$ for silicene (Fig. 10).

2.2 Experimental Realization of Epitaxial Silicene and Germanene

2.2.1 Epitaxial Silicene

Silicene, which has no parent lamellar crystal in nature from which it could be peeled off, like graphite for graphene or black phosphorous for phosphorene, had to be man made. As mentioned in Section 1, silicene was first synthesized in situ under ultra-high vacuum on Ag(111) (Vogt et al., 2012), then on other metallic templates, i.e., ZrB_2 (Fleurence et al., 2012), next Ir(111) (Meng et al., 2013), and, possibly, on semiconducting MoS_2 (Chiappe et al., 2014). On all-metallic substrates, silicene is compliant, forming reconstructed sheets coinciding with supercells of the substrates. We illustrate this later with silicene on silver (111), which forms, in optimum preparation conditions in-between 200 and 220°C, the archetype, highly ordered, unique 3×3 phase in coincidence with a 4×4 Ag(111) super cell (in short $3 \times 3/4 \times 4$), displaying the famous "flower pattern" in scanning tunneling microscopy (STM) imaging (Vogt et al., 2012), as illustrated in Fig. 11. The underlying honeycomb structure of silicene, hidden within the "flower pattern," was unraveled thanks to high-resolution synchrotron radiation angle-resolved photoelectron spectroscopy (SR-ARPES) measurements and DFT calculations based on an initial geometrical model inferred from the knowledge of the "magic" epitaxial relationship of the inverse system, upon growth of silver on Si(111) substrates (Le Lay et al., 2016).

Another interesting silicene phase, few percent extended, can be formed on Ag(111) at slightly higher temperatures (Resta et al., 2013). It exists in two symmetric $\sqrt{7} \times \sqrt{7}R \pm 19.1°$ structures on two rotated $\sqrt{13} \times \sqrt{13}R \pm 13.9°$ silver supercells (in short $\sqrt{7} \times \sqrt{7}/\sqrt{13} \times \sqrt{13}$), giving by symmetry a total of four equivalent domains, which have been simultaneously observed in the STM image shown in Fig. 12.

Upon STM imaging at low temperature a remarkable vortex structure is revealed (Liu et al., 2014d), as shown in Fig. 13. It is associated with different buckling arrangements within the silicene sheet, associated with the flexibility of the interatomic Si–Si bond angles due to intermediate sp^2/sp^3 hybridization states. We link the formation of such vortices to the inherent strain of the epitaxial silicene sheet, which has a negative thermal expansion coefficient (Huang et al., 2015), upon cooling, while its silver substrate behaves, instead, normally.

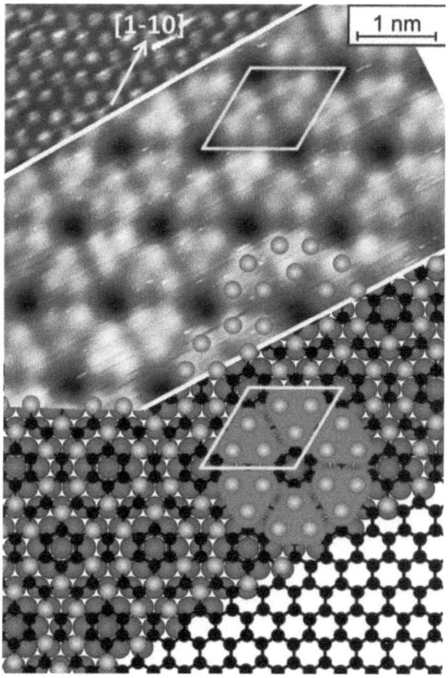

Fig. 11 Structural model for the archetype 3 × 3 reconstructed silicene phase coinciding with a 4 × 4 supercell on Ag(111). Filled-states STM images of the initial bare Ag(111) 1 × 1 surface (*top left corner*, in *gray*) and of the "flower pattern" formed by the 3 × 3 silicene sheet in *orange* (*light gray* in the print version). The atomic model of 3 × 3/4 × 4 silicene on Ag(111) is shown later, with the Si atoms sitting on *top* of Ag atoms highlighted as *orange* (*light gray* in the print version) *balls*, resembling the recorded STM image. The ball-and-stick model for the free-standing silicene layer is shown at the *bottom right corner*, with a Si–Si distance of ~0.22 nm. *Adapted from Vogt, P., De Padova, P., et al., 2012. Silicene: compelling experimental evidence for graphenelike two-dimensional silicon. Phys. Rev. Lett. 108, 155501.*

At significantly higher temperatures, typically above ~270°C, a $2\sqrt{3} \times 2\sqrt{3}R(30°)$ superstructure with reference to the Ag(111) surface, is observed (Le Lay et al., 2014). It appears inherently highly defective and inhomogeneous as a patchwork of fragmented silicene pieces with local $\sqrt{7} \times \sqrt{7}$ reconstructions (Liu et al., 2014e). Compelling observations and measurements reveal its sudden death, to end, in a dynamic fating process at ~300°C, on the one hand, in Si(111) islands through a dewetting mechanism (Acun et al., 2013), and, on the other hand, in the burial of an alloy, most probably confined below the surface (Liu et al., 2014e).

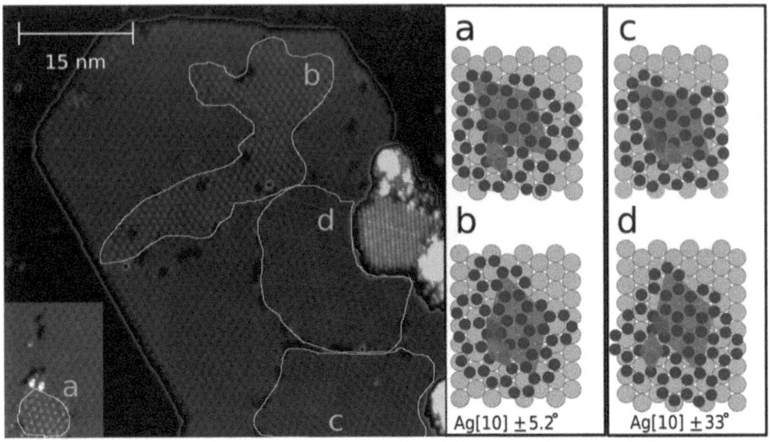

Fig. 12 (*Left*) Filled-states STM image acquired at room temperature of the four distinct possible domains a–c of $\sqrt{7}\times\sqrt{7}R\pm 19.1°$ reconstructed silicene on $\sqrt{13}\times\sqrt{13}R\pm 13.9°$ Ag(111) supercells. (*Right*) Ball model (Si atoms in *blue* (*black* in the print version), Ag atoms in *yellow* (*light gray* in the print version)) of the four A–C domains. The silicene $\sqrt{7}\times\sqrt{7}R\pm 19.1°$ cells in coincidence with the $\sqrt{13}\times\sqrt{13}R\pm 13.9°$ Ag(111) super cells are in *orange* (*dark gray* in the print version); the silicene 1 × 1 unit cells are in *magenta* (*gray* in the print version) and in *green* (*gray* in the print version) the Ag(111) 1 × 1 cells. *Adapted from Fig. 4 of Resta, A., Leoni, T., et al., 2013. Atomic structures of silicene layers grown on Ag(111): scanning tunneling microscopy and noncontact atomic force microscopy observations. Sci. Rep. 3, 2399.*

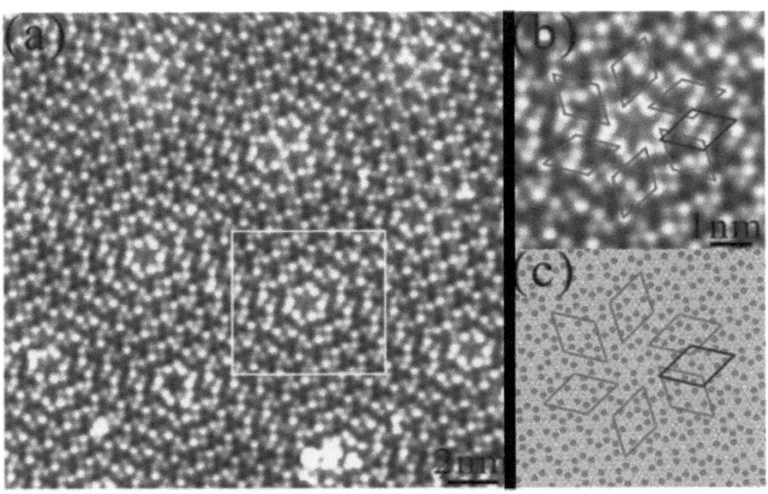

Fig. 13 (A) High-resolution STM image taken at 4 K on a $\sqrt{7}\times\sqrt{7}$ silicene domain showing a large hexagonal pattern. (B) Enlarged cut from the region delineated in *white* in (A). (C) Ball atomic model for the vortex pattern in (B), based on the structures depicted in Guo et al. (2013).

Concerning the electronic structure of single layer epitaxial silicene on Ag(111), most significant results have been obtained on the archetype 3×3 phase. While a linear dispersion of a silicene-related band has been originally measured at the $K_{1\times1}$ corner of the primitive silicene Brillouin zone (Vogt et al., 2012), the existence of Dirac cones is still in debate (Feng et al., 2015)

Beyond monolayer silicene, further Si deposition onto the $3 \times 3/4 \times 4$ or $\sqrt{7} \times \sqrt{7}/\sqrt{13} \times \sqrt{13}$ silicene sheet leads to the formation of double, and, next, multilayer silicene growing in successive flat terraces in a kind of Aztec pyramids (see Fig. 14) (Vogt et al., 2014). All flat terraces have the same $\sqrt{3} \times \sqrt{3}$ reconstruction with respect to silicene, which is contracted by a few percent compared to monolayer $3 \times 3/4 \times 4$, but their orientations follow those of the initial first layer, which acts as a kind of buffer layer (Salomon et al., 2014).

Multilayer silicene hosts, manifestly, Dirac fermions as proved by ARPES measurements (De Padova et al., 2013) and scanning tunneling spectroscopy (STS) ones (Wu, 2015). It is also robust toward oxidation, at least for 24 h in air, protected by an ultra-thin native oxide, which

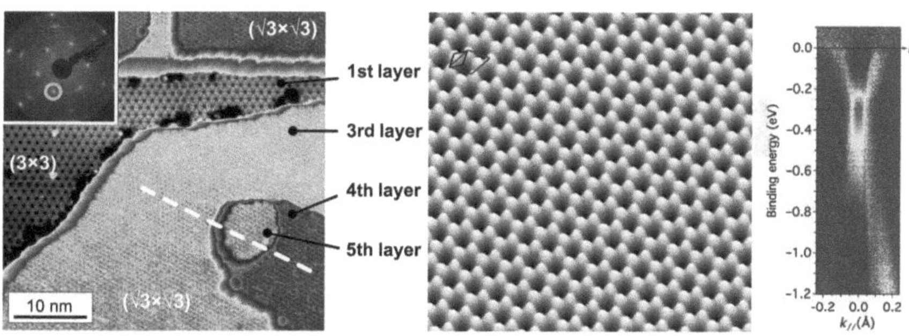

Fig. 14 (*Left*) Filled-states STM topograph after deposition of ~1.5 ML of Si onto the first layer 3×3 silicene phase, which is still visible. Well-ordered terraces can be seen showing the unique $\sqrt{3} \times \sqrt{3}$ reconstruction; (*inset*) LEED pattern of the surface where the (1/3,1/3) type superstructure spots of $\sqrt{3} \times \sqrt{3}$ multilayer silicene are circled in *blue* (*gray* in the print version). (*Middle*) 3D rendering of the $\sqrt{3} \times \sqrt{3}$ structure of multilayer silicene (filled-states 9 nm × 9 nm STM image). (*Right*) Evidence of Dirac fermions in multilayer silicene. A Dirac cone with π and π^* states, below, and above the Dirac point, respectively, is recorded in synchrotron radiation ARPES at the center of the Brillouin zone due to the band folding associated with the reconstruction. The Dirac point is situated at ~0.3 eV below the Fermi level ($E_F = 0.0$ eV). *Adapted from Fig. 2 of Vogt, P., Capiod, P., et al., 2014. Synthesis and electrical conductivity of multilayer silicene. Appl. Phys. Lett. 104, 021602 and Fig. 4 of Le Lay, G., Salomon, E., et al., 2016, Silicene: silicon conquers the 2D world. Europhys. News 47, 17–21.*

preserves the layers underneath (De Padova et al., 2014). This has allowed ex situ X-ray diffraction and Raman spectroscopy measurements performed in air without any protective capping, showing the (002) reflection and the G, D, and 2D Raman structures, which are unique fingerprints of multilayer silicene. This is why we think that the persisting debate on whether the observed $\sqrt{3}\times\sqrt{3}$ reconstruction is characteristic to multilayer silicene, or, instead, reflects diamond-like Si(111) thin islands, either intrinsically $\sqrt{3}\times\sqrt{3}$ reconstructed (Chen et al., 2015) or Ag terminated (Borensztein et al., 2015; Mahatha et al., 2015) will be soon closed (Solonenko et al., 2016). Indeed, single and multilayer silicene can been grown only in a narrow temperature window, which is often not well controlled by experimentalists, who work in reality at too high temperatures, where, Ag-terminated diamond-like Si(111) thin islands are likely to form.

Finally, for more detailed results and analysis, we refer the interested reader to a most recent review on emergent elemental 2D materials beyond graphene (Zhang et al., submitted).

2.2.2 Epitaxial Germanene

Two years after the first growths of silicene, the first claim of the synthesis of germanene on an Pt(111) substrate in 2014 (Li et al., 2014a) was immediately disputed as being, instead, a surface germanide (Švec et al., 2014). Soon after, still in 2014, germanene was synthesized on a gold (111) surface (see Fig. 15A–D) (Dávila et al., 2014). Among different surface structures, one forming large domains with a clear, nearly flat, honeycomb arrangement, as observed in STM imaging, was identified as a $\sqrt{3}\times\sqrt{3}R(30°)$ germanene layer in conjunction with a $\sqrt{7}\times\sqrt{7}R(\pm19.1°)$ Au(111) supercell, thanks to thorough synchrotron radiation core-level spectroscopy measurements and advanced DFT calculations.

A significant progress was achieved in 2015, upon growing 2×2 reconstructed germanene in a single phase at just 80°C on an aluminum (111) surface (see Fig. 15E and F) (Derivaz et al., 2015; Stephan et al., 2016). Now germanene, with 2D Dirac signature in STS, has been also demonstrated on Ge_2Pt clusters (Bampoulis et al., 2014; Zhang et al., 2015), yet this is not very practical. A challenge, the growth of germanene on a semiconducting or insulating substrate, has been partly attained upon evidence for germanene growth on epitaxial hexagonal (h)-AlN on

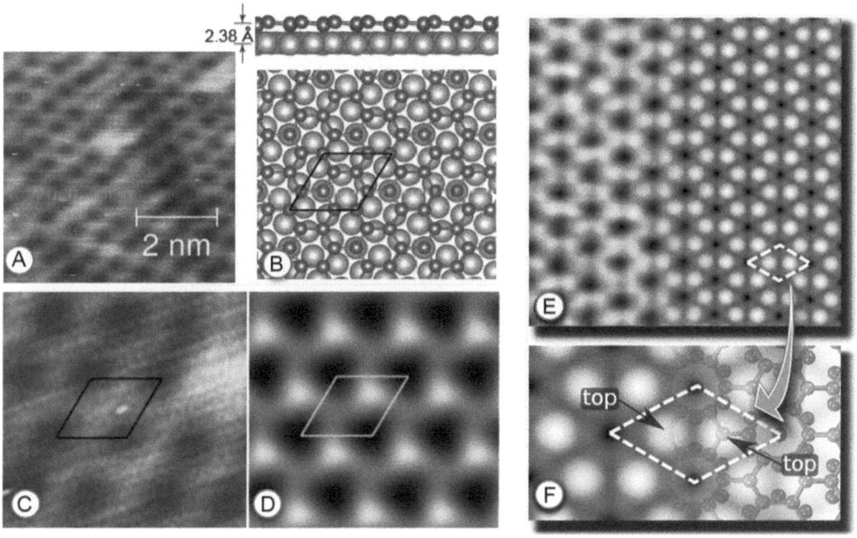

Fig. 15 (A) Filled-states STM image of the germanene sheet on Au(111). (B) Atomic model of the $\left(\sqrt{3}\times\sqrt{3}\right)$ reconstructed germanene phase coinciding with $\left(\sqrt{7}\times\sqrt{7}R(\pm19.1°)\right)$ Au(111) supercells. (C) close-up from (a). (D) Simulated STM image for the structure in (b) after smearing. (E) Empty-state experimental and calculated STM images of the single 2×2 germanene phase on a 3×3 Al(111) supercell. (F) Sketch of the atomic structure. *Panel D: After Fig. 3 of Dávila, M.E., Xian, L., et al., 2014. Germanene: a novel two-dimensional germanium allotrope akin to graphene and silicene. New J. Phys. 16, 095002. Panel F: After Fig. 4 of Derivaz, M., Dentel, D., et al., 2015. Continuous germanene layer on Al(111). Nano Lett. 15, 2510.*

Ag(111), where three germanene lattice parameters match four such parameters of the underlying AlN template (D'Acapito et al., 2016). If germanene can still be grown on thicker (h)-AlN films providing better isolation, a key step toward device fabrication could be reached.

Lastly, as previously with multilayer silicene, multilayer germanene has just been realized (Dávila and Le Lay, 2016). Thorough shallow core-level synchrotron radiation measurements have demonstrated the absence of gold atoms on top of the flat multilayer sheets, a point, which could be inferred from the absence of a simple $\sqrt{3}\times\sqrt{3}$ reconstruction that would be present in case of an Au-terminated Ge(111) growth (Le Lay, 1983). Despite the large Au(111) 8×8 coincidence cell and large inferred $3\sqrt{3}\times3\sqrt{3}$ germanene lattice parameter, clear signatures of Dirac cones have been evidenced in SR-ARPES, as shown in Fig. 16. Hence, multilayer

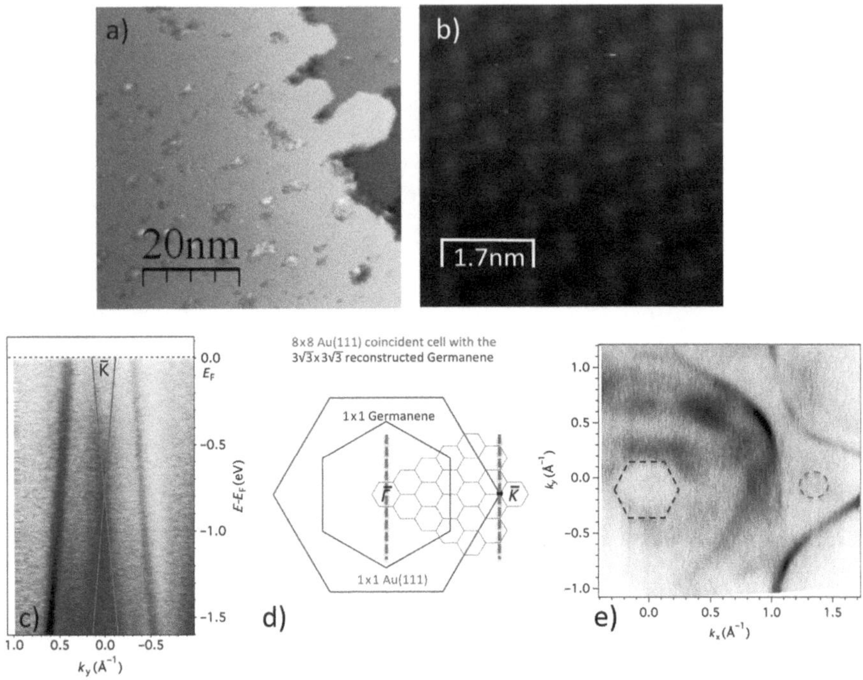

Fig. 16 (A) Large-scale filled-states STM image showing different flat terraces. (B) Higher resolution topograph showing large protrusions in a distorted hexagonal arrangement (image as acquired: no drift correction). (C) Cone-like dispersion, indicated by the *red* (*gray* in the print version) *lines*, along the *top dashed line* in panel (D) of a new state detected at the initial position of the K_{Au} point. (D) Scheme of the surface Brillouin zones: the small BZs in *red* (*light gray* in the print version) are common to the $\left(3\sqrt{3}\times 3\sqrt{3}\right)$ reconstructed germanene and the 8×8 Au(111) coincidence supercell; the K points of these small BZs coincide with the K_{Au} points, while the K and K' points of the germanene 1×1 surface BZ fold at Γ points of the small BZs and, hence, at the main zone center. (E) Fermi map of the multilayer germanene film: a new state around the zone center with hexagonal cross section (*dashed blue* (*black* in the print version) hexagon superposed) has replaced the initial Shockley surface state of the bare Au(111) substrate surface with circular cross section.

germanene appears a novel Dirac material, which could be very promising when referring to the predicted high mobilities of free-standing germanene (Ye et al., 2014), to the predicted robust 2D topological insulator character, nearly up to room temperature, resulting from the large effective spin–orbit coupling (Ezawa, 2015a), and, last but not least, to the possibility of very high Tc superconductivity, which should be searched for (Baskaran, 2013; Ezawa, 2015b).

2.3 Theoretical Studies of Silicene on Substrates

Motivated by the experimental success, the growth mechanism of silicene on Ag(111) substrate has been theoretically explored by atomistic first-principles calculations (Gao and Zhao, 2012; Shu et al., 2014). Using DFT calculations, Gao and Zhao (2012) explored the geometries and energies of small silicon clusters on Ag(111) surface to understand the early growth behavior of silicene. They showed that planar silicon clusters composed of six member rings as small patches of silicene clusters (Fig. 17) are more stable than 3D silicon clusters deposited on the Ag(111) surface. Upon relaxation, the planar honeycomb structures of small silicene patches are preserved on Ag(111), which in turn can act as the nucleation center during the growth of silicene. Indeed, small silicon clusters (~1.2 nm in diameter) as the nuclei of silicene growth were recognized as dotted spots in STM topography of 0.77 ML Si on Ag(111) in a recent experiment (Grazianetti et al., 2015). Due to a metal passivation effect, the formation energy of silicene-like 2D silicon clusters decreases by adsorption on the Ag(111) surface and also decreases as the number of Si atoms is increasing, suggesting that the 2D aggregation of silicon atom is energetically favorable. Shu et al. (2014) further demonstrated that the nucleation rate of silicene is very sensitive to both growth temperature and chemical potential ($\Delta\mu$) of the silicon atoms. For example, the nucleation rate shows a ~23 orders of magnitude

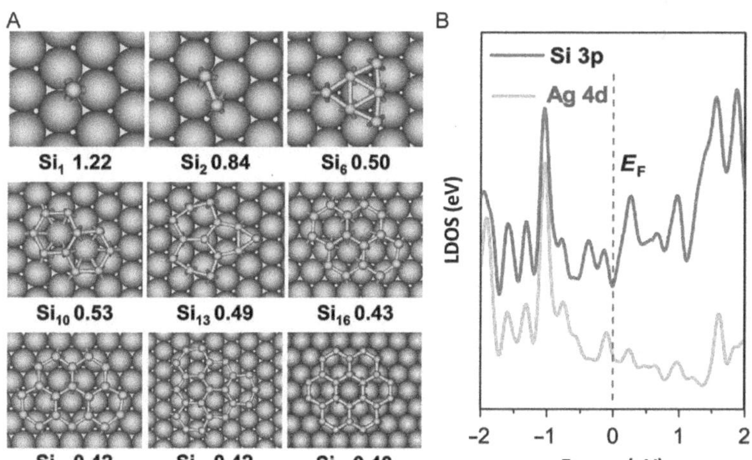

Fig. 17 (A) Atomic structures and formation energies (eV per Si atom) of 2D Si_n clusters on the Ag(111) surface. (B) local density of states for central six silicon atoms (3p orbital) and the Ag atoms (4d orbital) underneath the Si hexagon of the Si_{24} cluster on Ag(111).

increase (from $\sim 10^{-10}$ to $\sim 10^{13}$ cm^{-2} s^{-1}) as $\Delta\mu$ increases from 0.05 to 0.15 eV at a typical growth temperature of 500 K.

To date, there have been many theoretical calculations (Gao and Zhao, 2012; Guo et al., 2013; Liu et al., 2014c; Mahatha et al., 2014; Quhe et al., 2014; Wang and Cheng, 2013) on the interactions between silicene and Ag(111) surface, focusing on the electronic band structure and existence of Dirac cone. Using a planar Si_{24} cluster as a representative, Gao and Zhao (2012) revealed strong hybridization between Si $3p_z$ states and Ag $4d$ states in the vicinity of the Fermi level (see Fig. 17). Such strong Si–Ag hybridization, as evidenced by a recent Raman experiment Tao et al. (2015), not only stabilizes silicene sheets on the Ag(111) surface but also significantly affects the electronic band structure of silicene on the Ag substrate. As a consequence, the Dirac cone of monolayer silicene is usually destroyed on Ag(111) surfaces according to DFT calculations. Guo et al. (2013) considered several silicene superstructures with 4×4, $\sqrt{13}\times\sqrt{13}$, $2\sqrt{3}\times2\sqrt{3}$ periodicities (with respect to Ag(111)) and recognized two factors for the disappearance of Dirac electrons: (1) the substantial buckling of the silicene layers and (2) the mixing of the π state with the states of Ag atoms. Meanwhile, the linear dispersions observed in ARPES experiments were suggested to originate from either the *sp* band of the Ag(111) surface (Wang and Cheng, 2013), which, however, is at variance with the absence of dispersion with $k_\perp$ noted in Vogt et al. (2012), or the interface states between silicene and Ag(111) (Cahangirov et al., 2013).

In addition to Ag(111) substrates, there have been first-principle calculations about silicene on the surfaces of other metals, such as Al, Mg, Au, Cu, Pt, and Ir (Quhe et al., 2014). The strength of interaction between silicene and metal substrates ascends in the following order: Al < Mg < Ag < Au < Cu < Ir < Pt. Band structure calculations showed that the absence of the Dirac cone is a common feature for epitaxial silicene on metal substrates. However, intercalation of K atoms between silicene and these metal substrates would recover the Dirac cone of silicene, similar to the K intercalation in bilayer silicene (Liu et al., 2014c). The recovered Dirac cone is located at 0.40–0.78 eV below E_F, corresponding to an n-type doping of silicene.

For the real applications of silicene in electronic devices, silicene sheets have to be supported on some semiconducting or insulating substrate. Hence, there have been some theoretical efforts about the effect of semiconducting and insulating substrates on the electronic properties of silicene. As shown in Fig. 18, the Dirac cone in pristine silicene (highlighted by

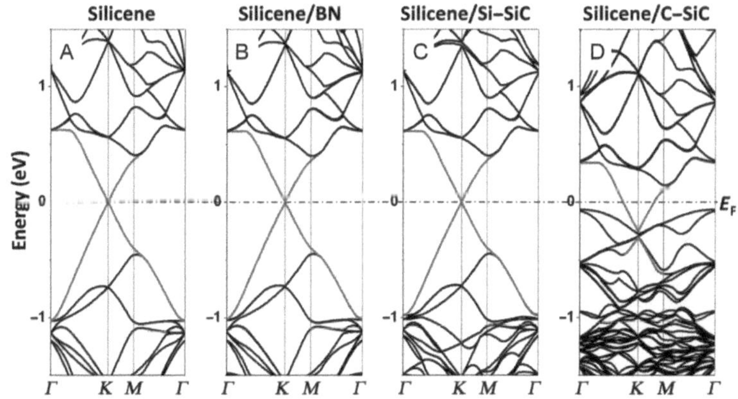

Fig. 18 Electronic band structures of (A) silicene, (B) silicene on BN, (C) silicene on Si-terminated SiC(0001), and (D) silicene on C–SiC. The *red* (*gray* in the print version) *lines* highlight the π-bands of silicene in different configurations; the Fermi level is set to zero (Liu et al., 2013).

red (*gray* in the print version) *lines*) can be well preserved after placed on h-BN monolayer and Si-terminated SiC(0001) surface, with small band gaps of several meV opened at the Fermi level (Liu et al., 2013). In contrast, the π bands of silicene are severely disturbed for the C-terminated SiC(0001) surface. Such different substrate effects can be interpreted by the relative locations of the valence band maxima and conduction band minima of silicene and substrates.

In experiment (Chiappe et al., 2014), high-buckled (HB) silicene has been reported to be synthesized on a semiconducting substrate, the MoS_2 surface. However, the prepared silicene is high-buckled and metallic. Accordingly, HB silicene on MoS_2 substrates has been theoretically investigated by DFT calculations (Li and Zhao, 2014). It was found that high-buckled silicene can be weakly adsorbed on MoS_2 substrate with binding energies around 80 meV per silicon atom and the metallic electronic band structure of HB silicene is preserved on MoS_2. In addition, many other insulating or semiconducting substrates, such as MoX_2 and GaX (X=S, Se, and Te), MgX_2 (X=Cl, Br, and I), WSe_2, β-Si_3N_4(0001)/Si(111), hydrogenated Si(111) and Ge(111) surfaces, have been considered as possible supports of silicene (Filippone, 2014; Kokott et al., 2013; Scalise et al., 2014; Zhu and Schwingenschlögl, 2014, 2015). First-principles calculations show that all these substrates can provide effective mechanical support without destroying the geometry and electronic structures of silicene and result in small band gaps from tens to more than 100 meV.

3. FUNCTIONALIZED SILICENE AND GERMANENE

3.1 Experimental Functionalization

As mentioned in Section 1, it was little known that functionalized group-IV sheets had been experimentally realized already 10 years ago by wet chemistry (Nakano et al., 2006). Now better control of, typically, germanane synthesis, i.e., of fully hydrogenated germanene (GeH), has been achieved by Bianco et al. (2013) from the topochemical deintercalation of the layered van der Waals solid calcium digermanide ($CaGe_2$) (Bianco et al., 2013); next, the stability of germanane was improved by replacing the H atom termination with a methyl group one (Jiang et al., 2014). However, as we focus in this chapter on structural and electronic properties, in the perspective of potential device applications, we will concentrate on experimental in situ realizations of functionalized silicene. At present, they are rather scarce.

A recent work (Du et al., 2014) reports a study of band-gap tuning in different silicene buckling structures on Ag(111) by controllable oxidation processes, using STM and in situ Raman spectroscopy, assisted by DFT calculations. The band gap can be modulated from semimetallic to semiconducting type, which can very well overcome the zero-gap disadvantage of pristine silicene. In fully monolayer oxidized silicene, the buckled silicene structure vanishes, with subsequent crumpling of the sample and exposure of bare Ag(111) surface areas. Hence, this does not look very practical. Instead, as indicated in Section 2.2.1, exposure of multilayer silicene to ambient air leads to the formation of an ultra-thin protective cap, which preserves the layers underneath (De Padova et al., 2014). Conversely, to hinder the oxidation of monolayer and multilayer silicene, ad hoc Al- and Al_2O_3-based nonreactive encapsulation of the silicene layers appears generally applicable to any silicene configuration, irrespective of the support substrate (Molle et al., 2013). It can lead to the fabrication of atomically sharp and chemically intact Al/silicene and Al_2O_3/silicene interfaces that can be functionally used for ex situ characterization as well as for gated device fabrication, as we will see later (Tao et al., 2015).

Other chemical modifications of silicene have been extensively reviewed in Wang et al. (2015a). We present, here, the challenge toward silicane, i.e., fully hydrogenated silicene (SiH), synthesis. The first experiments were carried out by Qiu et al. (2015a), who showed the surprisingly ordered adsorption of hydrogen atoms on the initial $3 \times 3/4 \times 4$ silicene phase, preserving, remarkably, the $3 \times 3/4 \times 4$ cell, but breaking the symmetry between the two half unit cells (HUC), as a result of a change in the buckling configuration, seemingly, but misleadingly, appearing as a lateral shift from the so-called α-(3 × 3) to β-(3 × 3) positions (see Fig. 19) (Qiu et al., 2015a). In a subsequent paper, Qiu et al. (2015b) has claimed to half-silicane fabrication (i.e., silicene complete hydrogenation on the exposed surface), starting, instead, from the $\left(2\sqrt{3} \times 2\sqrt{3}\right)R30°$ structure (noted with respect to Ag(111)). However, as this structure appears very disordered and defective in STM images (Liu et al., 2014e), and has been even assigned to a surface alloy (Rahman et al., 2015), this claim seems questionable.

Fig. 19 (A) A large-area STM image of a hydrogenated silicene-H(3 × 3) surface on Ag(111) showing an ordered 3 × 3/4 × 4 structure. (B) Enlarged image of the hydrogenated (3 × 3) phase. The *white rhombus marks* an apparent unit cell of the structure. There are six bright protrusions in one half unit cell and one protrusion in the other HUC. (C) Corresponding structural model; the *white* and *red* (*dark gray* in the print version) *rhombuses* correspond to the positions of apparent UCs of α-(3 × 3) and β-(3 × 3) phases, respectively, which are shifted relatively. *Adapted from Figs. 2 and 3 of Qiu, J., Fu, H., et al., 2015a. Ordered and reversible hydrogenation of silicene. Phys. Rev. Lett. 114, 126101.*

3.2 Functionalized Silicene: Theory

The zero-gap character of silicene and germanene hinders their applications in microelectronic and optoelectronic devices. Therefore, it is desirable to open a finite band gap in silicene and germanene by chemical functionalization. Hydrogenation can convert sp^2 hybridized silicon/germanium atoms into sp^3 and provide an effective way to fulfill the goal of gap opening (Houssa et al., 2011; Lew Yan Voon et al., 2010). Fully SiH and germanene are named as silicane and germanene, respectively. Two kinds of configurations for silicane or germanane, i.e., chair and boat, are found to be dynamically stable from phonon dispersion curves by first-principles calculations (Houssa et al., 2011), whereas chair configuration is slightly more stable than boat one for both silicane and germanane. As shown in Fig. 20, neighboring hydrogen atoms sit alternatively above and below the plane in the chair configuration, while the boat one has paired hydrogen atoms alternating.

The hydrogenated derivatives of silicene and germanene are all semiconductors, and their theoretical band gaps from DFT calculations (Houssa et al., 2011) are summarized in Table 5. The band gaps predicted from more accurate HSE (Heyd–Scuseria–Ernzerhof) or GW (the approximation is that the expansion of the self-energy Σ in terms of the single particle Green's function G and the screened Coulomb interaction W) methods are in the range of 2.9–4.0 eV. Chair-like silicane is an indirect semiconductor with larger band gap; while boat-like silicane is a direct semiconductor. In contrast, germanane is always a direct semiconductor irrelevant to its atomic configuration, with less magnitude of band gap than silicane.

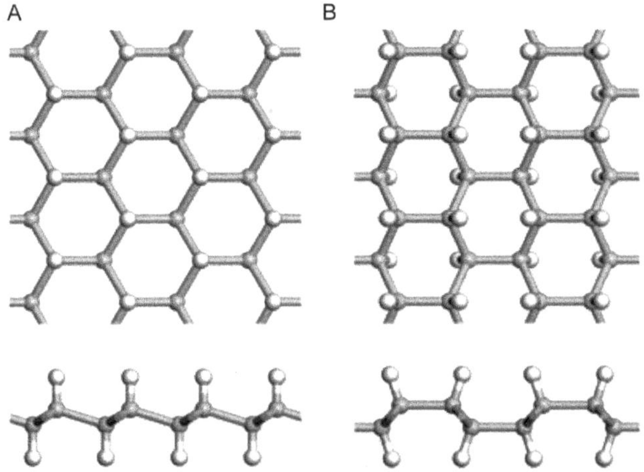

Fig. 20 Top view (*upper*) and side view (*lower*) of silicane or germanane in (A) chair and (B) boat configurations. The *yellow* (*gray* in the print version) *balls* represent Si/Ge atoms and the *white balls* represent H atoms.

Table 5 Energy Gaps (eV) of Silicane and Germanane in Chair and Boat Configurations, Computed Using Local-Density Approximations (LDA) Functional, Heyd–Scuseria–Ernzerhof (HSE) Hybrid Functional, and GW Approximation (Houssa et al., 2011)

	Silicane		Germanane	
Method	**Chair**	**Boat**	**Chair**	**Boat**
LDA	2.0	1.6	1.4	1.3
HSE	4.0	3.3	3.6	2.9
GW	3.8	2.9	3.5	2.9

In addition to two-side hydrogenation, semihydrogenation on one side of silicene or germanene have also been investigated by DFT calculations (Zhang et al., 2012; Wang et al., 2012) and found to be dynamically stable from the phonon dispersion curves. One-side semihydrogenated silicene and germanene are both semiconductors with direct band gaps. The band gaps of semihydrogenated silicene from HSE06 calculations of 1.79 eV (chair) and 1.14 eV (boat) (Zhang et al., 2012), are only about half of the fully SiH with the same configuration. Interestingly, ferromagnetism is induced in silicene and germanene by one-side semihydrogenation, since semihydrogenation breaks the extended π-bonding network in silicene and germanene (Wang et al., 2012). As a result, every unsaturated Si (Ge) atom has an unpaired localized electron and carries a magnetic moment of 1 μ_B.

Similar to the case of hydrogenation, chemisorption of halogen elements can also open a band gap in silicene (or germanene), and chair configuration is still more stable than boat configuration for X-silicene (X = F, Cl, Br, and I). The band structures of the chair-like X-silicene systems are shown in Fig. 21. From sX-LDA calculations, all X-silicene are direct semiconductors with band gaps of 1.469, 1.979, 1.950, and 1.194 eV for F, Cl, Br, and I, respectively (Gao et al., 2012). The variation trend of band gap can be interpreted by the competition of Si–Si and Si–X bond strengths.

Chemical stability of silicene exposed in air and oxidation of silicene are critical concerns to its device applications. The stability of free-standing silicene sheets under oxygen environment has also been investigated by first-principles calculations (Liu et al., 2014b). It was found that an O_2 molecule can be easily adsorbed on silicene and dissociates into two O atoms spontaneously without overcoming any energy barrier. The dissociated O atoms form strong Si–O bonds, lowering the energy by 4.046–5.355 eV per O_2. Oxidization also provides an effective way to tune band gap of silicene in

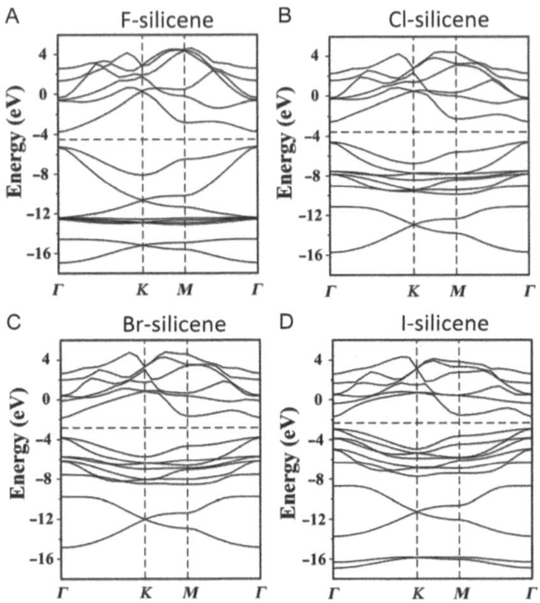

Fig. 21 Energy band structures for (A) F-silicene, (B) Cl-silicene, (C) Br-silicene, and (D) I-silicene. The Fermi level is labeled with a *dashed line* (Gao et al., 2012). *Reproduced from Gao, N., Zheng W.T., et al., 2012. Density functional theory calculations for two-dimensional silicene with halogen functionalization. Phys. Chem. Chem. Phys. 14, 257, with permission from the PCCP Owner Societies.*

a wide range. Wang et al. (2013) investigated the structures and electronic properties of silicene oxides by incorporating atomic oxygen (O) and hydroxyl (OH) onto silicene lattice in a variety of bonding configurations. Depending on the bonding of O and OH, the partially and fully oxidized silicene could be metallic, semimetallic, semiconducting, and insulating.

4. FIRST SILICENE FIELD EFFECT TRANSISTORS

Although there were proposals for silicene-based transistors in the literature, it was generally considered that silicene would not withstand isolation from its metallic substrate or exposure to ambient air (Geim and Grigorieva, 2013). Hence, the report in 2015 of the first fabrication of a field effect transistor (FET) operating at room temperature and exhibiting an ambipolar behavior, in which silicene was used as the channel material, as illustrated in Fig. 22, came as a surprise (Le Lay, 2015; Tao et al., 2015). Even if the carrier mobilities of about 100 cm^2/Vs at room temperature are significantly less than what could be expected in principle, and if the ON/OFF

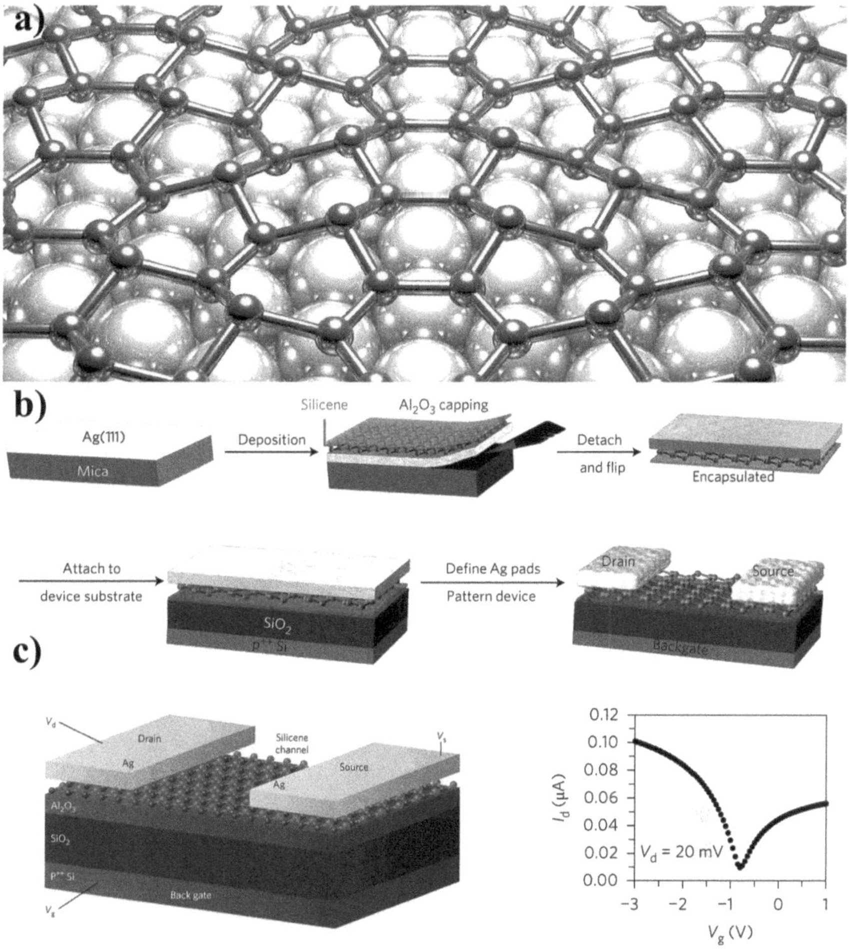

Fig. 22 (A) Perspective view of silicene in the $3 \times 3/4 \times 4$ structure on Ag(111). (B) The synthesis–transfer–fabrication process of a silicene transistor with following key steps: epitaxial growth of silicene on a silver (111) epitaxial thin film on mica, in situ Al_2O_3 capping, encapsulated delamination transfer of silicene, and native contact electrode formation to enable back-gated silicene transistors. (C) The fabricated transistor and its drain current vs V_g displaying the ambipolar electron–hole symmetry expected from silicene (V_s, V_d, and V_g are the source, drain, and gate voltages, I_d is the drain current). *Adapted from Tao, L., Cinquanta, E., et al., 2015. Silicene field-effect transistors operating at room temperature. Nat. Nanotechnol. 10, 227–231; Le Lay, G., 2015. 2D materials: silicene transistors. Nat. Nanotechnol. 10, 202, Nature Publishing Group.*

current ratio is inferior to 10, undoubtedly, this represents a remarkable breakthrough. Indeed, the direct compatibility with the current Si-based technologies is the key point, while plenty of room exists for improving the performance. This could be achieved by optimizing the silicene growth and engineering process, and tuning and enlarging the band gap, typically, by surface adsorption, e.g., using for example half silicane (Pan et al., 2014), fabricating tunneling FETs (Ni et al., 2014), and protecting the silicene gate (which is exposed to air in the present devices). Multilayer silicene, which is, as we have seen earlier, robust in air (De Padova et al., 2014), could be possibly used to solve this last problem.

Furthermore, this breakthrough opens a novel route to push the down-scaling of the FET toward transistors with gate lengths of 5 nm or less, as shown in Fig. 23, typically to overcome the so-called ON/OFF and energy dissipation problems, i.e., detrimental short-channel effects (Dimoulas, 2015).

To this end, the use of 2D materials in the active channel of the transistor might be essential (Fischetti et al., 2013) and novel materials such as silicene, germanene, and their derivatives could be a source of innovation and performance gain. A large ON/OFF ratio (>10) is needed for digital devices, and a good drain current saturation is needed for power gain in radio frequency FETs (Schwierz et al., 2015). In this respect, although intensive efforts have made remarkable progresses in graphene device technology, still, graphene fares poorly on both aspects due to the absence of a band gap (Schwierz, 2010).

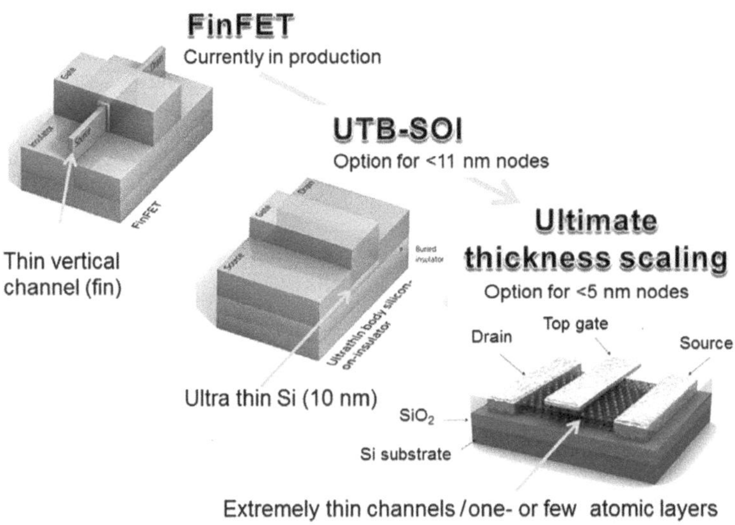

Fig. 23 Reduction of channel thickness is at the core of the nanoelectronic device scaling scenario. Ultimate thickness scaling with atomically thin layers offering excellent electrostatic control could be a key option for scaling beyond the 5 nm technology node. *Adapted from Fig. 1 of Dimoulas, A., 2015. Silicene and germanene: silicon and germanium in the "flatland". Microelectr. Eng. 131, 68.*

As we have seen earlier, the reactive surfaces of silicene and germanene enable chemical modification to tune their band structures, which appears highly promising.

5. FUTURE APPLICATIONS

Owing to the prominent properties, silicene and germanene are expected to find a variety of applications in future 2D nanoelectronic devices, just like their brother, graphene. In addition to the prototypical FET discussed in Section 4, there have been many theoretical proposals of other kinds of silicene/germanene-based FETs, such as dual-gated FET (Ni et al., 2012), tunneling FET (Ni et al., 2014), and all-metallic FET (Wang et al., 2015b). Spintronic (Pan et al., 2014; Tsai et al., 2013) and thermoelectric devices (Yang et al., 2014b) can be also constructed, for instance, spin filters and spin switches using semihydrogenated silicene (Pan et al., 2014), and a spin filter based on a quantum point contact in a silicene sheet (Tsai et al., 2013). In recent theoretical studies, other potential applications including highly sensitive gas sensors (Hu et al., 2014), DNA nucleobase sensors (Amorim and Scheicher, 2015), cathode materials for Li–air batteries (Hwang et al., 2015), anode materials for Li–ion batteries (Tritsaris et al., 2013), and Na–ion batteries (Kulish et al., 2014), have been proposed and await experimental validation. Metal decorated silicene or silicane might be also utilized as hydrogen storage materials with the maximum gravimetric density of H_2 up to 9.40 wt.% (Wang et al., 2014b).

6. SUMMARY

The study of elemental 2D materials from group IV is likely to remain the most studied group of elements from the periodic table not only because of the popularity of graphene, but also because of the challenges in growth for the other materials and the wide array of properties. The progression from the perfectly flat sp^2 graphene to the buckled sp^2/sp^3 bonded structure of the other elements makes this group a unique one to study.

In this chapter, we have presented at length the properties of silicene and germanene, both predicted and measured. Even though they were both considered in the seminal work by Takeda and Shiraishi (1994), it is not surprising that silicene has to date received most of the attention (Zhao et al., 2016) and was next fabricated, in 2012, after graphene. Germanene was next to be fabricated, in 2014. It has not taken much longer, though, for the first report of the fabrication of stanene on $Bi_2Te_3(111)$ substrates in 2015 (Zhu et al., 2015). Because, this was the only experimental report at the time we wrote this chapter, we thought it was premature to extend it with the study of stanene, despite its great potential for hosting the quantum spin Hall effect at room temperature and even above, and, consequently, the potential for achieving dissipation-less spintronic devices. However, we must ultimately note, just before going for press, that a new realization of buckled primitive stanene on noble metal substrates (Ag(111) and Au(111)) has been presented by the group of J. Yuhara (with one of the present author, G.L.L.) at the 71st Annual Meeting of the Physical Society of Japan (Sendai, March 2016), (Yuhara et al., 2016), which confirms large advances in the synthesis of this new 2D tin graphene-like allotrope. In this respect, it is also worth-mentioning the discovery of a new 2D-related materials system, namely, 2D tin monoxide (SnO), accompanied by the fabrication of FETs (Saji et al., 2016).

Finally, the gold rush for novel artificial 2D elemental materials continues: a first report on the synthesis of borophene, i.e., 2D column III boron sheets, has been published in late 2015 (Mannix et al., 2015), rapidly followed by a second report in March 2016 (Feng et al., 2016).

Finally, coming back to column IV, if leadene, that is lead films with honeycomb structure and tantalizing topological properties (Lu et al., 2016) has not yet been synthesized, we are pretty convinced that its realization is now at hand. Theoretically, all of the earlier and many other elemental 2D materials have been proposed; however, we leave out consideration of them here since they have not yet been realized.

ACKNOWLEDGMENTS

L.C.L.Y.C.'s contribution was supported by The Citadel Foundation. M.E.D. acknowledges support from the Projects Nos. AYA2012-39832-C02-01/02 and ESP2015-67842-P of MINECO of Spain.

REFERENCES

Acun, A., Poelsema, B., et al., 2013. The instability of silicene on Ag(111). Appl. Phys. Lett. 103, 263119.

Amorim, R.G., Scheicher, R.H., 2015. Silicene as a new potential DNA sequencing device. Nanotechnology 26, 154002.

Balandin, A.A., Ghosh, S., et al., 2008. Superior thermal conductivity of single-layer graphene. Nano Lett. 8, 902–907.

Bampoulis, P., Zhang, L., et al., 2014. Germanene termination of Ge_2Pt crystals on Ge(110). J. Phys. Condens. Matter 26, 442001.

Baroni, S., de Gironcoli, S., et al., 2001. Phonons and related crystal properties from density-functional perturbation theory. Rev. Mod. Phys. 73, 515.
Baskaran, G., 2013. Room temperature superconductivity, spin liquid and mott insulator: silicene and germanene as prospective playgrounds. http://arxiv.org/abs/1309.2242. arXiv:1309.2242v2.
Bianco, E., Butler, S., et al., 2013. Stability and exfoliation of germanane: a germanium graphane analogue. ACS Nano 7, 4414.
Borensztein, Y., Curcella, A., et al., 2015. Silicene multilayers on Ag(111) display a cubic diamondlike structure and a $\sqrt{3}\times\sqrt{3}$ reconstruction induced by surfactant Ag atoms. Phys. Rev. B 92, 155407.
Brumfiel, G., 2013. Sticky problem snares wonder material. Nature 495, 152–153.
Cahangirov, S., Topsakal, M., et al., 2009. Two- and one-dimensional honeycomb structures of silicon and germanium. Phys. Rev. Lett. 102, 236804.
Cahangirov, S., Audiffred, M., et al., 2013. Electronic structure of silicene on Ag(111): strong hybridization effects. Phys. Rev. B 88, 035432.
Chen, J., Du, Y., et al., 2015. Delocalized surface state in epitaxial Si(111) film with spontaneous $\sqrt{3}\times\sqrt{3}$ superstructure. Sci. Rep. 5, 13590.
Chiappe, D., Scalise, E., et al., 2014. Two-dimensional Si nanosheets with local hexagonal structure on a MoS_2 surface. Adv. Mater. 26, 2096–2101.
D'Acapito, F., Torrengo, S., et al., 2016. Evidence for germanene growth on epitaxial hexagonal (h)-AlN on Ag(1 1 1). J. Phys. Condens. Matter 28, 045002.
Dávila, M.E., Le Lay, G., 2016. Few layer epitaxial germanene: a novel two-dimensional Dirac material. Sci. Rep. 6, 20714.
Dávila, M.E., Xian, L., et al., 2014. Germanene: a novel two-dimensional germanium allotrope akin to graphene and silicene. New J. Phys. 16, 095002.
Day, C., 2015. The dayside: hot physics. Phys. Today (25 September).
De Padova, P., Avila, J., et al., 2013. The quasiparticle band dispersion in epitaxial multilayer silicene. J. Phys. Condens. Matter 25, 382202.
De Padova, P., Ottaviani, C., et al., 2014. 24 h stability of thick multilayer silicene in air. 2D Mater. 1, 021003.
Derivaz, M., Dentel, D., et al., 2015. Continuous germanene layer on Al(111). Nano Lett. 15, 2510.
Dimoulas, A., 2015. Silicene and germanene: silicon and germanium in the "flatland". Microelectron. Eng. 131, 68.
Drummond, N.D., Zólyomi, V., et al., 2012. Electrically tunable band gap in silicene. Phys. Rev. B 85, 075423.
Du, Y., Zhuang, J., et al., 2014. Tuning the band gap in silicene by oxidation. ACS Nano 8, 10019.
Durajski, A.P., Szczęśniak, D., et al., 2014. Study of the superconducting phase in silicene under biaxial tensile strain. Solid State Commun. 200, 17–21.
Durgun, E., Tongay, S., et al., 2005. Silicon and III-V compound nanotubes: structural and electronic properties. Phys. Rev. B 72, 075420.
Ezawa, M., 2015a. Monolayer topological insulators: silicene, germanene, and stanene. J. Phys. Soc. Jpn. 84, 121003.
Ezawa, M., 2015b. Photo-induced topological superconductor in silicene, germanene, and stanene. J. Supercond. Nov. Magn. 28, 1249–1253.
Feng, Y., Liu, D., et al., 2015. Direct evidence of interaction-induced Dirac cones in monolayer silicene/Ag(111) system. http://arxiv.org/abs/1503.06278. arXiv:1503.06278.
Feng, B., Zhang, J., et al., 2016. Experimental realization of two-dimensional boron sheets. Nat. Chem.. http://dx.doi.org/10.1038/NCHEM.2491.
Ferrari, A.C., Meyer, J.C., et al., 2006. Raman spectrum of graphene and graphene layers. Phys. Rev. Lett. 97, 187401.

Filippone, F., 2014. Interaction of silicene with beta-Si3N4(0001)/Si(111) substrate; energetics and electronic properties. J. Phys. Condens. Matter 26, 395009.
Fischetti, M.V., Fu, B., et al., 2013. Theoretical study of the gate leakage current in sub-10-nm field-effect transistors. IEEE Trans. Electron Devices 60, 3862.
Fleurence, A., Friedlein, R., et al., 2012. Experimental evidence for epitaxial silicene on diboride thin films. Phys. Rev. Lett. 108, 245501.
Gao, J., Zhao, J., 2012. Initial geometries, interaction mechanism and high stability of silicene on Ag(111) surface. Sci. Rep. 2, 861.
Gao, N., Zheng, W.T., et al., 2012. Density functional theory calculations for two-dimensional silicene with halogen functionalization. Phys. Chem. Chem. Phys. 14, 257.
Geim, A.K., Grigorieva, I.V., 2013. Van der Waals heterostructures. Nature 499, 419–425.
Geim, A.K., Novoselov, K.S., 2007. The rise of graphene. Nat. Mater. 6, 183–191.
Grazianetti, C., Chiappe, D., et al., 2015. Nucleation and temperature-driven phase transitions of silicene superstructures on Ag(111). J. Phys. Condens. Matter 27, 255005.
Guo, Z.-X., Furuya, S., et al., 2013. Absence and presence of Dirac electrons in silicene on substrates. Phys. Rev. B 87, 235435.
Guzmán-Verri, G.G., Lew Yan Voon, L.C., 2007. Electronic structure of silicon-based nanostructures. Phys. Rev. B 76, 075131.
Hoffmann, R., 2013. Small but strong lessons from chemistry for nanoscience. Angew. Chem. Int. Ed. 52, 93–103.
Houssa, M., Scalise, E., et al., 2011. Electronic properties of hydrogenated silicene and germanene. Appl. Phys. Lett. 98, 223107.
Hu, M., Zhang, X., et al., 2013. Anomalous thermal response of silicene to uniaxial stretching. Phys. Rev. B 87, 195417.
Hu, W., Xia, N., et al., 2014. Silicene as a highly sensitive molecule sensor for NH_3, NO and NO_2. Phys. Chem. Chem. Phys. 16, 6957.
Huang, L.-F., Gong, P.-L., et al., 2015. Phonon properties, thermal expansion, and thermomechanics of silicene and germanene. Phys. Rev. B 91, 205433.
Hwang, Y., Yun, K.H., et al., 2015. Carbon-free and two-dimensional cathode structure based on silicene for lithium–oxygen batteries: a first-principles calculation. Power Sources 275, 32–37.
Jiang, S., Butler, S., et al., 2014. Monolithic graphene oxide sheets with controllable composition. Nat. Commun. 5, 3383.
Kaloni, T.P., Schwingenschlögl, U., 2013. Stability of germanene under tensile strain. Chem. Phys. Lett. 583, 137–140.
Kaloni, T.P., Cheng, Y.C., et al., 2013. Hole doped Dirac states in silicene by biaxial tensile strain. J. Appl. Phys. 113, 104305.
Kaltsas, D., Tsetseris, L., 2013. Stability and electronic properties of ultrathin films of silicon and germanium. Phys. Chem. Chem. Phys. 15, 9710–9715.
Kamatagi, M.D., Sankeshwar, N.S., 2015. Thermal conductivity of silicene—a molecular dynamics study. AIP Conf. Proc. 1665, 110036.
Kokott, S., Matthes, L., et al., 2013. Silicene on hydrogen-passivated Si(111) and Ge(111) substrates. Phys. Status Solidi RRL 7, 538–541.
Kulish, V.V., Malyi, O.I., et al., 2014. Controlling Na diffusion by rational design of Si-based layered architectures. Phys. Chem. Chem. Phys. 16, 4260.
Le Lay, G., 1983. Physics and electronics of the noble-metal/elemental-semiconductor interface formation: a status report. Surface Sci. 132, 169–204.
Le Lay, G., 2015. 2D materials: silicene transistors. Nat. Nanotechnol. 10, 202.
Le Lay, G., Cahangirov, S., et al., 2014. Silicene phases on Ag(111). In: IEEE Xplore, Int. Conf. on Manipulation, Manufacturing and Measurement on the Nanoscale (3M-NANO), pp. 108–111.

Le Lay, G., Salomon, E., et al., 2016. Silicene: silicon conquers the 2D world. Europhys. News 47, 17–21.

Lew Yan Voon, L.C., Sandberg, E., et al., 2010. Hydrogen compounds of group-IV nanosheets. Appl. Phys. Lett. 97, 163114.

Lew Yan Voon, L.C., López-Bezanilla, A., et al., 2015. Effective Hamiltonians for phosphorene and silicene. New J. Phys. 17, 025004.

Li, H.-P., Zhang, R.-Q., 2012. Vacancy-defect-induced diminution of thermal conductivity in silicene. EPL (Europhys. Lett.) 99, 36001.

Li, L., Zhao, M., 2014. Structures, energetics, and electronic properties of multifarious stacking patterns for high-buckled and low-buckled silicene on the MoS_2 substrate. J. Phys. Chem. C 118, 19129–19138.

Li, L., Lu, S.-Z., et al., 2014. Buckled germanene formation on Pt(111). Adv. Mater. 26, 4820–4824.

Lin, X., Ni, J., 2015. Dirac points and van Hove singularities of silicene under uniaxial strain. J. Appl. Phys. 117, 164305.

Liu, C.-C., Feng, W., Yao, Y., 2011. Quantum spin Hall effect in silicene and two-dimensional germanium. Phys. Rev. Lett. 107, 076802.

Liu, G., Wu, S., et al., 2012. Strain-induced semimetal-metal transition in silicene. EPL (Europhys. Lett.) 99, 17010.

Liu, H., Gao, J., et al., 2013. Silicene on substrates: a way to preserve or tune its electronic properties. J. Phys. Chem. C 117, 10353–10359.

Liu, B., Reddy, C.D., et al., 2014a. Thermal conductivity of silicene nanosheets and the effect of isotopic doping. J. Phys. D. Appl. Phys. 47, 165301.

Liu, G., Lei, X.L., et al., 2014b. Is silicene stable in O_2? First-principles study of O_2 dissociation and O_2-dissociation-induced oxygen atoms adsorption on free-standing silicene. Europhys. Lett. 106, 47001.

Liu, H., Han, N., et al., 2014c. Band gap opening in bilayer silicene by alkali metal intercalation. J. Phys. Condens. Matter 26, 475303.

Liu, Z.-L., Wang, M.-X., et al., 2014d. Various atomic structures of monolayer silicene fabricated on Ag(111). New J. Phys. 16, 075006.

Liu, Z.-L., Wang, M.-X., et al., 2014e. The fate of the $2\sqrt{3}\times2\sqrt{3}R(30°)$ silicene phase on Ag(111). APL Mater. 2, 092513.

Lu, Y.H., Zhou, D., et al., 2016. Topological properties of atomic lead film with honeycomb structure. Sci. Rep. 6, 21723.

Mahatha, S.K., Moras, P., et al., 2014. Silicene on Ag(111): a honeycomb lattice without Dirac bands. Phys. Rev. B 89, 201416.

Mahatha, S.K., Moras, P., et al., 2015. Evidence for a diamondlike electronic band structure of Si multilayers on Ag(111). Phys. Rev. B 92, 245127.

Manjanath, A., Kumar, V., et al., 2014. Mechanical and electronic properties of pristine and Ni-doped Si, Ge, and Sn sheets. Phys. Chem. Chem. Phys. 16, 1667–1671.

Mannix, A.J., Zhou, X.-F., et al., 2015. Synthesis of borophenes: anisotropic, two-dimensional boron polymorphs. Science 350, 1513.

Meng, L., Wang, Y., et al., 2013. Buckled silicene formation on Ir(111). Nano Lett. 13, 685–690.

Mohan, B., Kumar, A., et al., 2014. Electronic and optical properties of silicene under uni-axial and bi-axial mechanical strains: a first principle study. Physica E Low Dimens. Syst. Nanostruct. 61, 40–47.

Molle, A., Grazianetti, C., et al., 2013. Hindering the oxidation of silicene with non-reactive encapsulation. Adv. Funct. Mater. 23, 4340.

Nakano, H., Mitsuoka, T., et al., 2006. Soft synthesis of single-crystal silicon monolayer sheets. Angew. Chem. 118, 6451–6454.

Ng, T.Y., Yeo, J., et al., 2013. Molecular dynamics simulation of the thermal conductivity of shorts strips of graphene and silicene: a comparative study. Int. J. Mech. Mater. Des. 9, 105–114.

Ni, Z., Liu, Q., et al., 2012. Tunable bandgap in silicene and germanene. Nano Lett. 12, 113.

Ni, Z., Zhong, H., et al., 2014. Tunable band gap and doping type in silicene by surface adsorption: towards tunneling transistors. Nanoscale 6, 7609.

Pan, F., Quhe, R., et al., 2014. Gate-induced half-metallicity in semihydrogenated silicene. Physica E 56, 43–47.

Pei, Q.-X., Zhang, Y.-W., et al., 2013. Tuning the thermal conductivity of silicene with tensile strain and isotopic doping: a molecular dynamics study. J. Appl. Phys. 114, 033526.

Qin, R., Wang, H., et al., 2012. First-principles calculations of mechanical and electronic properties of silicene under strain. AIP Adv. 2, 022159.

Qin, R., Zhu, W., et al., 2014. Uniaxial strain-induced mechanical and electronic property modulation of silicene. Nanoscale Res. Lett. 9, 521.

Qiu, J., Fu, H., et al., 2015a. Ordered and reversible hydrogenation of silicene. Phys. Rev. Lett. 114, 126101.

Qiu, J., Fu, H., et al., 2015b. From silicene to half-silicane by hydrogenation. ACS Nano 9, 11192.

Quhe, R., Yuan, Y., et al., 2014. Does the Dirac cone exist in silicene on metal substrates? Sci. Rep. 4, 5476.

Rahman, M.S., Nakagawa, T., et al., 2015. Growth of Si on Ag(111) and determination of large commensurate unit cell of high-temperature phase. Jpn. J. Appl. Phys. 54, 015502.

Resta, A., Leoni, T., et al., 2013. Atomic structures of silicene layers grown on Ag(111): scanning tunneling microscopy and noncontact atomic force microscopy observations. Sci. Rep. 3, 2399.

Roome, N.J., Carey, J.D., 2014. Beyond graphene: stable elemental monolayers of silicene and germanene. ACS Appl. Mater. Interfaces 6, 7743–7750.

Şahin, H., Cahangirov, S., et al., 2009. Monolayer honeycomb structures of group-IV elements and III-V binary compounds: first-principles calculations. Phys. Rev. B 80, 155453.

Saji, K.J., Tian, K., et al., 2016. 2D Tin monoxide—an unexplored p-type van der waals semiconductor: material characteristics and field effect transistors. Adv. Electron. Mater., 2, 1500453.

Salomon, E., El Ajjouri, R., et al., 2014. Growth and structural properties of silicene at multilayer coverage. J. Phys. Condens. Matter 26, 185003.

Scalise, E., Houssa, M., et al., 2014. Engineering the electronic properties of silicene by tuning the composition of MoX2 and GaX (X = S, Se, Te) chalchogenide templates. 2D Mater. 1, 011010.

Schwierz, F., 2010. Graphene transistors. Nat. Nanotechnol. 5, 487.

Schwierz, F., Pezoldt, J., et al., 2015. Two-dimensional materials and their prospects in transistor electronics. Nanoscale 7, 8261.

Shu, H., Cao, D., et al., 2014. Two-dimensional silicene nucleation on a Ag(111) surface: structural evolution and the role of surface diffusion. Phys. Chem. Chem. Phys. 16, 304–310.

Solonenko, D., Gordan, O.D., et al., 2016. 2D vibrational properties of epitaxial silicene on Ag(111) probed by in situ Raman spectroscopy. https://arxiv.org/abs/1605.01593. arXiv:1605.01593.

Stephan, R., Hanf, M.C., et al., 2016. Germanene on Al(111): interface electronic states and charge transfer. J. Phys. Chem. C 2016 (120), 1580.

Švec, M., Hapala, P., et al., 2014. Silicene versus two-dimensional ordered silicide: atomic and electronic structure of Si-($\sqrt{19}\times\sqrt{19}$)R23.4°/Pt(111). Phys. Rev. B 89, 201412(R).

Takeda, K., Shiraishi, K., 1994. Theoretical possibility of stage corrugation in Si and Ge analogs of graphite. Phys. Rev. B 50, 14916–14922.
Tao, L., Cinquanta, E., et al., 2015. Silicene field-effect transistors operating at room temperature. Nat. Nanotechnol. 10, 227–231.
Tritsaris, G.A., Kaxiras, E., et al., 2013. Adsorption and diffusion of lithium on layered silicon for Li-ion storage. Nano Lett. 13, 2258.
Tsai, W.F., Huang, C.Y., et al., 2013. Gated silicene as a tunable source of nearly 100% spin polarized electrons. Nat. Commun. 4, 1500.
Vogt, P., De Padova, P., et al., 2012. Silicene: compelling experimental evidence for graphene like two-dimensional silicon. Phys. Rev. Lett. 108, 155501.
Vogt, P., Capiod, P., et al., 2014. Synthesis and electrical conductivity of multilayer silicene. Appl. Phys. Lett. 104, 021602.
Wang, Y.-P., Cheng, H.-P., 2013. Absence of a Dirac cone in silicene on Ag(111): first-principles density functional calculations with a modified effective band structure technique. Phys. Rev. B 87, 245430.
Wang, Y., Ding, Y., 2013. Strain-induced self-doping in silicene and germanene from first-principles. Solid State Commun. 155, 6–11.
Wang, L., Sun, H., 2012. Thermal conductivity of silicon and carbon hybrid monolayers: a molecular dynamics study. J. Mol. Model. 18, 4811–4818.
Wang, X.Q., Li, H.D., et al., 2012. Induced ferromagnetism in one-side semihydrogenated silicene and germanene. Phys. Chem. Chem. Phys. 14, 3031.
Wang, R., Pi, X., et al., 2013. Silicene oxides: formation, structures and electronic properties. Sci. Rep. 3, 3507.
Wang, B., Wu, J., et al., 2014a. Stable planar single-layer hexagonal silicene under tensile strain and its anomalous Poisson's ratio. Appl. Phys. Lett. 104, 081902.
Wang, Y., Zheng, R., et al., 2014b. Metal adatoms-decorated silicene as hydrogen storage media. Int. J. Hydrog. Energy 39, 14027.
Wang, R., Xu, M.-S., et al., 2015a. Chemical modification of silicene. Chin. Phys. B 24, 086807.
Wang, Y., Ni, Z., et al., 2015b. All-metallic vertical transistors based on stacked Dirac materials. Adv. Funct. Mater. 25, 68.
Winkler, R., Zülicke, U., 2010. Invariant expansion for the trigonal band structure of graphene. Phys. Rev. B 82, 245313.
Wu, K.-H., 2015. A review of the growth and structures of silicene on Ag(111). Chin. Phys. B 24, 086802.
Xie, H., Hu, M., et al., 2014. Thermal conductivity of silicene from first-principles. Appl. Phys. Lett. 104, 131906.
Yang, X., Ni, J., 2005. Electronic properties of single-walled silicon nanotubes compared to carbon nanotubes. Phys. Rev. B 72, 195426.
Yang, C.-H., Yu, Z.-Y., et al., 2014a. Phonon instability and ideal strength of silicene under tension. Comput. Mater. Sci. 95, 420.
Yang, K., Cahangirov, S., et al., 2014b. Thermoelectric properties of atomically thin silicene and germanene nanostructures. Phys. Rev. B 89, 125403.
Ye, X.-S., Shao, Z.-G., et al., 2014. Intrinsic carrier mobility of germanene is larger than graphene's: first-principle calculations. RSC Adv. 4, 21216–21220.
Yeo, J.J., Ng, T.Y., et al., 2014. Molecular dynamics analysis of the thermal conductivity of graphene and silicene monolayers of different lengths. J. Comput. Theor. Nanosci. 11, 1790–1796.
Yuhara J. et al., 2016, in preparation.
Zhang, Y., Rubio, A., et al., submitted.
Zhang, P., Li, X.D., et al., 2012. First-principles studies of the hydrogenation effects in silicene sheets. Phys. Lett. A 376, 1230.

Zhang, X., Xie, H., et al., 2014. Thermal conductivity of silicene calculated using an optimized Stillinger–Weber potential. Phys. Rev. B 89, 054310.
Zhang, L., Bampoulis, P., et al., 2015. Two-dimensional Dirac signature of germanene. Appl. Phys. Lett. 107, 111605.
Zhao, H., 2012. Strain and chirality effects on the mechanical and electronic properties of silicene and silicane under uniaxial tension. Phys. Lett. A 376, 3546–3550.
Zhao, J., Liu, H., et al., 2016. Rise of silicene: a competitive 2D material. Prog. Mater. Sci. 83, 24–151.
Zhu, J., Schwingenschlögl, U., 2014. Structural and electronic properties of silicene on MgX (2) (X = Cl, Br, and I). ACS Appl. Mater. Interfaces 6, 11675–11681.
Zhu, J., Schwingenschlögl, U., 2015. Stability and electronic properties of silicene on WSe2. J. Mater. Chem. C 3, 3946–3953.
Zhu, F.-F., Chen, W.-J., et al., 2015. Epitaxial growth of two-dimensional stanene. Nat. Mater. 14, 1020.

CHAPTER FIVE

Synthesis, Properties, and Stacking of Two-Dimensional Transition Metal Dichalcogenides

K. Zhang, Y.-C. Lin, J.A. Robinson[1]
The Center for 2-Dimensional and Layered Materials, The Pennsylvania State University, University Park, PA, United States
The Center for Atomically Thin Multifunctional Coatings, The Pennsylvania State University, University Park, PA, United States
[1]Corresponding author: e-mail address: jrobinson@psu.edu

Contents

1. SYNTHESIS OF TRANSITION METAL DICHALCOGENIDES

Exfoliation has become the most popular method to prepare monolayer materials after the discovery of exfoliated graphene (Novoselov et al., 2004). Mechanical exfoliation has enabled many promising electronic and optoelectronic applications of two-dimensional (2D) materials. However, the nonuniformity and small domain size of mechanical exfoliated 2D materials is a long-term problem. During the past 10 years, a large variety of

Semiconductors and Semimetals, Volume 95
ISSN 0080-8784
http://dx.doi.org/10.1016/bs.semsem.2016.04.005

synthesis methods for TMDs including chemical exfoliation (Eda et al., 2011), ion exchange (Browning et al., 2015), powder vaporization (PV) (Najmaei et al., 2013), chemical vapor deposition (CVD) (Eichfeld et al., 2015), atomic layer deposition (ALD) (Song et al., 2013), and molecular beam epitaxy (MBE) (Liu et al., 2015) have been developed.

2. CHEMICAL EXFOLIATION

Chemical exfoliation of TMDs is a solution-based synthesis method. Solid TMD precursors (typically powder) are dissolved in certain solutions with specific surface tension, followed by ultra-sonication. Traditionally, dimethylformamide and *N*-methyl-2-pyrrolidone have been employed as the solvent. Recently, this method has been applied to the synthesis of MoS_2, WS_2, $MoSe_2$, $MoTe_2$, $TaSe_2$, $NbSe_2$, $NiTe_2$, BN, and Bi_2Te_3 (Coleman et al., 2011). It requires the solvent to overcome the cohesive energy between each layer in order to prevent layer reaggregation. As a result, this method needs matched surface energy between the solvent and layered materials. Although the chemical exfoliation is able to provide large quantity of mono-few layer TMDs on the same substrate, the difficulty to maintain the lateral and integrate film, due to the ultra-sonication, is detrimental.

In order to overcome the difficulties of traditional chemical exfoliation, the lithium intercalation method has been developed (Eda et al., 2011). *N*-Butyl lithium is employed as the lithiation agent, and hexane is used as the solvent to remove the residuals. Lithium intercalation is able to control the yield of TMDs monolayers by tuning the formation of the lithiation compound: Li_xMX_2. Compared to traditional exfoliation, the lithium intercalation yields nearly 100% monolayers. The other important contribution of the lithium intercalation is the achievement of a 1T phase TMDs (Fan et al., 2015). In the lithiation process, the lithium ions contain extra charges. To accommodate this extra charge, the 2H phase transforms to a 1T phase, which passes through the extra charge. However, a long reaction time ($\sim$3 days) and the presence of nanoparticles are still the challenge for the lithium intercalation. Most recently, Zhang's group designed a simple electrolysis system to effectively limit the presence of nanoparticles in the lithium intercalation process (Zeng et al., 2011). The electrolysis system is shown in Fig. 1A (Zeng et al., 2011). The lithium foil provides lithium ions under a 0.05 mA current, followed by the lithium ion insertion into layers of TMDs. After the complete insertion, the lithium ions extend the distance between each layer and therefore weaken the van der Waals (vdW)

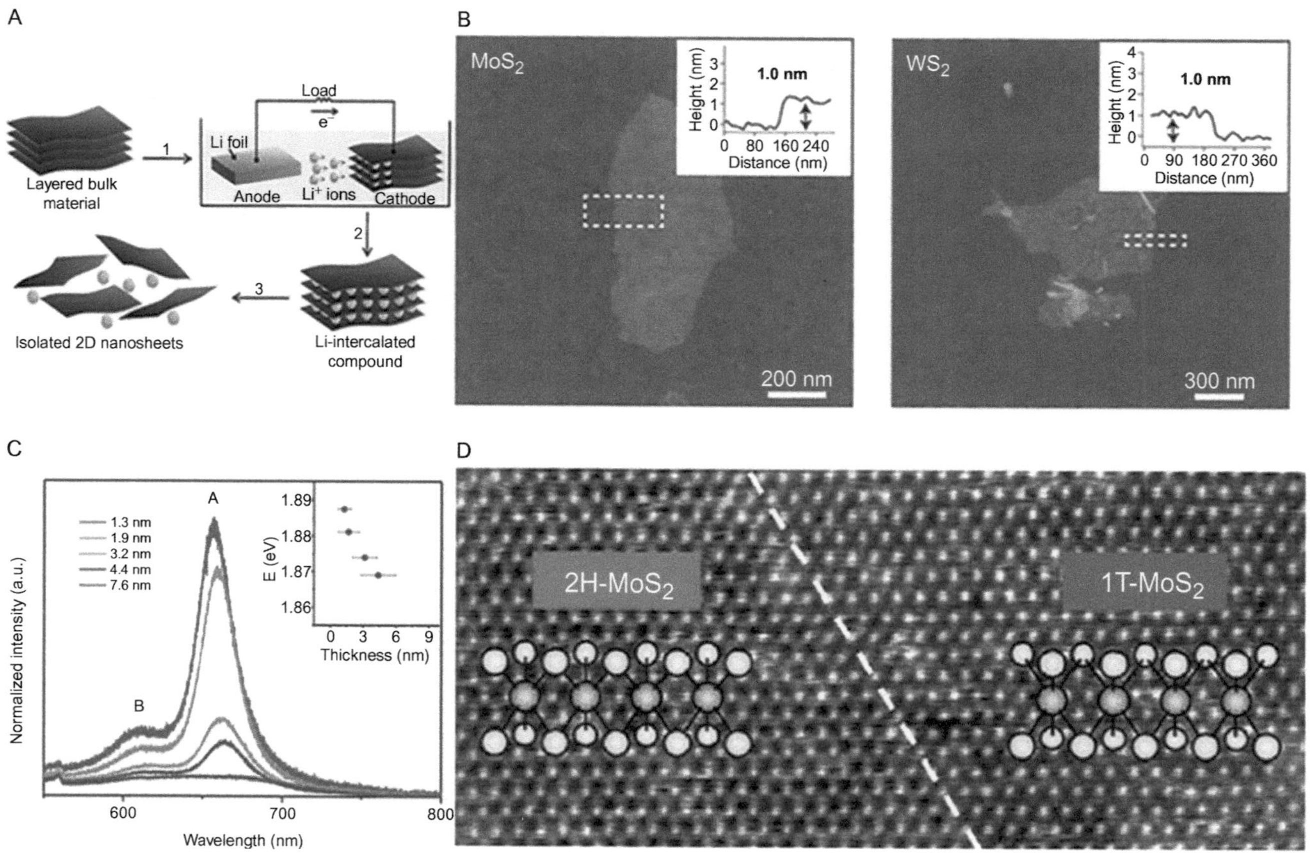

Fig. 1 See legend on next page.

interaction between layers. At the same time, the Li^+ reacts with water and the induced H_2 bubbles help the separation of each layer. The yield of this method is proved at 92%.

The fundamental properties of chemically exfoliated TMDs are shown in Fig. 1B–D. The thickness of the TMDs is identified as ~1 nm, which is similar to the mechanically exfoliated samples (Fig. 1B) (Chhowalla et al., 2013). It exhibits a thickness-dependent photoluminescence (PL). As the thickness increases, the PL intensity reduces. It is clear that the PL energy is at 1.89 eV for 1.3 nm (two layers) MoS_2 (Fig. 1C). Additionally, the PL energy strongly indicates that the chemical exfoliated TMDs on SiO_2 are free-standing (minimum interaction with substrates). In Fig. 1D, it is clear that two atomic arrangements are characterized by high-resolution scanning transmission electron microscope (HRSTEM), evidenting the 1T phase formation (Eda et al., 2012).

3. PHYSICAL VAPOR DEPOSITION AND ION EXCHANGE

In order to synthesize the wafer-scale uniform TMD thin films, the chalcogenization method has been widely used (Browning et al., 2015; Lin et al., 2012; Zhan et al., 2012). The thickness of the TMD films is determined by the thickness of predeposited films. The schematic of chalcogenization is shown in Fig. 2A. Metal films were deposited on insulating substrates via thermal evaporation or e-beam deposition. The deposited metal films are subsequently chalcogenized. Zhan et al. reported the sulfurization of Mo thin films in sulfur-rich condition (Zhan et al., 2012).

Fig. 1 (A) The schematic of the lithium insertion process; (B) the AFM images of chemically exfoliated MoS_2 and WS_2. The thickness is ~1 nm and the single-crystal domain size is ~1μm; (C) the thickness-dependent PL spectra of chemical exfoliation. It is clear that the PL intensity is enhanced as the thickness reduces due to the bandgap transition; and (D) the HRSTEM image of the lithium intercalated MoS_2. *Panel (A): Reproduced with permission from Zeng, Z., et al., 2011. Single-layer semiconducting nanosheets: high-yield preparation and device fabrication. Angew. Chem. Int. Ed. Engl. 50, 11093–11097. Copyright 2011 John Wiley and Sons. Panel (B): Reproduced with permission from Chhowalla, M., et al., 2013. The chemistry of two-dimensional layered transition metal dichalcogenide nanosheets. Nat. Chem. 5, 263–275. Copyright 2013 Nature Publishing Group. Panel (C): Reproduced with permission from Eda, G., et al., 2011. Photoluminescence from chemically exfoliated MoS 2. Nano Lett. 11, 5111–5116. Copyright 2011 American Chemical Society. Panel (D): Reproduced with permission from Eda, G., et al., 2012. Coherent atomic and electronic heterostructures of single-layer MoS2. ACS Nano 6, 7311–7317. Copyright 2012 American Chemical Society.*

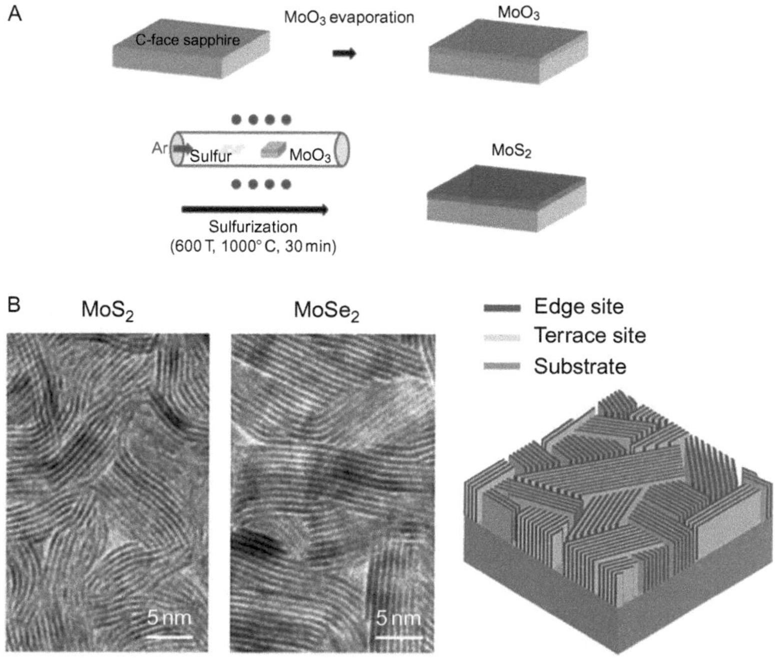

Fig. 2 (A) The schematic of physical vapor deposition and ion exchange method to synthesize TMD thin films. Metal or metal oxide thin film is first deposited on substrates followed by the chalcognization; (B) the high-resolution planar TEM image of vertical aligned MoS_2 and $MoSe_2$ and the schematic to demonstrate the large area edge exposure of edge for vertical aligned TMDs. *Panel (A): Reproduced with permission from Lin, Y.-C., et al., 2012. Wafer-scale MoS(2) thin layers prepared by MoO(3) sulfurization. Nanoscale 4, 6637–6641. doi:10.1039/c2nr31833d. Copyright 2011 Royal Society of Chemistry. Panel (B): Reproduced with permission from Kong, D., et al., 2013. Synthesis of MoS_2 and $MoSe_2$ films with vertically aligned layers. Nano Lett. 13, 1341–1347. Copyright 2013 American Chemical Society.*

It is suggested that the sulfurization of the Mo film yields nonuniform layer thickness due to nonuniformity of Mo deposition. Also, sulfurization of Mo film seems to be incomplete due to the observance of the metallic-like transport behavior, which is due to the unsulfurized Mo metal. To solve this problem, metal oxides is a good alternate. For example, MoO_3 thin film is sulfurized in a two-step procedure. MoO_3 is reduced as MoO_2 at 500° C, and MoO_2 is sulfurized at 1000°C. Wafer-scale semiconducting MoS_2 is achieved by this method (Lin et al., 2012). Similarly, this method is applied in the synthesis of WSe_2 thin films (Browning et al., 2015). Interestingly, the rapid thermal annealing of WO_3 results in microsize domain of WO_3 thin film. After the selenization, the WSe_2 thin films exhibit the same topography of WO_3 with ~500 nm domain size and perfect layer structure.

Other than the lateral layer structure, the chalcogenization method is currently utilized to produce vertical aligned TMDs for hydrogen evolution reaction (HER) (Kong et al., 2013). The vertical structure with dominating edge exposure exhibits high surface energy but is metastable (Fig 2B). The vertical structure exposes the maximum edges on the surface and it can be applied in diverse catalytic reactions. In Cui's study, the HER catalytic activity is directly correlated to the exposed edge sites (Kong et al., 2013). It is suggested that the vertical alignment of TMDs is a kinetic-controlled process. The metal thin film converts to sulfides/selenides much faster than the sulfur/selenium diffusion at high temperature. As a result, the sulfur/selenium diffusion is the controlling process for the sulfurization and selenization. Due to the natural anisotropic structure of TMDs, the sulfur/selenium is expected to diffuse faster along the vdW gap between layers than diffusing across the layers. Consequently, the film is vertically aligned.

The chalcogenization method realizes the preparation of few layer wafer-scale TMDs. However, there are still some limits in this method. For instance, the thickness of the predeposited film is hard to be controlled accurately. Song et al. proposes the hybrid of ALD and chalcogenization (Song et al., 2013). In their study, the WO_3 film thickness can be well controlled by controlling the number of cycle of ALD WO_3 films and subsequently control the thickness of the TMDs. The long-standing challenge of the chalcogenization is the small domain size and polycrystallinity of the films. To synthesize large area high-quality TMD film, efforts were made on thermolysis of precursors containing transition metal and chalcogen elements.

4. THERMOLYSIS AND PV

The first dedication of thermolysis was done by Li's group (Liu et al., 2012). Two steps of annealing are involved in their study. The substrate is dip-coated in ammonium thiomolybdate ($(NH_4)_2MoS_4$), followed by the first annealing at 500°C in order to convert $(NH_4)_2MoS_4$ to MoS_2. The second annealing is employed at 1000°C in H_2/S condition in order to remove the oxygen-related defects introduced by the first annealing (conversion). MoS_2 bilayer and trilayer synthesized by this method exhibits good electrical properties (>10^5 on/off ratio 6 cm^2/V s carrier mobility). However, this method is not able to provide the large area single-crystal monolayer TMDs.

Vapor phase reaction between transition metal oxides and chalcogens is first reported by Li's group (Lee et al., 2012). For example, in MoS_2 case, MoO_3 and sulfur powder are used as precursors. It is suggested that the MoO_3 will be reduced to MoO_{3-x} at the elevated temperature (~400°C), and MoO_{3-x} diffuses on the substrate as an intermediate phase and then be sulfurized by sulfur vapor at 600–800°C. The monolayer MoS_2 synthesized by this method exhibits 10^4 current on/off ratio and 0.02 cm^2/V s FET mobility in ambient. It is a valuable measurement to bring the society's attention to the mobility that is questionable for vapor phase grown monolayer TMDs. The key of this method is the control of the defect density on the substrate surface. It is believed that the defect density on the substrate determines the nucleation density of MoS_2 and subsequently determines the domain size of the monolayer MoS_2. It is required that the substrate should be ultra clean before the growth.

In order to decrease the nucleation density, many precursors have been demonstrated as preseed and they reduce the nucleation density significantly (Ling et al., 2014). The different film quality on SiO_2 grown with different preseed materials is summarized in Fig. 3A. It has been demonstrated that the aromatic molecules such as $F_{16}CuPc$, CuPc, and DBP can reduce the nucleation density due to the enhanced surface wettability by these chemicals. At the same time, the seed is able to promote the lateral growth. As a result, the high-quality, large area, and continuous MoS_2 monolayer film is synthesized with the assistance of preseeds. Perylene-3,4,9,10-tetracarboxylic acid tetrapotassium salt (PTAS) works best for hydrophilic substrates, while $F_{16}CuPc$ is promising for hydrophobic surfaces. Interestingly, the inorganic seeds such as Al_2O_3 and HfO_2 do not result in monolayer MoS_2.

Other than the organic small-molecule seeds, Najmaei et al. used MoO_3 nanoribbons as the precursor and nucleation agent (Najmaei et al., 2013). It is reported that the MoS_2 monolayer triangular domains are concentrated along the step edges of the nanoribbons (Fig. 4A). It is suggested that the nucleation energy along the step edge is lower than in the flat area due to the presence of extra dangling bonds. It is found that the coalescence between different triangular domains results in the continuous films. The continuous film may or may not show chemical bonded grain boundaries. In their study, the grain boundary is systematic studied by HRSTEM. Along the grain boundary of monolayer MoS_2, it is preferable to form five to seven rings (Fig. 4B) and S substitution of the sharing Mo atom is also found (Fig. 4B). The average FET carrier mobility including few layer and monolayer is ~4.3 cm^2/V s (Fig. 4C). It has been demonstrated that there might be up to an order of magnitude variation for the on/off ratio for monolayer MoS_2. The highest on/off ratio achieved in this study is 10^6 (Fig. 4D).

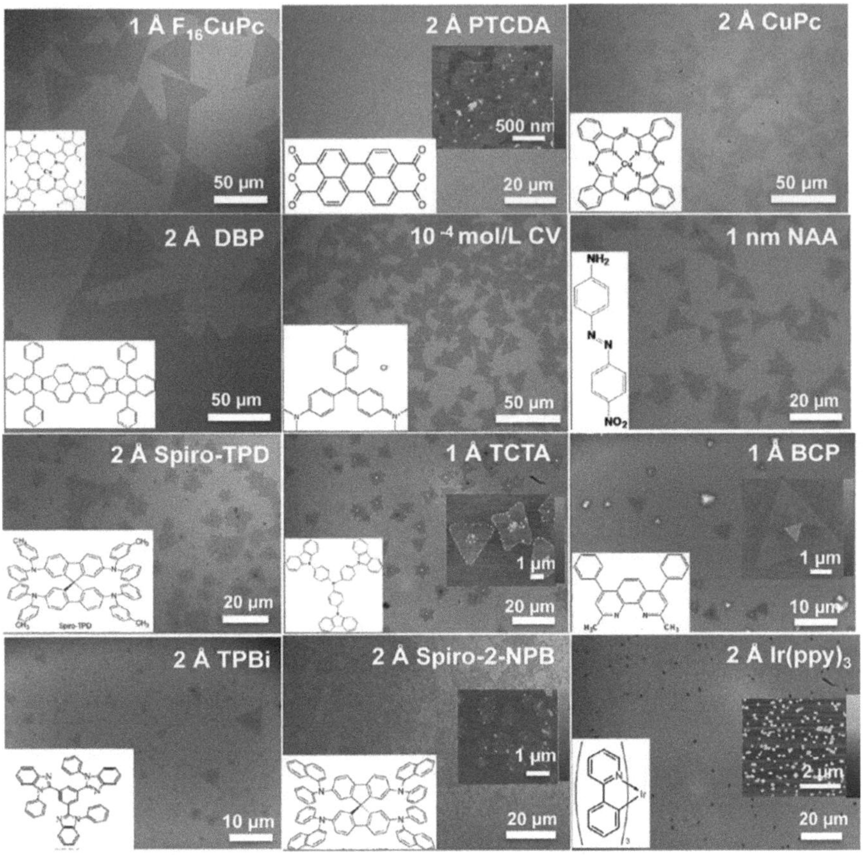

Fig. 3 Optical microscope images show the different film quality of MoS_2 synthesized on SiO_2 substrates with different nucleation agents. The *left bottom inset* shows the molecular formula of the nucleation agent. Some AFM images are also shown as the *inset. Reproduced with permission from Ling, X., et al., 2014. Role of the seeding promoter in MoS_2 growth by chemical vapor deposition. Nano Lett. 14, 464–472. Copyright 2014 American Chemical Society.*

One method has been reported to produce large area (>100μm) triangular domain of MoS_2 without any step edges and preseeding molecules by van der Zande et al. They used the ultra-cleaned substrate and fresh precursor in order to ensure the low defect density on the substrate surface and subsequently control the nucleation density (van der Zande et al., 2013). The anisotropic optical properties along the grain boundary are revealed. Fifty percent intensity quench and 100% intensity enhancement are observed along the mirror twin grain boundary and the tilt grain boundary,

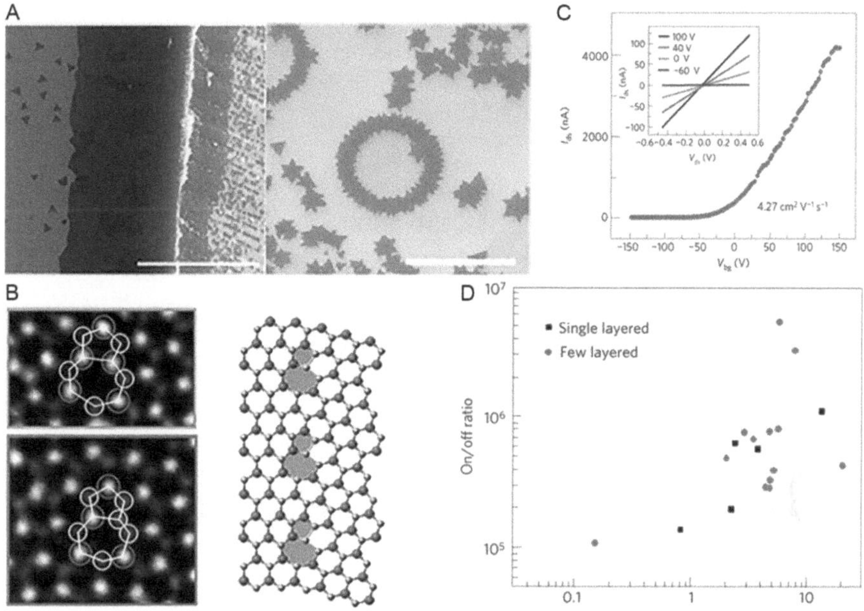

Fig. 4 (A) The SEM images of the MoS_2 preferably grow along the step edge of the MoO_3 nanoribbons. (B) The HRSTEM images of the grain boundary of MoS_2. It is clear that the five to seven rings are found (*upper*) and S substitution of sharing Mo atom is also found (*down*). The schematic at *right* shows the atomic structure of the defects along the grain boundary. (C) The *I–V* curve of the MoS_2-based FET device at the room temperature. (D) A summary of mobility and on/off ratio of MoS_2 FET devices, demonstrating that order of magnitude variation in mobility and on/off ratio exists between device and device. *Reproduced with permission from Najmaei, S., et al., 2013. Vapour phase growth and grain boundary structure of molybdenum disulphide atomic layers. Nat. Mater. 12, 754–759. Copyright 2013 Nature Publishing Group.*

respectively. (Fig. 5B and F). The peak position and the peak width exhibit a homogeneous increase in both cases (Fig. 5C, D, G, and H). It is believed that the variation of optical properties is due to the additional density of state (DOS) along the grain boundary (Fig. 5I and J).

Another deviation of the PV technique is the vapor–solid method. In this method, no metal oxides and chalcogens are applied. Pristine TMD powders are used as the precursor. In Wu et al.' s study, MoS_2 powder is put in high temperature (~900°C, slightly lower than the melting point of MoS_2 (1185°C)). The substrate is put in a lower temperature zone (650°C). MoS_2 vapor is transferred from the powder to the substrate by inert gas. Surprisingly, the MoS_2 flakes synthesized by this method exhibit superb optical quality and high crystallinity (Wu et al., 2013).

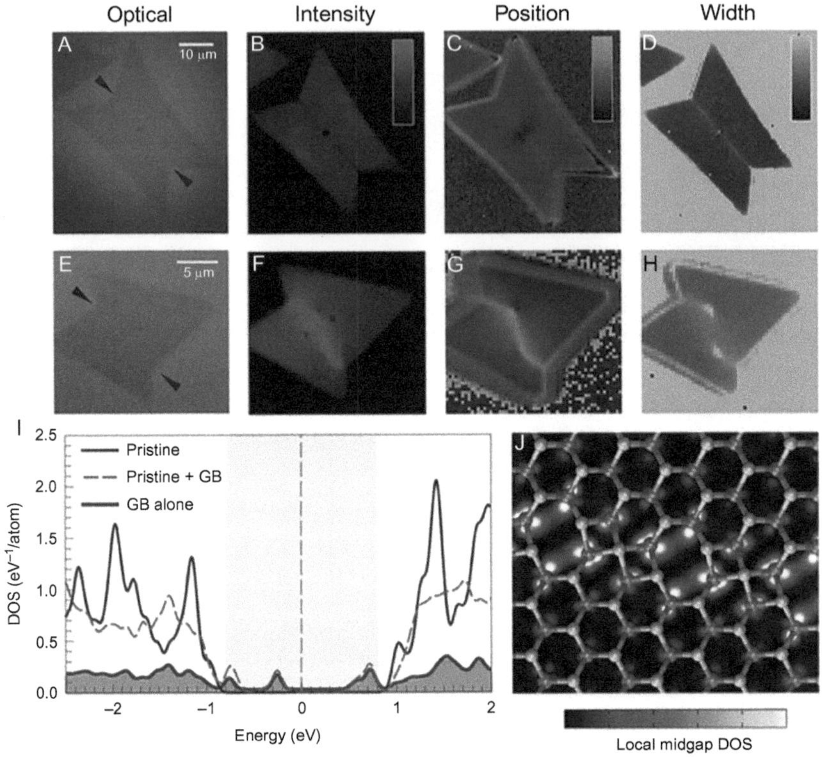

Fig. 5 (A and E) The optical image of the MoS_2 monolayer with twin mirror (A) and typical tilt grain boundary. (B and F) The correlating intensity PL map of the MoS_2 flakes, the scale bar is 0–1100 a.u. (C and G) The correlating peak position map, the scale bar is 1.82–1.87 eV. There is 8 meV upshift for twin mirror grain boundary and 26 meV upshift for tilt grain boundary. (D and H) The peak width map of the correlating flakes, the scale bar is 55–62 meV. About 6 meV broaden of peak width along the grain boundary is found in both cases; (I) The total DOS of pristine (*black, solid*), grain boundary included (*red* (*gray* in the print version), *dash*), and grain boundary along (*blue* (*dark gray* in the print version), *solid*). Additional DOS is observed with the existence of grain boundary. (J) A 2D DOS map of the lattice of MoS_2 grain boundary. *Reproduced with permission from van der Zande, A.M., et al., 2013. Grains and grain boundaries in highly crystalline monolayer molybdenum disulphide. Nat. Mater. 12, 554–561. Copyright 2013 Nature Publishing Group.*

The forementioned methods have been applied on the synthesis of various TMDs such as WS_2 (Elías et al., 2013), $MoSe_2$ (Wang et al., 2014a), and WSe_2 (Huang et al., 2014). The biggest challenge of this method is the uniformity. There are also some modified methods to improve the uniformity such as angled growth (Zhang et al., 2015).

5. METAL ORGANIC CHEMICAL VAPOR DEPOSITION

Metal organic chemical vapor deposition (MOCVD) is a well-known, controllable synthesis method. It features in the all vapor phase reaction, which allows the control of precursor ratio, growth time, and growth rate. It also features in the capability of low-temperature growth (Mun et al., 2016). MOCVD is the best method to synthesize wafer-scale uniform TMDs. A simple reactor schematic of MOCVD is shown in Fig. 6. Transition metal hexacarbonyl ($W(CO)_6$, $Mo(CO)_6$) is used as the transition metal precursor and stored in highly sealed bubblers with H_2 (Eichfeld et al., 2015; Kang et al., 2015). Both liquid (Kang et al., 2015) and gasoline (Kranthi Kumar et al., 2015) chalcogen precursors have been used as chalcogen source.

The reaction temperature ranges from 300 to 900°C. It is suggested that high temperature results in higher growth rate and facilitates to eliminate the carbon residuals from the precursor. Pressure plays an important role on the domain size and morphology of the TMD film. It is suggested that higher pressure facilitates the large single-crystal domain, while lower pressure drives coalescence few layer films (Fig. 7A). It is noticeable that the metal/chalcogen ratio has a significant impact on the growth mode of TMDs. In WSe_2 case, higher Se/W ratio improves the lateral growth of WSe_2, while lower Se/W ratio results in the growth of 3D domains and clusters. Another important factor of the domain size is the salt, and the single-crystal domains are $>10\times$ larger than the growth without salt (Fig. 7B). It is believed that the salt acts as a desiccant that removes the water residuals. The continuous wafer-scale TMDs (MoS_2, WS_2) have been synthesized, and it enables the wafer-scale fabrication of FET devices. The film exhibits highly homogenous transport characteristics with high mobility $>10\ cm^2/V\ s$ for 99% devices (Kang et al., 2015; Fig. 7C).

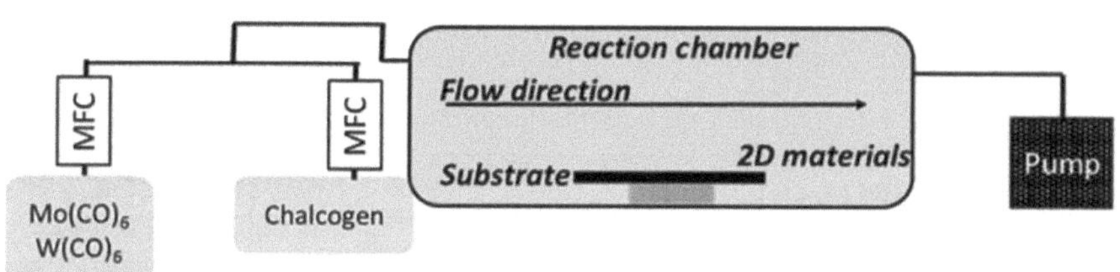

Fig. 6 The schematic of an MOCVD reactor for TMD synthesis; in this figure, only a Mo- and W-based precursor is sketched.

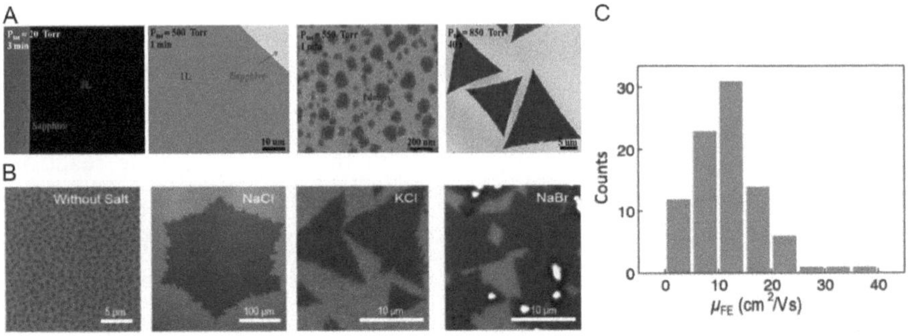

Fig. 7 (A) The pressure dependence of MOCVD grown MoS_2 domain morphology. Higher pressure facilitates large single-crystal domains. (B) The impact of salt on the domain size of MOCVD grown MoS_2. It is clear that the presence of salt (NaCl, KCl, and NaBr) increases the domain size. (C) The statistics of the FET mobility of a batch of MoS_2 FET devices on a 4-in. SiO_2 wafer. *Reproduced with permission from Kang, K., et al., 2015. High-mobility three-atom-thick semiconducting films with wafer-scale homogeneity. Nature 520, 656–660. Copyright 2015 Nature Publishing Group.*

6. MOLECULAR BEAM EPITAXY

MBE, known as one of the most advanced and controllable growth methods, has also been utilized for the TMD synthesis. Its ultra-high vacuum environment produces the highest achievable purity. High-quality $MoSe_2$ is synthesized by the Knudsen cell-evaporated Mo and Se source with a Mo:Se flux ratio at 1:8 (Zhang et al., 2014). The transition from the indirect to direct band gap is directly observed by ARPES with the MBE synthesized $MoSe_2$ thin films with accurate layer thickness. By tuning the flux ratio, it is able to control the deficiency in the film. It has been demonstrated that the Se deficiency in the MBE synthesized $MoSe_2$ films shows a dense network along the twin mirror grain boundary (Mose et al., 2015). Additionally, due to the epitaxial nature of the MBE, it has been used for the growth of vdW heterostructure. The $MoSe_2$ and WSe_2 exhibit a same orientation as the graphene template (Liu et al., 2015).

7. HETEROSTRUCTURES BUILT ON vdW CRYSTALS

A heterostructure consists of different electronic materials and has a varied energy gap. In principle, a heterostructure utilizes its energy gap variations to control electrons and holes in terms of their flow and distribution, in addition to electrical fields. Since the central of design principle of heterostructure devices was proposed by Kroemer (1957), it has been an essential requirement for high-performance transistors, semiconducting lasers, and optical devices made out of conventional semiconductors (Kroemer, 1982; Sze and Kwok, 2006).

Graphene and beyond-graphene-layered materials, especially atomically thin TMDs, have created a vast field that generates more than a thousand of publications on the study of their fundamentals and material applications each year (Bhimanapati et al., 2015; Geim and Grigorieva, 2013). These publications provide throughput fundamental understandings on every aspect of each layered material and enable people to select specific layered materials for their needs. While new opportunities of discovering exotic phenomena in one layered material itself are running low, a new focus going beyond this field has been initiated. Various isolated monolayers of TMDs and other vdW crystals are assembled into a sophisticated structure made into a layer-by-layer sequence that is purposely designed. These vdW heterostructures have been synthesized and investigated extensively since 2010, and already revealed new properties and exotic phenomena yet presented in their constituent layers (Dean et al., 2010; Geim and Grigorieva, 2013). While most of ultra thin-layered crystals have been explored and demonstrated in optoelectronics, the emerging vdW heterostructure is raising a "layered renaissance" for the next-generation devices (Akinwande et al., 2014; Bonaccorso et al., 2015; Das et al., 2015; Wang et al., 2014b).

vdW heterostructures are consisted of a variety of 2D layered crystals that have strong in-plane covalent bonds and weak out-of-plane vdW forces (Fig. 8) (Das et al., 2015; Lotsch, 2015). Van der Waals gap between constituent layers is the key feature that enables ones distinguish vdW heterostructures (Fig. 8A and C; Koma, 1999) from conventional heterostructures (Fig. 8B and D; Koma, 1999). While the convenient heterostructures derived from 3D solids, such as III–V compounds, SiGe epitaxy layers, and oxides (ie, perovskites, spinels, and dielectrics) (Schlom et al., 2008), involve covalent bonds bridging their constituent materials (Lotsch, 2015; Schlom et al., 2008; Sze and Kwok, 2006), the vdW heterostructures bridge their constituent layers with weak vdW forces. Without physical bonds involved, their interfaces can tolerate a highly lattice mismatch combination (Fig. 8C).

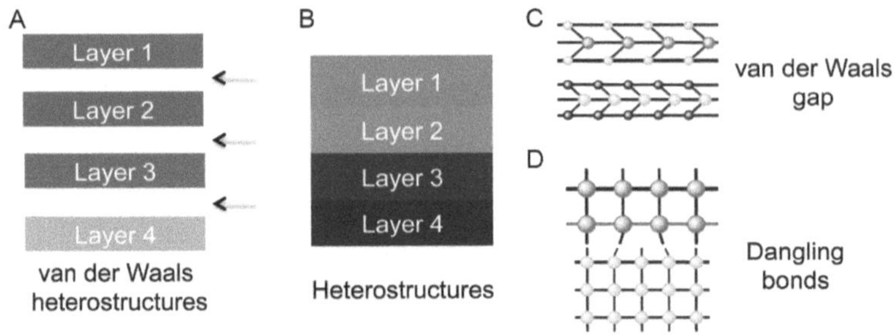

Fig. 8 (A) Schematic illustration of "van der Waals" (vdW) heterostructures and (B) "conventional" heterostructures. The symbolic feature of a vdW heterostructure is the presence of vdW gaps (*arrows* in A) in between constituent 2D layers, which attract to their adjacent layers by a weak vdW force, shown in (C). (D) On the other hand, the heterostructures derived from ionic and covalent compounds have physical bonds at their interfaces connecting each constituent 3D building block. Dangling bonds would be caused, if there is a large lattice mismatch between grown material and growth template in the 3D cases. *Reproduced with permission from Koma, A., 1999. Waals epitaxy for highly lattice-mismatched systems. J. Cryst. Growth 201–202, 236–241. Copyright 1999 Elsevier.*

8. MECHANICALLY EXFOLIATED vdW HETEROSTRUCTURES

When reliable techniques to synthesis high-quality vdW heterostructures are on development still, the simplest fabrication technique is by mechanically transfer of one 2D crystal onto another in a step-by-step manipulation (Lee et al., 2011; Lotsch, 2015). The easiest report of this route is from Dean et al. (Dean et al., 2010; Weitz and Yacoby, 2010) on graphene and hexagonal boron nitride (hBN) stacks, in which a micromanipulator was used, under an optical microscopy, to precisely deposit graphene that is closely aligned to a hBN flake (Fig. 9A). The electrical transport measurement on the graphene integrated with hBN flakes shows a significant improvement in the field-effect mobility of graphene (Fig. 9B). These results indicate that hBN serves as a substrate better than SiO_2 for graphene electronics due to their close lattices, an atomically flat surface, and lack of dangling bonds. The stacking methods can apply to layered materials that are not structurally compatible or unlikely can be grown on each other. This method had inspired numerous works that built exotic vdW heterostructures to discover the new properties and applications (Britnell et al., 2012; Geim and Grigorieva, 2013; Yankowitz et al., 2012). There have been many new prototypes of vertical devices made from stacked exfoliated

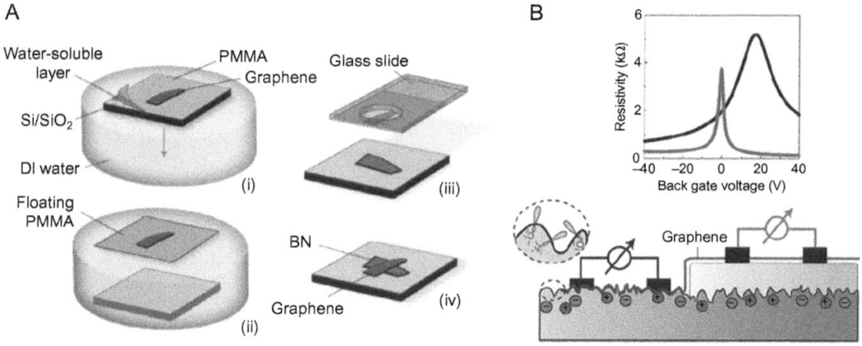

Fig. 9 (A) Schematic flow of the transfer process used to deposit exfoliated graphene on hBN flakes. (B) *Top*: The field-effect mobility of graphene–hBN devices achieved 60,000–80,000 cm^2/V s, a significant improvement compared with graphene–SiO_2 devices. *Bottom*: The atomically flat and dangling bond-free surface of hBN attribute to the success of the high mobility. *Reproduced with permission from Dean, C.R., et al., 2010. Boron nitride substrates for high-quality graphene electronics. Nat. Nanotechnol. 5, 722–726; Weitz, R.T., Yacoby, A., 2010. Nanomaterials: graphene rests easy. Nat. Nanotechnol. 5, 699–700. Copyright 2010 Nature Publishing Group.*

layers, while the synthesis techniques for ultra high-quality heterostructures are under development.

There are significant amounts of reported devices demonstrated on vdW heterostructures utilizing a variety of 2D crystal categories including conductors, insulators, and semiconductors. The groundbreaking work by Britnell et al. (Britnell et al., 2012; Lim et al., 2014) utilized hBN flakes ranging from five to seven layers as a tunneling barrier between two sheets of graphene serving as the top and bottom electrodes in the vertical field-effect tunneling transistors in hBN–Gr–hBN–Gr–hBN (Gr: graphene) vertical heterostructures (Fig. 10A–C). The amount of tunneling current density of the vertical devices can be tuned by controlling finite doping density and applied bias (Fig. 10D–G). The transistors show a tunneling *I*–*V* characteristics and orders of the on/off ratio, which address the weakness of planar graphene field-effect transistors due to lack of on/off ratio. In addition to rigid devices, integrated flexible technology that benefits from 2D crystals with high flexibility and stretchability is also emerging. The strain limit of thin-film devices made of TMDs and other monolayers possess a value three to five imes greater than that made from III to V compounds, metal oxides, and crystalline silicon. Combining with their high transparency, it is fair to say that 2D vdW heterostructures may be attractive candidates for flexible

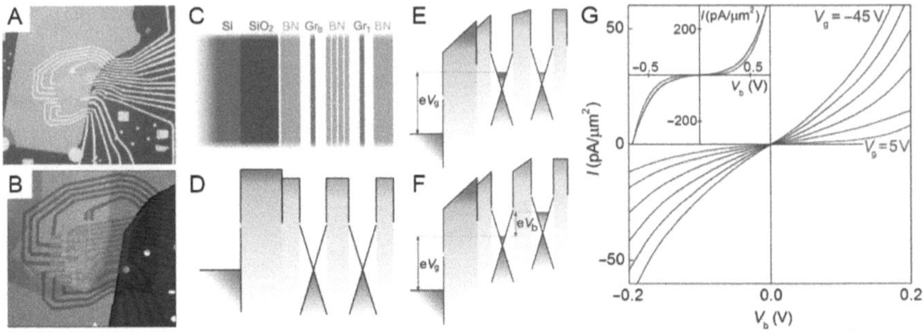

Fig. 10 Field-effect tunneling transistors on vertical hBN–graphene–hBN–graphene–hBN heterostructure. (A) Optical image of the final device. (B) Electron microscopic image captured prior to evaporation of the Au electrodes shows two Hall bars made from graphene are shaded in *green* (*light gray* in the print version) and *orange* (*gray* in the print version). (C) Schematic structure of the experimental vertical devices. (D) The corresponding band structure without applied gate voltage. (E) The same band structure subjected to a finite gate voltage (V_g) and zero bias (V_b). (F) Both of V_g and V_b are applied. (Only the tunnel barrier for electrons is considered in the illustrations.) (G) Tunneling characteristics for the vertical tunneling device with five to seven layers of hBN as the tunnel barrier. *I–V* curves for different V_g, in a 10-V step. Due to finite doping, the minimum tunneling conductivity is achieved at V_g around 3 V. The inset compares the experimental *I–V* curves at $V_g = 5$ V (*red* (*gray* in the print version) *curve*) with theory (*dark curve*), which takes the linear density of state in the two graphene layers into consideration and assumes no momentum conservations. *Reproduced with permission from Lim, H., Yoon, S.I., Kim, G., Jang, A.-R., Shin, H.S., 2014. Stacking of two-dimensional materials in lateral and vertical directions. Chem. Mater. 26, 4891–4903. Copyright 2014 American Chemical Society.*

and transparent electronics. Similar to the first prototype of Gr–hBN–Gr devices, Georgiou et al. (2013) prepared a Gr–WS_2–Gr vertical heterostructure fabricated on a flexible polyethylene terephthalate films using the same transfer methods (Fig. 11A and B). The device exhibits tunneling characteristics under applied characteristics (Fig. 11C) and maintains its electrical performance subjected to a strain up to 5% (Fig. 11D) (Georgiou et al., 2013). The flexible structures also demonstrate a transistor effect (inset, Fig. 11D) and may have improved performances with the optimal thickness of dielectrics.

Besides the vertical tunneling devices, engineering the band structures of heterostructures can also lead to practical lightning devices using TMDs (MoS_2, WS_2, WSe_2) as a light emitter (Withers et al., 2015a,b). The work by Withers et al. (2015a) fabricated single quantum well (SQW) emitters

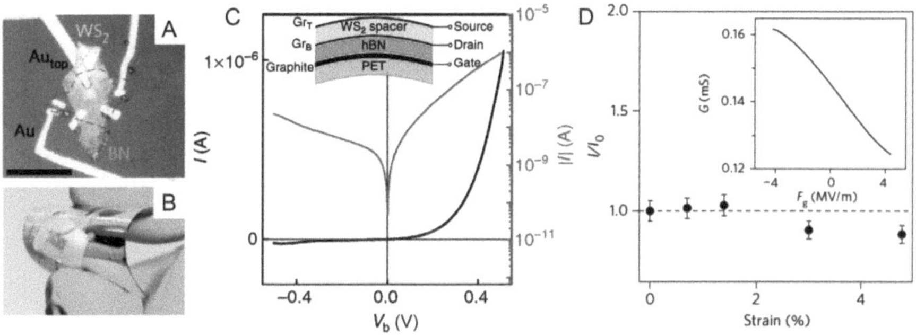

Fig. 11 A transistor Gr–WS_2–Gr built on a flexible polymer substrate: (A) Optical image and (B) image of the device under bending. (C) *I–V* plot at $T=300$ K for the bended device with $V_g=0$. Curvature is 0.05 mm^{-1}. (D) Relative current variation vs applied strain. Standard variations for several consecutive measurements are shown in error bars. *Inset* is the gating transport of the device under strain. *Panels (A, B, and D): Reproduced with permission from Georgiou, T., et al., 2013. Vertical field-effect transistor based on graphene-WS_2 heterostructures for flexible and transparent electronics. Nat. Nanotechnol. 8, 100–103. Copyright 2013 Nature Publishing Group. (C) Reproduced with permission from Akinwande, D., Petrone, N., Hone, J., 2014. Two-dimensional flexible nanoelectronics. Nat. Commun. 5, 5678. Copyright 2014 Nature Publishing Group.*

made from a stack of hBN/Gr_{Bottom}/3L hBN/1L MoS_2/3L hBN/Gr_{Top}/hBN (Fig. 12A and B), in which 1L MoS_2 serves as a light emitter excited by an applied bias and operates at 300 K (Fig. 12C–G). One of the demonstrated devices in this work achieved extrinsic quantum efficiency nearly to 10%, and the emission can be tuned over a wide range of wavelength by choosing different types and thickness of 2D semiconductors. By stacking more repetitive SQW, a multiple quantum well (MQW) vdW heterostructure with enhancement emitting intensity was realized.

Combining different TMD monolayers can lead to a new class of vdW solids that exhibit new optical and electrical properties. The theoretical work by Terrones et al. (2013) predicted that stacking MoS_2–WSe_2 heterostructures (Fig. 13A) will yield electronic properties that are entirely different from their constituent layers, such as a significantly reduced bandgap energy (Fig. 13B and C) (Lv et al., 2015). This exciting theoretical work got support from many experimental works aiming to realize these theoretical results through manual stacking of different TMD layers (Chiu et al., 2014; Fang et al., 2014; Rivera et al., 2015). For example, a manually stacked MoS_2–WSe_2 heterostructure made by Fang et al. (2014) exhibits an interlayer excitons at 1.55–1.59 eV (Fig. 14) (Lim et al., 2014), in addition to

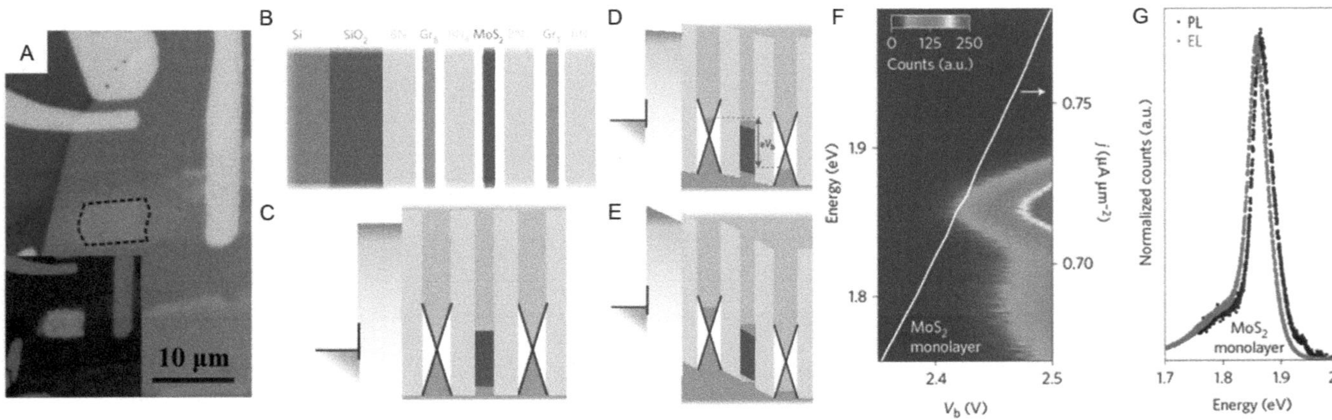

Fig. 12 Heterostructure devices with a single quantum well (SQW), made from hBN/Gr_{Bottom}/3L hBN/1L MoS_2/3L hBN/Gr_{Top}/hBN, shown in (A), the optical image; the inset of (A) is electroluminescence (EL) image from the same device, under $V_b = 2.5$ V, $T = 300$ K. (B) The schematic structure of the same SQW device. (C–E) Band alignment for the case of zero applied bias (C), intermediate applied bias (D), and high applied bias (E), for the heterostructure presented in (B). (F) EL spectra as a function of applied bias (V_b) for the SQW device made from MoS_2. *White curve* is its current density vs applied bias characteristics (j–V_b). and (G) comparison of the PL and EL spectra for the same devices. *Reproduced with permission from Withers, F., et al., 2015. Light-emitting diodes by band-structure engineering in van der Waals heterostructures. Nat. Mater. 14, 301–306. Copyright 2015 Nature Publishing Group.*

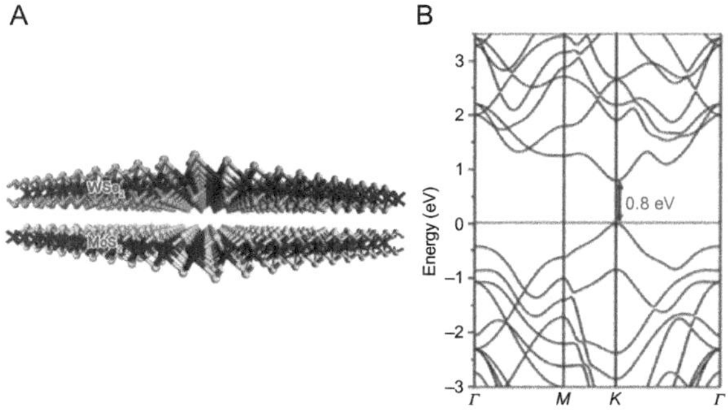

C

TDMC	Band Gap (eV) Monolayer	Bulk
MoS_2	**1.8**	1.2
WS_2	**1.9**	1.4
$MoSe_2$	1.5	1.1
WSe_2	**1.7**	1.2

Heterostructure	Bilayer Band Gap (eV) Direct	Indirect
WS_2–MoS_2	1.71	1.19
WSe_2–WS_2	**1.07**	1.31
WSe_2–MoS_2	**0.9**	1.14
$MoSe_2$–WS_2	1.18	1.05
WSe_2–$MoSe_2$	1.47	1.11
$MoSe_2$–MoS_2	1.01	0.9

Fig. 13 (A) Simulated MoS_2–WSe_2 heterostructure yields a new direct band gap, which is shown in (B). (C) Band gap of monolayer and bulk TMDs and their heterostructures. *Reproduced with permission from Lv, R. et al., 2015.Transition metal dichalcogenides and beyond: synthesis, properties, and applications of single- and few-layer nanosheets. Acc. Chem. Res., 48, 56–64 Copyright 2015 American Chemical Society.*

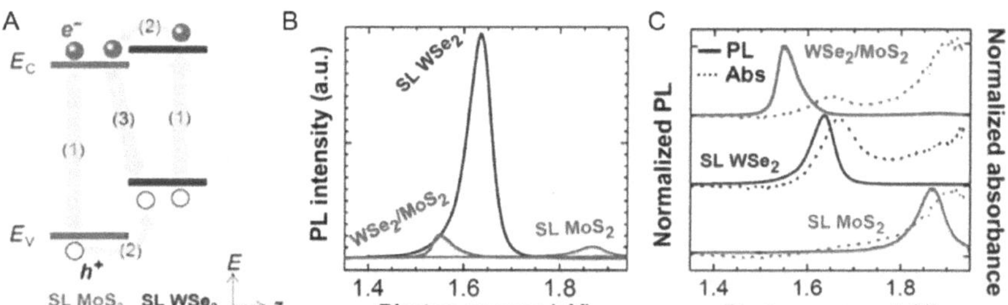

Fig. 14 (A) Band diagram of WSe_2/MoS_2 heterobilayer under photoexcitation, illustrating (1) exciton generation in WSe_2 and MoS_2 single layers (SL), (2) relaxation of excitons at the MoS_2–WSe_2 interface where band has been offset, and (3) radiative recombination of spatially indirect excitons; (B) SL MoS_2, WSe_2, and their heterolayer exhibit PL spectra created by radiative recombination of intralayer and interlayer excitons; (C) Normalized PL (*solid lines*) and absorbance (*dashed lines*) spectra of SL WSe_2, MoS_2, and their corresponding heterolayers, where the spectra are normalized to the height of the strongest PL/absorbance peak. *Reproduced with permission from Lim, H., Yoon, S.I., Kim, G., Jang, A.-R., Shin, H.S., 2014. Stacking of two-dimensional materials in lateral and vertical directions. Chem. Mater. 26, 4891–4903. Copyright 2014 American Chemical Society.*

intralayer excitons of (1.87 eV) MoS_2 and (1.65 eV) WSe_2 monolayers (Fang et al., 2014). More interestingly, adding electrically insulating hBN monolayers into MoS_2–WSe_2 heterostructures can modify the strength of the interlayer coupling and result in decoupling of the layers, as evident by a decreased intensity of the interlayer excitons (Fang et al., 2014).

9. ROTATIONALLY DEPENDENT OPTOELECTRICAL PROPERTIES IN vdW HETEROSTRUCTURES

Transition metal dichalcogenide (TMD)-based heterostructures can "feel" different force fields by different patterns of their superlattices; rotational angles between two TMD monolayered crystals would determine the features of the optical properties in their vdW heterostructures (Rivera et al., 2015; van der Zande et al., 2014). For example, van der Zande et al. (2014) stacked two MoS_2 monolayers grown by CVD techniques (Fig. 15).

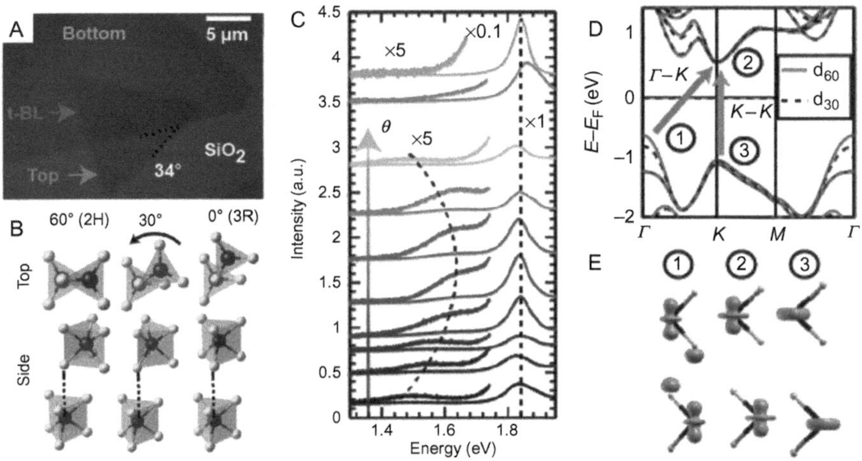

Fig. 15 (A) Optical microscope image of two overlapping CVD MoS_2 monolayers, forming a bilayer with a twisted angle. The *dashed lines* are plotted to indicate the different orientations of the triangle edges. (B) *Top* and *side view* of the aligned bilayer MoS_2 with rotation angles of 0 degree, 30 degree, and 60 degree. The 0 degree and 60 degree alignments match the 2H and 3R phase of nature MoS_2 crystals, respectively. (C) Photoluminescence spectra for the *bottom* (as-grown, *gray line*, one at the most top in (C)) and *top* (transferred, *green* (*gray* in the print version) *line*) MoS_2 monolayers, and twisted bilayers at different twist angles. The spectra are normalized and offset for clarity and color-labeled by the angle with *black* and *yellow* (*light gray* in the print version) corresponding to 0 degree and 60 degree, respectively. The *black dashed* line highlights relevant tuning features of the MoS_2 indirect band gap, alone with increased twisted angles; (D) calculated electronic structures of 60 degree and 30 degree bilayers are presented in *solid pick lines* and *broken black lines*, respectively; (E) The partial charge density of states corresponding to (1)–(3) in (D) shows that only the inner sulfur atoms in the structures and their separation are the main factors involved in the indirect transition in (D). The twisted angle dictates the separation of sulfur atoms, which is the key parameter for this optical modulation. *Reproduced with permission from van der Zande, A.M., et al., 2014. Tailoring the electronic structure in bilayer molybdenum disulfide via interlayer twist. Nano Lett. 14, 3869–3875. Copyright 2014 American Chemical Society.*

A serious of artificial MoS_2 bilayers were made with the twisted angle ranging from 0 to 60 degree (Fig. 15A and B; van der Zande et al., 2014), and the intensity and the peak position of their indirect transition (indirect band gap) are highly tunable in PL measurements for 532 nm excited wavelength (Fig. 15C; van der Zande et al., 2014). Theoretical calculations confirmed the changed distance between inner sulfur atoms in MoS_2 bilayer would modify the indirect transition in their band structures due to a changed interaction between the *p* orbitals form sulfur atoms (Fig. 15D and E). Given by these examples, one can say that the rotation angle between stacked 2D lattices is an additional degree of freedom for people to engineer the properties of vdW heterostructures.

10. INTERFACE IMPERFECTION IN MECHANICALLY EXFOLIATED vdW HETEROSTRUCTURES

The manual stacking process is indeed a practically useful for integrating various 2D materials to create a variety of proof-of-concept vdW heterostructures. However, the process requires multiple steps to complete an assembly, including isolating a 2D crystal in μm size, transferring it onto polymer-supporting films, stacking 2D crystals repeatedly, repeating standard clean room procedure in terms of cleaning, dissolving, resist spinning, and so on, and a precision-demanding alignment under a microscope with a micromanipulator (Geim and Grigorieva, 2013; Lotsch, 2015). These multiple steps, carried out in ambient, unavoidably introduce contaminations at the interfaces of constituent layers (Fig. 16A) (Georgiou et al., 2013), which could be due to the presence of adsorbents and the usage of polymer films in the transfer process. Although a clean and sharp interface in these heterostructures is still obtained by confining trapped residues into micrometer-sized "bubbles" with vdW forces that bond adjacent constituent layers (Geim and Grigorieva, 2013; Haigh et al., 2012), a sophisticated process is still needed in order to fabricate a useful device. The visual example of the bubbles is shown in Fig. 18B, where the top gate contact was deposited in a shape that the "bubbles" would be avoided (Robinson, 2016). In order to make vdW heterostructures practically useful for digital industries, an alternative for the synthetic vdW heterostructures with clean interfaces needs to come out.

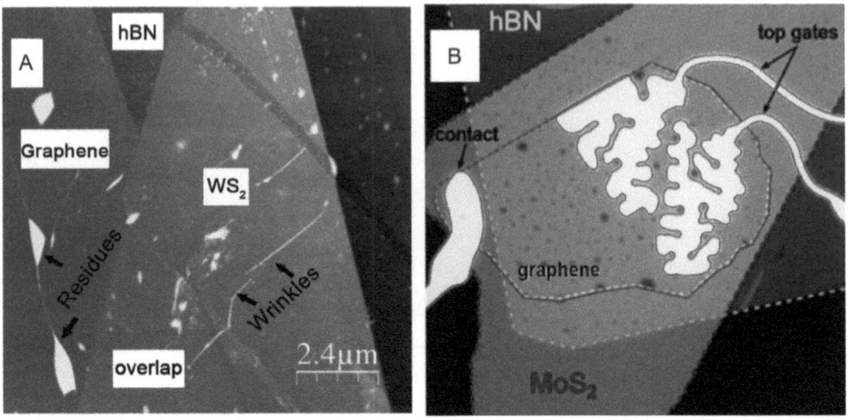

Fig. 16 (A) "Bubbles" and wrinkles in manually stacked van der Waals heterostructures formed due to segregated residues at their interfaces after the transfer process. Each stacked layer and their overlap are highlighted, and none of them are completely free of imperfection. (B) While fabricating devices on the heterostructures, the top contacts are shaped irregularly to avoid the polymer residue (*black dots* in figure) that is common in this technique. *Panel (A): Reproduced with permission from Georgiou, T., et al., 2013. Vertical field-effect transistor based on graphene-WS_2 heterostructures for flexible and transparent electronics. Nat. Nanotechnol. 8, 100–103. Copyright 2013 Nature Publishing Group. Panel (B): Reproduced with permission from Robinson, J.A., 2016. Growing vertical in the flatland. ACS Nano 10, 42–45. Copyright 2016 American Chemical Society.*

11. SYNTHETIC vdW HETEROSTRUCTURES

The practically useful heterostructures made of III–V compounds, such as heterostructure bipolar transistors, phototransistor with wide-gap emitters, and double-heterostructure lasers, had not appeared until the growth technologies of MOCVD and MBE were developed in the early 1970s (Kroemer, 1982). Similar to the early development on their conventional counterparts, vdW heterostructures have not been practical since its demonstration in 2010 due to limited size of clean interface obtained and absence of techniques for the large area growth. Although the vdW epitaxy, growing one vdW solid on another, has already been recognized in the 1980s, many were by Koma (1999); it did not get much attention from research societies until the breakthrough results exploited in manually stacked vdW heterostructures. Recently, the emergence of direct synthesis of vdW solids, utilizing CVD, MOCVD, and MBE (Das et al., 2015), also made impressive progress in synthetic vdW heterostructures like graphene–hBN transistors (Yang et al., 2013), graphene–TMD photosensors (Ago et al., 2015; Lin et al., 2014a), and TMD p–n junctions and tunneling diodes grown on graphene (Lin et al., 2014b, 2015) and insulating substrates (Gong et al., 2014; Li et al., 2015). In view of these recent results, synthetic vdW heterostructures appear to revolutionize the digital electronics and their industries.

In order to synthesize crystalline TMD layers, lattice of the selected substrate is critical for the epigrowth of vdW heterostructures. Shi et al. (2012) initiatively used CVD graphene grown on copper foils as the template for MoS_2 growth (Fig. 17A). The reported process utilizes $(NH_4)_2MoS_4$ precursors that were thermally decomposed into MoS_2 in the vapor phase and then subsequently deposited on CVD graphene/Cu foil. The as-grown MoS_2 domains on graphene adopted the same orientation of underlying graphene. This experiment indicated that an epitaxial vdW heterostructure can be realized still, even the lattice mismatch can be 20–23% (Das et al., 2015; Shi et al., 2012). Similarly, Lin et al. (2014a) used epitaxial graphene (EG)/SiC as the growth template for monolayer MoS_2 made via PV (Fig. 17B). This study by Lin et al. also found that morphology and defects of EG/SiC can significantly impact the nucleation density and the thickness of MoS_2 layers (Lin et al., 2014a). Scanning transmission electron microscopy (STEM) images show that the atomically sharp interface is possible to achieve through vapor deposition techniques (bottom, Fig. 17B). In addition, it is also possible to grow larger domain of WSe_2 monolayers on EG/SiC via vdW epitaxy (Lin et al., 2014b). Following vdW epitaxy, monolayered WSe_2 domains grown on graphene consistently align at either the same direction or 180 degree rotated and thus achieve a commensurability between WSe_2 and graphene (scanning electron microscopy image in Fig. 17C), as evident by low-energy electron diffraction (LEED) patterns, which roughly shows that four transition metal atoms can align with nine carbon atoms in a long-range order (inset, Fig. 17C). The vdW heterostructures can be put on a more sophisticated level by stacking other types of TMDs layers. Lin et al. (2015) performed two-step growth of MoS_2 and WSe_2 to create MoS_2–WSe_2–graphene and WSe_2–$MoSe_2$–graphene that have clean and sharp interfaces without Mo–W and Se–Se alloys, as evident in STEM images (Fig. 17D–F). The success of the trilayered vdW heterostructures is the first step toward synthetic heterostructures beyond the trilayered stacks. Besides the techniques of pyrolysis, PV, CVD, and MOCVD, MBE is also emerging for synthetic 2D crystals. Bradley et al. (2015) synthesized 1–3L $MoSe_2$ on bilayer graphene through MBE. Albeit the domain size of the $MoSe_2$ is typically less than 1 μm in scanning tunneling microscopy (STM), performing scanning tunneling spectroscopy (STS)

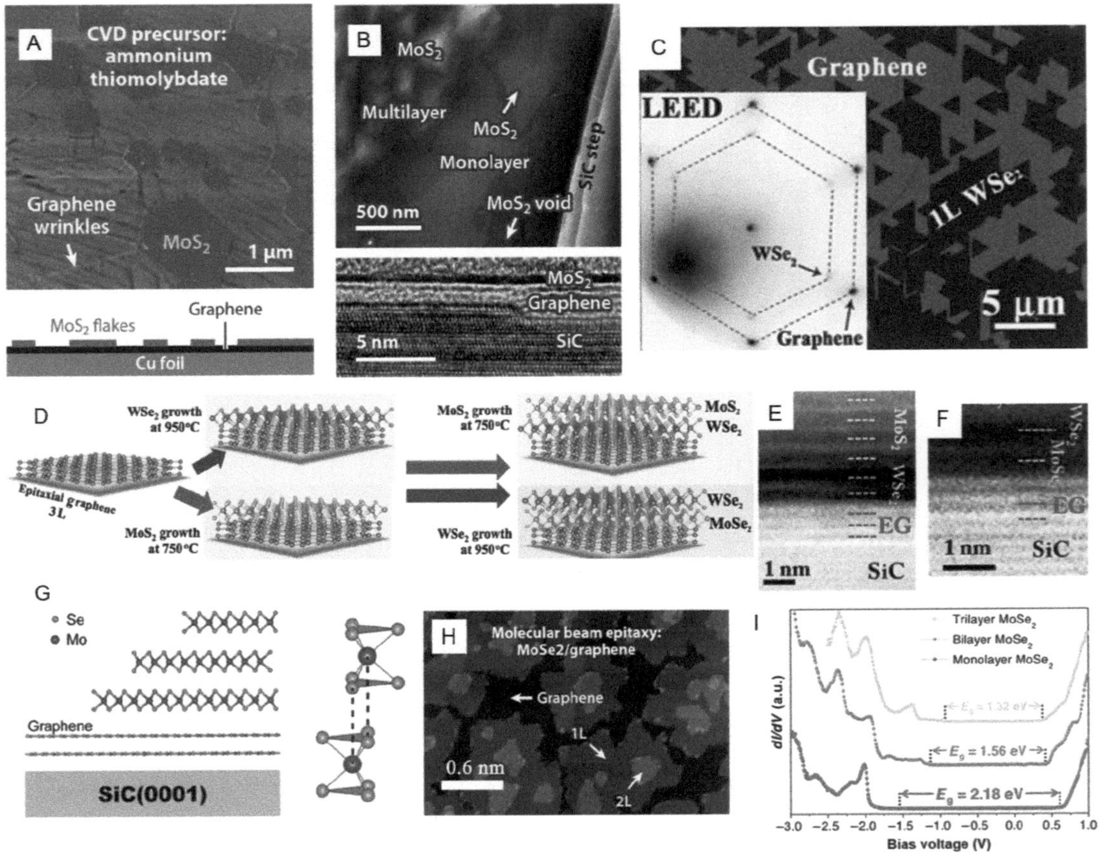

Fig. 17 See legend on opposite page.

on these films is able to obtain the quasiparticle band gaps and exciton binding energy of 1–3L $MoSe_2$.

Although TMD–graphene heterostructures, in which graphene is utilized as a bottom electrode, can be used as electrical diodes (Das et al., 2015; Tan and Zhang, 2015), photosensors (Das et al., 2015), and studied materials for STM/STS measurements (Ugeda et al., 2014), majority of device research is focusing on metal–oxide–semiconductor device geometry, such as TMDs deposited on SiO_2/Si. To fulfill this need, many efforts had been made to grow high-quality and large size TMD-based vdW heterostructures on SiO_2/Si, sapphire, and other insulating substrates, mainly through a CVD process. Among insulating growth templates, the most popular one is SiO_2/Si since it is easy to prepare and immediately makes a metal–oxide–semiconductor devices after the material growth. Gong et al. (2014) used Te-assisted PV involving the reaction of MoO_3, W, and S powders to grow both of the lateral and vertical MoS_2–WS_2 heterostructures in an in situ process. The role of Te powders involved is for lowering the melting point of W powders via forming metastable Te–W alloys during the reaction (Gong et al., 2014). The lateral MoS_2–WS_2 grows at 650°C (Fig. 18A and B), while the vertical one grows at a higher temperature, at 850°C (Fig. 18C and D)

Fig. 17 (A) Multilayer MoS_2 grown on CVD graphene/Cu through the thermal decomposition of ammonium thiomolybdate. (B) Monolayer and few layer MoS_2 grown on EG from powder vaporization (PV) process. (C) WSe_2 monolayers grown EG through PV process or metallic–organic CVD (MOCVD). The WSe_2 lattices are fully registered to the graphene lattices, as evident by low-energy electron diffraction pattern (LEED, inset). (D) The flow for growing "trilayer" van der Waals heterostructures. MoS_2–EG was converted into $MoSe_2$–EG during the growth of WSe_2 layers via a Se–S ionic exchange occurring in high temperatures. (E and F) STEM image confirms that these trilayer stacks exhibit pristine interfaces without W–Mo or S–Se formation. (G–I) 1L–3L $MoSe_2$ layers grown on 2L EG through molecular beam epitaxy. EG serves as bottom electrodes for STS in (I), which measures the quasiparticle band gap of 1L–3L $MoSe_2$. *(A–C) Reproduced with permission from Lin, Y.-C., et al., 2014. Direct synthesis of van der Waals solids. ACS Nano 8, 3715–3723; Lin, Y.-C., et al., 2014. Atomically thin heterostructures based on single-layer tungsten diselenide and graphene. Nano Lett. 14, 6936–6941; Shi, Y., et al., 2012. van der Waals epitaxy of MoS_2 layers using graphene as growth templates. Nano Lett. 12, 2784–2791, American Chemical Society. (D–F) Reproduced with permission from Lin, Y.-C., et al., 2015. Atomically thin resonant tunnel diodes built from synthetic van der Waals heterostructures. Nat. Commun. 6, 7311. Copyright 2015 Natural Publish Group. (G–I) Reproduced with permission from Bradley, A.J., et al., 2015. Probing the role of interlayer coupling and coulomb interactions on electronic structure in few-layer $MoSe_2$ nanostructures. Nano Lett. 15, 2594–2599. Copyright 2015 American Chemical Society.*

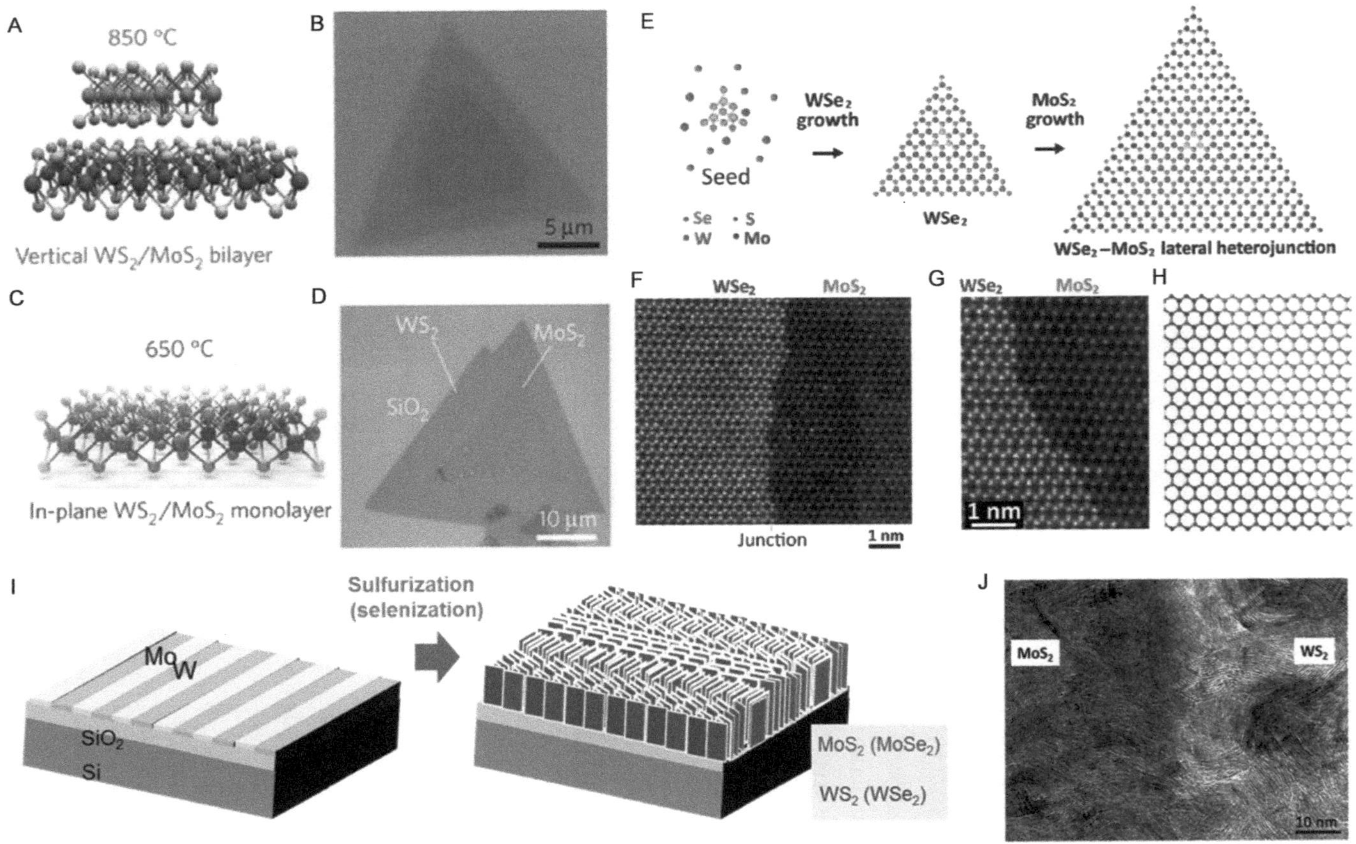

Fig. 18 See legend on opposite page.

(Gong et al., 2014). Besides the heterostructure that uses single chalcogen atom, Li et al. (2015) also developed a two-step ex situ process using the edges of the WSe_2 monolayers pregrown at 950°C as nucleation sites and then growing MoS_2 monolayers epitaxially around the MoS_2 monolayers at 700°C to obtain MoS_2–WSe_2 lateral heterostructures (Fig. 18E). The order for material growth, that is WSe_2 first and MoS_2 second, is deliberately decided to avoid the ionic exchange of Se–S occurring above 800°C. The STEM performed on MoS_2–WSe_2 confirmed that the lateral interface is atomically abrupt and no sign of Mo–W and Se–S formation in a micrometer range in parallel to the junction (Fig. 18F–H) (Li et al., 2015). Besides the above "flat" cases, vdW heterostructures can also exist in a vertically aligned fashion. Jung et al. (2014) sulfurized (selenized) patterned Mo/W arrays to synthesize MoS_2–WS_2 ($MoSe_2$–WSe_2) heterostructures in large area (Fig. 18I). Functionality and properties of this type of structures may be completely controllable because the dimension and the thickness of Mo/W arrays can be controlled by the lithography and sputtering time, respectively. Although electrical transports do not favors the vertical formation, as evident in STEM images (Fig. 18J) (the measured mobility is <0.01 cm^2/V s in both of the direction vertical and parallel to the array), preferably exposed dangling bonds on the edge sites are useful for HER (Jung et al., 2014).

Fig. 18 (A and B) Schematic illustration and optical images of the vertical WS_2–MoS_2 heterostructures. (C and D) Schematic illustration and optical images of the lateral WS_2–MoS_2 heterostructures. (E) Schematic illustration of the growth process of the monolayer WSe_2–MoS_2 lateral heterostructure on sapphire. (F and G) HRSTEM images of the WSe_2–MoS_2 lateral heterostructure at the interface regions. (H) Atomic model shows the interface structure of the WSe_2–MoS_2 lateral heterostructure. (I) Sputtered Mo–W strips patterned on SiO_2 were converted into MoS_2–WS_2 ($MoSe_2$–WSe_2) vertically aligned heterostructures after sulfurization (selenization). (J) STEM shows the interface of vertically aligned MoS_2–WS_2 heterostructures. A higher Z number of W atoms makes WS_2 region brighter than the other in STEM. *(A–H) Reproduced with permission from Tan, C., Zhang, H., 2015. Epitaxial growth of hetero-nanostructures based on ultrathin twodimensional nanosheets. J. Am. Chem. Soc. 137, 12162–12174. Copyright 2015 American Chemical Society. (I and J) Reproduced with permission from Bradley, A.J., et al., 2015. Probing the role of interlayer coupling and coulomb interactions on electronic structure in few-layer $MoSe_2$ nanostructures. Nano Lett. 15, 2594–2599. Copyright 2014 American Chemical Society.*

12. CONCLUSIONS

In this chapter, various synthesis methods of TMDs are reviewed. The structural, electronic, and optoelectronic properties are highly dependent on the synthesis techniques. Many challenges still exist and they have to be addressed in order to achieve the superb properties of TMDs and high performance of TMD-based devices. A heterostructure made of various semiconducting materials is an important remark toward advanced optoelectronics. Two-dimensional TMD layers represent a new class of building blocks for engineering a heterostructure. Thanks to rapid development of synthesis techniques for TMDs, novel vdW heterostructures that are mainly made via transfer techniques could one day be realized in a large scale and large area through a direct synthesis.

REFERENCES

Ago, H., et al., 2015. Controlled van der Waals epitaxy of monolayer MoS_2 triangular domains on graphene. ACS Appl. Mater. Interfaces 7, 5265–5273.
Akinwande, D., Petrone, N., Hone, J., 2014. Two-dimensional flexible nanoelectronics. Nat. Commun. 5, 5678.
Bhimanapati, G.R., et al., 2015. Recent advances in two-dimensional materials beyond graphene. ACS Nano 9, 11509–11539.
Bonaccorso, F., et al., 2015. Graphene, related two-dimensional crystals, and hybrid systems for energy conversion and storage. Science 347, 1246501.
Bradley, A.J., et al., 2015. Probing the role of interlayer coupling and coulomb interactions on electronic structure in few-layer $MoSe_2$ nanostructures. Nano Lett. 15, 2594–2599.
Britnell, L., et al., 2012. Field-effect tunneling transistor based on vertical graphene heterostructures. Science 335, 947–950.
Browning, P., et al., 2015. Large-area synthesis of WSe_2 from WO_3 by selenium–oxygen ion exchange. 2D Mater. 2, 014003.
Chhowalla, M., et al., 2013. The chemistry of two-dimensional layered transition metal dichalcogenide nanosheets. Nat. Chem. 5, 263–275.
Chiu, M.-H., et al., 2014. Spectroscopic signatures for interlayer coupling in MoS_2-WSe_2 van der Waals stacking. ACS Nano 8, 9649–9656.
Chow, P.K., et al., 2015. Defect-induced photoluminescence in monolayer semiconducting transition metal dichalcogenides. ACS Nano 9, 1520–1527.
Coleman, J.N., et al., 2011. Two-dimensional nanosheets produced by liquid exfoliation of layered materials. Science 331, 568–571.
Das, S., Robinson, J.A., Dubey, M., Terrones, H., Terrones, M., 2015. Beyond graphene: progress in novel two-dimensional materials and van der Waals solids. Annu. Rev. Mater. Res. 45, 1–27.
Dean, C.R., et al., 2010. Boron nitride substrates for high-quality graphene electronics. Nat. Nanotechnol. 5, 722–726.
Eda, G., et al., 2011. Photoluminescence from chemically exfoliated MoS_2. Nano Lett. 11, 5111–5116.

Eda, G., et al., 2012. Coherent atomic and electronic heterostructures of single-layer MoS_2. ACS Nano 6, 7311–7317.
Eichfeld, S.M., et al., 2015. Highly scalable, atomically thin WSe_2 grown via metal-organic chemical vapor deposition. ACS Nano 9, 2080–2087.
Elías, A.L., et al., 2013. Controlled synthesis and transfer of large-area WS_2 sheets: from single layer to few layers. ACS Nano 7, 5235–5242.
Fan, X., et al., 2015. Fast and efficient preparation of exfoliated 2H MoS_2 nanosheets by sonication-assisted lithium intercalation and infrared laser-induced 1 T to 2H phase reversion. Nano Lett. 15, 5956–5960.
Fang, H., et al., 2014. Strong interlayer coupling in van der Waals heterostructures built from single-layer chalcogenides. Proc. Natl. Acad. Sci. U.S.A 111, 6198–6202.
Geim, A.K., Grigorieva, I.V., 2013. Van der Waals heterostructures. Nature 499, 419–425.
Georgiou, T., et al., 2013. Vertical field-effect transistor based on graphene-WS_2 heterostructures for flexible and transparent electronics. Nat. Nanotechnol. 8, 100–103.
Gong, Y., et al., 2014. Vertical and in-plane heterostructures from WS_2/MoS_2 monolayers. Nat. Mater. 13, 1135–1142.
Haigh, S.J., et al., 2012. Cross-sectional imaging of individual layers and buried interfaces of graphene-based heterostructures and superlattices. Nat. Mater. 11, 764–767.
Huang, J.-K., et al., 2014. Large-area synthesis of highly crystalline WSe_2 monolayers and device applications. ACS Nano 8, 923–930.
Jung, Y., Shen, J., Sun, Y., Cha, J.J., 2014. Chemically synthesized heterostructures of two-dimensional molybdenum/tungsten-based dichalcogenides with vertically aligned layers. ACS Nano 8, 9550–9557.
Kang, K., et al., 2015. High-mobility three-atom-thick semiconducting films with wafer-scale homogeneity. Nature 520, 656–660.
Koma, A., 1999. Waals epitaxy for highly lattice-mismatched systems. J. Cryst. Growth 201–202, 236–241.
Kong, D., et al., 2013. Synthesis of MoS_2 and $MoSe_2$ films with vertically aligned layers. Nano Lett. 13, 1341–1347.
Kranthi Kumar, V., Dhar, S., Choudhury, T.H., Shivashankar, S.A., Raghavan, S., 2015. A predictive approach to CVD of crystalline layers of TMDs: the case of MoS_2. Nanoscale 7, 7802–7810.
Kroemer, H., 1957. Theory of a wide-gap emitter for transistors. Proc. IRE 45, 1535–1537.
Kroemer, H., 1982. Heterostructure bipolar transistors and integrated circuits. Proc. IEEE 70, 13–25.
Lee, G.-H., et al., 2011. Electron tunneling through atomically flat and ultrathin hexagonal boron nitride. Appl. Phys. Lett. 99, 243114.
Lee, Y.-H., et al., 2012. Synthesis of large-area MoS_2 atomic layers with chemical vapor deposition. Adv. Mater. 24, 2320–2325.
Li, M.-Y., et al., 2015. Epitaxial growth of a monolayer WSe_2-MoS_2 lateral p-n junction with an atomically sharp interface. Science 349, 524–528.
Lim, H., Yoon, S.I., Kim, G., Jang, A.-R., Shin, H.S., 2014. Stacking of two-dimensional materials in lateral and vertical directions. Chem. Mater. 26, 4891–4903.
Lin, Y.-C., et al., 2012. Wafer-scale MoS_2 thin layers prepared by MoO_3 sulfurization. Nanoscale 4, 6637–6641.
Lin, Y.-C., et al., 2014a. Direct synthesis of van der Waals solids. ACS Nano 8, 3715–3723.
Lin, Y.-C., et al., 2014b. Atomically thin heterostructures based on single-layer tungsten diselenide and graphene. Nano Lett. 14, 6936–6941.
Lin, Y.-C., et al., 2015. Atomically thin resonant tunnel diodes built from synthetic van der Waals heterostructures. Nat. Commun. 6, 7311.
Ling, X., et al., 2014. Role of the seeding promoter in MoS_2 growth by chemical vapor deposition. Nano Lett. 14, 464–472.

Liu, K.-K., et al., 2012. Growth of large-area and highly crystalline MoS_2 thin layers on insulating substrates. Nano Lett. 12, 1538–1544.
Liu, H.J., et al., 2015. Molecular-beam epitaxy of monolayer and bilayer WSe_2: a scanning tunneling microscopy/spectroscopy study and deduction of exciton binding energy. 2D Mater. 2, 034004.
Lotsch, B.V., 2015. Vertical 2D heterostructures. Annu. Rev. Mater. Res. 45, 85–109.
Lv, R., et al., 2015. Transition metal dichalcogenides and beyond: synthesis, properties, and applications of single- and few-layer nanosheets. Acc. Chem. Res. 48, 56–64.
Mose, T., et al., 2015. Atomic scale microstructure and atomic scale microstructure and properties of Se-deficient two-dimensional $MoSe_2$. ACS Nano 9, 3274–3283.
Mun, J., et al., 2016. Low-temperature growth of layered molybdenum disulphide with controlled clusters. Sci. Rep. 6, 21854.
Najmaei, S., et al., 2013. Vapour phase growth and grain boundary structure of molybdenum disulphide atomic layers. Nat. Mater. 12, 754–759.
Novoselov, K.S., et al., 2004. Electric field effect in atomically thin carbon films. Science 306, 666–669.
Rivera, P., et al., 2015. Observation of long-lived interlayer excitons in monolayer $MoSe_2$-WSe_2 heterostructures. Nat. Commun. 6, 6242.
Robinson, J.A., 2016. Growing vertical in the flatland. ACS Nano 10, 42–45.
Schlom, D.G., Chen, L.-Q., Pan, X., Schmehl, A., Zurbuchen, M.A., 2008. A thin film approach to engineering functionality into oxides. J. Am. Ceram. Soc. 91, 2429–2454.
Shi, Y., et al., 2012. van der Waals epitaxy of MoS_2 layers using graphene as growth templates. Nano Lett. 12, 2784–2791.
Song, J.-G., et al., 2013. Layer-controlled, wafer-scale, and conformal synthesis of tungsten disulfide nanosheets using atomic layer deposition. ACS Nano 7, 11333–11340.
Sze, S.M., Kwok, K.N., 2006. Physics of Semiconductor Devices. John Wiley & Sons, Hoboken, NJ.
Tan, C., Zhang, H., 2015. Epitaxial growth of hetero-nanostructures based on ultrathin two-dimensional nanosheets. J. Am. Chem. Soc. 137, 12162–12174.
Terrones, H., López-Urías, F., Terrones, M., 2013. Novel hetero-layered materials with tunable direct band gaps by sandwiching different metal disulfides and diselenides. Sci. Rep. 3, 1549.
Ugeda, M.M., et al., 2014. Giant bandgap renormalization and excitonic effects in a monolayer transition metal dichalcogenide semiconductor. Nat. Mater. 13, 1091–1095.
van der Zande, A.M., et al., 2013. Grains and grain boundaries in highly crystalline monolayer molybdenum disulphide. Nat. Mater. 12, 554–561.
van der Zande, A.M., et al., 2014. Tailoring the electronic structure in bilayer molybdenum disulfide via interlayer twist. Nano Lett. 14, 3869–3875.
Wang, X., et al., 2014a. Chemical vapor deposition growth of crystalline monolayer $MoSe_2$. ACS Nano 8, 5125–5131.
Wang, H., et al., 2014b. Two-dimensional heterostructures: fabrication, characterization, and application. Nanoscale 6, 12250–12272.
Weitz, R.T., Yacoby, A., 2010. Nanomaterials: graphene rests easy. Nat. Nanotechnol. 5, 699–700.
Withers, F., et al., 2015a. Light-emitting diodes by band-structure engineering in van der Waals heterostructures. Nat. Mater. 14, 301–306.
Withers, F., et al., 2015b. WSe_2 light-emitting tunneling transistors with enhanced brightness at room temperature. Nano Lett. 15, 8223–8228.
Wu, S., Huang, C., Aivazian, G., Ross, J.S., Cobden, D.H., 2013. Vapor–solid growth of high optical quality MoS_2 monolayers with near-unity valley polarization. ACS Nano 7, 2768–2772.

Yang, W., et al., 2013. Epitaxial growth of single-domain graphene on hexagonal boron nitride. Nat. Mater. 12, 792–797.
Yankowitz, M., et al., 2012. Emergence of superlattice Dirac points in graphene on hexagonal boron nitride. Nat. Phys. 8, 382–386.
Zeng, Z., et al., 2011. Single-layer semiconducting nanosheets: high-yield preparation and device fabrication. Angew. Chem. Int. Ed. Engl. 50, 11093–11097.
Zhan, Y., Liu, Z., Najmaei, S., Ajayan, P.M., Lou, J., 2012. Large area vapor phase growth and characterization of MoS_2 atomic layers on SiO_2 substrate. Small 8, 966–971.
Zhang, Y., et al., 2014. Direct observation of the transition from indirect to direct bandgap in atomically thin epitaxial $MoSe_2$. Nat. Nanotechnol. 9, 111–115.
Zhang, K., et al., 2015. Synthesis of two-dimensional materials for beyond graphene devices. Proc. SPIE 9467.

CHAPTER SIX

Advances in 2D Materials for Electronic Devices

B.M. Nichols, A.L. Mazzoni, M.L. Chin, P.B. Shah, S. Najmaei, R.A. Burke, M. Dubey[1]
US Army Research Laboratory, Sensors and Electron Devices Directorate, Adelphi, MD, United States
[1]Corresponding author: e-mail address: madan.dubey.civ@mail.mil

Contents

1. INTRODUCTION

Since the isolation of graphene in 2004 and the demonstration of a field-effect transistor (FET) using this material (Novoselov et al., 2004), there has been a great deal of interest in the electrical behavior of this

Semiconductors and Semimetals, Volume 95
ISSN 0080-8784
http://dx.doi.org/10.1016/bs.semsem.2016.03.001

two-dimensional (2D) material. Coupled with the miniaturization of the metal–oxide–semiconductor field-effect transistor (MOSFET) that has dominated the microelectronic industry over the last 50 years, the promise of 2D materials has excited many both within and outside of the field. Since the initial demonstration, a search for additional 2D materials for electronic applications has yielded a wide variety of potential candidates—layered compounds such as transitional metal dichalcogenides (TMDs) and single/few layer single element "X-enes" such as silicene and phosphorene (Miró et al., 2014).

In this chapter, we review the advancements made using 2D materials as the active channel in electrical transistors and devices. First, we examine why 2D materials are considered for electronic applications, particularly for achieving sub-5 nm device sizes. Next, the use of 2D materials in simple FETs is explored, including a discussion of experimental issues to consider when fabricating and testing these devices. Then, efforts made to expand the use of 2D materials from simple transistors to complex logic devices and circuits are given. Consideration of 2D materials for radio frequency (RF) applications is also detailed. Finally, the use of 2D materials in nonconventional electronic devices, such as tunneling and valleytronics devices, is discussed.

1.1 Why 2D?

Conventional scaling down of microelectronic transistors has relied on the steady reduction of the minimum feature size of traditional integrated circuit (IC) components (Ieong et al., 2004). In particular, the MOSFET gate length has been shrinking, with state-of-the-art commercially available devices reaching the 14-nm fabrication node. However, achieving sub-5 nm gate lengths by 2020, as predicted by the 2013 International Technology Roadmap for Semiconductors, using silicon (Si) will be challenging. As transistor sizes are scaled down, device operational issues, such as increased current leakage due to electron tunneling, arise which can hinder performance. Scaling factors based on electrostatics and device design can be defined to predict the minimum gate length of a FET device necessary to avoid degraded subthreshold characteristics. The FET scaling length parameter λ is derived as $\lambda = \sqrt{t_s t_{ox} \varepsilon_s / \varepsilon_{ox}}$, where t_s and t_{ox} are the thickness of the semiconductor and gate dielectric/oxide and ε_s and ε_{ox} are the permittivity of the semiconductor and gate dielectric/oxide, respectively (Yan et al., 1992). As shown in the equation, λ is proportional to the square root of the semiconductor thickness t_s meaning a thinner channel enables a shorter device. However, thinning Si comes with issues, such as an increase of surface roughness scattering, which causes mobility degradation (Gold, 1987;

Uchida and Takagi, 2003). Layered 2D materials do not suffer from this degradation with thin channels like Si and other bulk materials and can be the ideal material for this scaling requirement. 2D materials also have the advantage of being substrate independent (Franklin, 2015). Most 2D materials can be transferred to different substrates—rigid, flexible, semiconductors, insulators, or amorphous materials—and still maintain their crystalline structure and unique properties. This has opened up a wide range of applications, including flexible displays and electronics, sensors, and even composite materials.

Although 2D materials cover a wide range of compounds across the periodic table, only a handful of material groups have been explored for traditional electronic applications. Semiconductors and insulators are all needed for creating electronic devices, and this holds for 2D-based electrical devices, as well. Table 1 lists the significant 2D materials used for fabricating electrical devices and a few of their electronic properties. They

Table 1 Room Temperature Solid-State Properties for Select 2D Materials

Material	Bandgap (eV)	Mobility ($cm^2/V \cdot s$)
Graphene[a]	0	$500–2 \times 10^5$
Silicene[b]	0–0.4	33–129
Phosphorene[c] SL[d]	2	286 – 1000
Phosphorene bulk	0.3	e-: 220, h+: 350
MoS_2 SL	1.8	1–80
MoS_2 bulk	1.2	200 – 500
$MoSe_2$ SL	1.49	50 – 200
$MoSe_2$ bulk	1.1	25 – 160
$MoTe_2$ SL[d]	1.13	11 – 40
$MoTe_2$ bulk	Semimetal	
WS_2 SL	1.93	0.01 – 234
WS_2 bulk	1.35	205
WSe_2 SL	1.7	140–200
WSe_2 bulk	1.2	120 – 500
Boron nitride[e]	4.69	N/A

[a]Mobility values highly dependent upon graphene deposition method and substrate.
[b]Predicted bandgap values.
[c]Bandgap is tunable by layer number.
[d]Mobility values reported taken from single to nm-thick material.
[e]Optical bandgap is 5.9 eV.
SL, single layer.
Mobility values taken primarily from Akinwande et al. (2014), Kam (1982), and Li et al. (2016).

range from graphene and its zero bandgap to small/mid-bandgap materials, such as phosphorene and the TMDs, and to wide bandgap hexagonal boron nitride (h-BN), which is used exclusively as a topologically smooth insulator. As discussed below, each material's combination of unique properties lends itself to specific electronic applications.

2. 2D MATERIALS IN FETs

This section focuses on FET devices with the active channels made from graphene, common TMDs, and phosphorene, as they represent the most prolific 2D materials found in literature. However, it is noted that other 2D materials are being investigated for their electronic properties as well (Bhimanapati et al., 2015; Miró et al., 2014). First, basic electrical performance metrics and characterization methods are introduced and explained. Then, FET device performance and important results are discussed for each material system. Recent experimental advances exploring the electrical characterization of silicene, the silicon counterpart to graphene, will also be covered. Potential sources of error in measurement testing and metric calculations are addressed.

2.1 FETs and Basic Electrical Characterization

The first step in creating beneficial devices and circuits from 2D materials is to determine their fundamental electrical properties. This is typically accomplished by fabricating simple devices, such as FETs, and characterizing their electrical performance. In a FET, current flows from one metallic contact through a channel made of semiconducting material to a second metallic contact (the metal contacts are called the source and the drain). When the semiconductor channel is highly conductive, current can easily flow and the FET is considered "on." Conversely, when the semiconductor is highly resistive, very little current can flow and the FET is considered "off." The conductivity of the semiconducting channel is controlled by the field effect from a third metallic contact (called the gate) that is separated from the channel by an insulator (often called the gate dielectric or gate oxide). Essentially, the FET is a voltage-controlled current source, where the gate voltage determines the current flowing from drain to source.

For fabrication simplicity, many researchers choose to create 2D material-based FETs that have a back-gated configuration (see Fig. 1A). A standard back-gated FET fabrication process consists of placing or growing the 2D material on an oxidized Si wafer, defining the channel

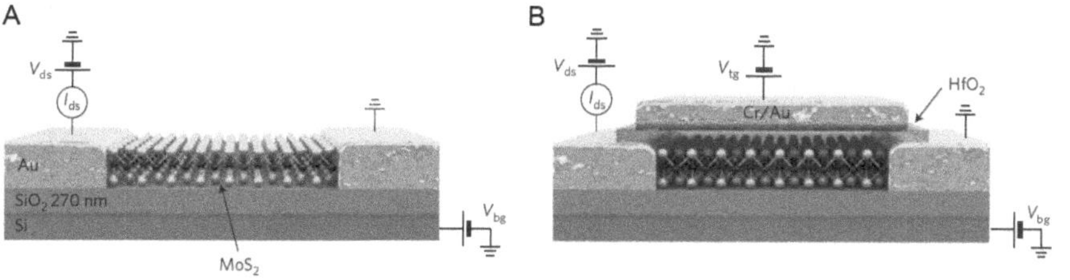

Fig. 1 (A) Cross-sectional drawing of a back-gated 2D FET. (B) Cross-sectional drawing of a dual-gated 2D FET (Radisavljevic and Kis, 2013b). *Reproduced with permission from Nature Publishing Group.*

dimensions with an etch process, and depositing metal contacts. The Si wafer is highly doped and acts as a global back gate. The thermal silicon oxide layer is used as the gate dielectric.

A slightly more complicated but practical configuration is the top-gated FET. In this setup, in addition to defining the channel area and depositing source/drain contacts, a dielectric is deposited on top of the channel, followed by a metal deposition for the local top gate contact. This top gate can be in sole control of the channel or implemented in a dual gate configuration with a back gate (see Fig. 1B).

Three important parameters that can be extracted from FET performance include the on/off current ratio (I_{on}/I_{off}), contact resistance (R_c), and carrier mobility (μ). Given a constant drain–source bias, the current flowing in a FET is controlled by the gate voltage. I_{on}/I_{off} is simply the ratio of the current that flows through a transistor in the on state divided by the current in the off state. The on/off current ratio is an important metric when choosing the application space for a material or device. For instance, in digital logic, a high I_{on}/I_{off} is required to create two discrete states and limit power consumption. To save power, the FET should leak as little current as possible in the off state; yet, when turned on, the FET should be highly conductive to allow for the power savings gained by decreasing voltage supply levels.

A parasitic resistance, called the contact resistance, is formed at the interface of the semiconductor and metallic contact. This contact resistance can severely limit the on-state current of a transistor and must be satisfactorily small for many device applications. To measure contact resistance, the transfer length method device geometry is commonly used (Schroder, 1990, p. 146). In this geometry, total device resistance is plotted against channel length. One can extract the contact resistance by noting that the y-intercept represents the resistance of a device with zero channel length (ie, the resistance of two contacts, $2R_c$). Since the contact resistance is dependent on the width of the contact, values of contact resistance are typically reported as $R_c W$, with units of Ω μm.

Carrier mobility reflects how easily an electron or hole can flow through a material. High carrier mobility enables higher transistor speed and efficiency. Two different techniques are typically employed to extract carrier mobility: field-effect mobility measurement and Hall mobility measurement. Field-effect mobility measurements are commonly performed by applying a constant drain–source bias (V_{ds}), sweeping the gate voltage (V_g), and measuring the drain–source current (I_{ds}) in the linear region of operation. Field-effect mobility is then extracted using the following equation (in units of $cm^2/V \cdot s$):

$$\mu_{FE} = \frac{d\frac{I_{ds}}{V_{ds}}}{dV_g} \times \frac{1}{C_{gate}} \times \frac{l}{w} \tag{1}$$

C_{gate} represents the gate-channel capacitance per area and l and w are the channel length and width, respectively (Fuhrer and Hone, 2013; Radisavljevic and Kis, 2013a). This measurement is based upon the formula for conductivity (shown below), where the conductivity, σ, is the product of the carrier density, the elementary charge value, and the carrier mobility, symbolized as n, e, and μ, respectively.

$$\sigma = ne\mu \tag{2}$$

In Eq. (1), the gate voltage changes the charge density by a known amount, $dV_g \times C_{gate}$, via the capacitive coupling of the gate and channel, and the measured change in sheet conductance, $d\frac{I_{ds}}{V_{ds}} \times \frac{l}{w}$, enables calculation of the mobility. Eq. (1) also assumes that the voltage drop across the channel is equal to V_{ds}, which is not always accurate due to the effects of contact resistance. Therefore, extracting material and device performance is more accurate when implementing additional electrical contacts. Voltages are commonly measured using the four point probe (4PP) configuration, where the inner two probes have high input impedance and act as voltage sensing probes. Performing voltage measurements in a 4PP configuration negates the effects of contact resistance that can underestimate mobility for two point probe (2PP) measurements. When measuring field-effect mobility with a 4PP configuration, the voltage drop between the two inner probes (represented below as V_{diff}) replaces V_{ds} in Eq. (1), and the channel dimensions change as well, denoted by l_{4PP} and w_{4PP}:

$$\mu_{FE\,4PP} = \frac{d\dfrac{I_{ds}}{V_{diff}}}{dV_g} \times \frac{1}{C_{gate}} \times \frac{l_{4PP}}{w_{4PP}} \tag{3}$$

To measure the Hall mobility, a magnetic field is applied orthogonal to the direction of current flow. Carriers will deflect due to the Lorentz force and build up a voltage (called the Hall voltage) that will oppose and equal the Lorentz force. Carriers with a higher mobility are more easily deflected by the magnetic field, which creates a larger Hall voltage.

Hall mobility measurements require a strong magnetic field and are more difficult to set up, but they have a few advantages over field-effect mobility measurements. Hall measurements are inherently done in a 4PP configuration; therefore, contact resistance is not a factor. Additionally, the measurements do not depend on a calculation of carrier concentration from a capacitance value. Rather, carrier concentration is determined directly from the measurements.

2.2 Comparison of Performance Metrics

Fig. 2 compares the mobility and I_{on}/I_{off} values for graphene, the semiconducting TMDs, and phosphorene. These three 2D materials systems cover a wide range of the mobility-on/off current ratio spectrum. On one end of the spectrum, graphene has extremely high mobility but a very low on/off current ratio. As a semimetal with zero bandgap, graphene cannot be used as the

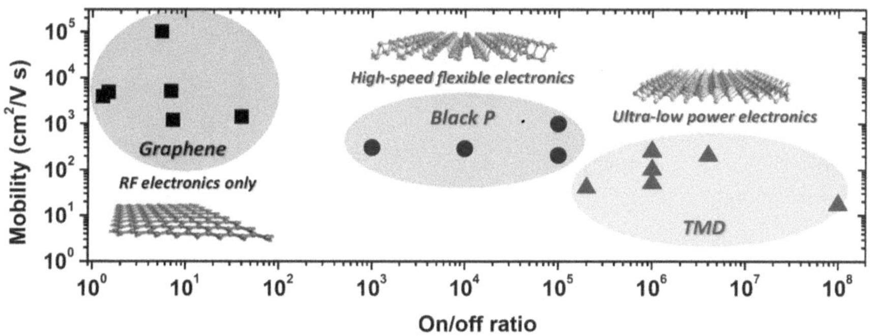

Fig. 2 Distribution of the basic electrical properties of graphene, the TMDs, and phosphorene (Ling et al., 2015). *Reproduced with permission from the National Academy of Sciences of the United States of America.*

channel material for digital logic applications since it does not have an off state. However, some RF electronics applications could potentially use graphene (Schwierz, 2013). The RF performance of graphene and other 2D materials will be covered in a later section. For an exhaustive source on the prospects of graphene, the article written by Ferrari et al. (2015) is recommended.

On the other side of the mobility vs on/off ratio spectrum, the common TMDs, such as molybdenum disulfide (MoS_2), have comparatively low mobility, but a very high on/off ratio. As a result of their large bandgaps, many TMDs have a very low I_{off}, which makes them well suited for low power applications. A collection of experimental mobility measurements on TMDs, including fabrication and testing parameters, is available from Li et al. (2016). Bridging the gap between graphene and MoS_2 is phosphorene, with mobility values well in the 100s of $cm^2/V \cdot s$ and an on/off current ratio around 10^4. These properties have excited many researchers and sparked a burgeoning interest in this 2D material (Samuel Reich, 2014).

Often the performance of 2D materials is compared to Si and other well-established material systems, serving as a reference for the progress made and potential that exists. Fiori et al. (2014) compared the performance of 2D materials to traditional semiconductor materials (Fig. 3A) and materials used for flexible electronics (Fig. 3B).

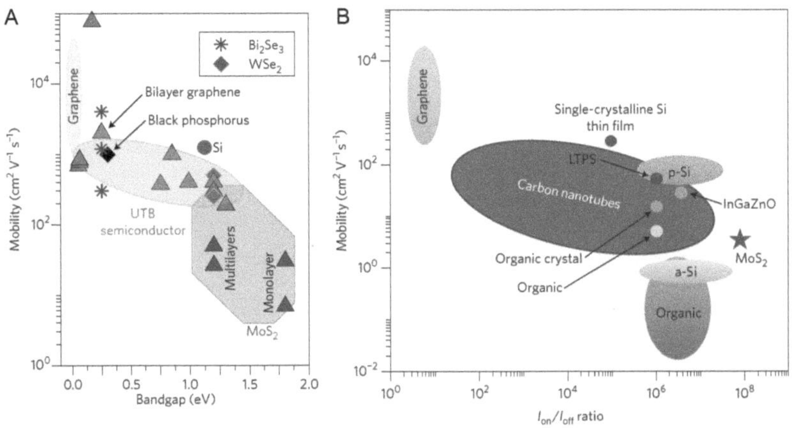

Fig. 3 (A) Comparison of common 2D materials and traditional semiconducting materials, plotted on a mobility vs bandgap graph. (B) Comparison of common 2D materials and traditional flexible electronics materials, plotted on a mobility vs I_{on}/I_{off} ratio graph (Fiori et al., 2014). *Reproduced with permission from the Nature Publishing Group.*

Table 2 Main Advantages (Pros), Drawbacks (Cons), and Potential Applications of Select 2D Materials (Schwierz et al., 2015)

Material	Main Pros	Main Cons	Potential for	Not Suited for
Graphene	High μ	Zero gap	Flex, mp RF	Logic, hp RF
Phosphorene	Reasonable μ	Stability	Flex, mp RF and logic hp logic $L > 5$ nm	Logic at $L < 5$ nm
Mo-based TMDCs	Reasonable E_g	Moderate μ	Flex, mp RF and logic hp logic $L < 5$ nm	hp RF
W-based TMDCs	Reasonable E_g	Moderate μ	Flex, mp RF and logic	hp RF

mp, medium performance; *hp*, high performance; *flex*, flexible, E_g, bandgap; *m*, mobility; *L*, gate length.
Reprinted with permission from the Royal Society of Chemistry.

Based on the fundamental properties measured (ie, mobility and on/off ratio), Schwierz et al. (2015) identified possible applications of various 2D materials, as well as their main pros and cons. As shown in Table 2, graphene shows potential for flexible electronics and some RF applications due to its high mobility, but its lack of a bandgap limits the number of potential applications. Phosphorene could be used for flexible electronics and RF, as well as digital logic, as it has a bandgap. A distinction is made between logic applications with gate lengths greater than and less than 5 nm because tunneling between the source and drain contacts limits the scaling of transistors by increasing the off-state current and degrading the subthreshold characteristics (Wang and Lundstrom, 2002). At gate lengths less than 5 nm, direct source–drain tunneling is an issue and materials with carriers having large effective mass perform better (Majumdar et al., 2014). The carriers in Mo-based TMDs in particular have large effective masses, which mean they could be used for logic with gates less than 5 nm. Phosphorene, with its lighter carriers, would be applicable to longer gate length devices.

As discussed earlier, regardless of device application, contact resistance is a major parasitic effect that needs to be minimized for acceptable device performance. Table 3 shows the contact resistances achieved in 2D material FETs compared to the contact resistance in traditional Si or III–V materials. As the oldest and most established 2D material, graphene devices have the lowest contact resistances, almost comparable to the contact resistance in silicon devices. Compared to graphene, contact engineering in TMDs and phosphorene is less developed due to the relative youth of the materials, particularly for the latter.

Table 3 Contact Resistance of Different FET Structures (Schwierz et al., 2015)

Transistor Channel Material	R_cW (Ω mm)	Contact Material	Layer Count
Graphene	0.011 – 0.08	Ni, Ti	BL
	0.1 – 0.2	Ti, Ni, Pd/Au, Cr/Au	SL
MoS_2	0.2 – 1.6	Ni/Au, Ti/Au, Au	FL
	2	Au	SL
WSe_2	1.4	Graphene	BL
	2	Graphene	FL
Phosphorene	1.75	Ni/Au, Pd/Au	FL
Silicon	<0.01		
InP	0.03		
GaAs	0.02		

SL, single layer; *BL*, bilayer; *FL*, few-layer.
Reprinted with permission from the Royal Society of Chemistry.

2.3 Important Results/Findings

2.3.1 Graphene

Theoretically, graphene possesses a room temperature electron mobility of 250,000 $cm^2/V \cdot s$ (Chen et al., 2008); however, the transport behavior is extremely dependent on the local environment and processing done to the material. When placed on a substrate, charged impurity scattering and remote interfacial phonon (RIP) scattering reduce mobility. On SiO_2, RIP scattering limits the graphene mobility to 40,000 $cm^2/V \cdot s$ (Chen et al., 2008). Atmospheric exposure and processing contaminants such as resist residue, water, and metallic impurities also act as scattering sources and degrade the mobility (Chan et al., 2012).

One method to approach graphene's incredible intrinsic electronic properties is to suspend the graphene channel. In a method used by Bolotin et al. (2008), the SiO_2 under the graphene channel is etched away using buffered oxide etchant (1:6, BOE). The sample is transferred from BOE to ethanol, and then critical-point dried to keep the graphene sheet suspended and intact. Current annealing is used to clean the suspended devices, which improved the performance substantially. Only devices that were both suspended and current annealed show the high performance. A mobility above 60,000 $cm^2/V \cdot s$ was measured for every device, with a peak mobility of 230,000 $cm^2/V \cdot s$ reported.

Another important demonstration of improved performance is through the replacement of the usual SiO_2 substrate with h-BN, a 2D insulator material (Dean et al., 2010). Compared to SiO_2, the improved performance is attributed to the h-BN layer being atomically flat, free of dangling bonds, and possessing higher energy surface optical phonon modes. For these reasons, h-BN has established itself as the substrate and gate dielectric of choice when fabricating 2D devices and heterostructures. When encapsulated with h-BN, hole mobility of 100,000 $cm^2/V \cdot s$ and electron mobility of 70,000 $cm^2/V \cdot s$ were achieved at room temperature, as well as the observation of ballistic transport on the micrometer scale. Importantly, the h-BN encapsulation also provided protection from the ambient environment (Mayorov et al., 2011).

The previous two examples (ie, graphene suspension and h-BN encapsulation) were achieved using mechanical exfoliation of graphene from bulk graphite crystals. Exfoliation of individual flakes from bulk crystals may enable fruitful characterization and innovation, but for manufacturability, large-area growth is a necessity. It is worthwhile to briefly address the multitude of methods demonstrated to isolate or grow graphene and compare the resulting electrical properties. Fig. 4 shows the mobility measured for different growth synthesis techniques. With over a decade of extensive research, numerous ways to grow graphene have been developed. For large-area, high-mobility material, two promising growth methods are the

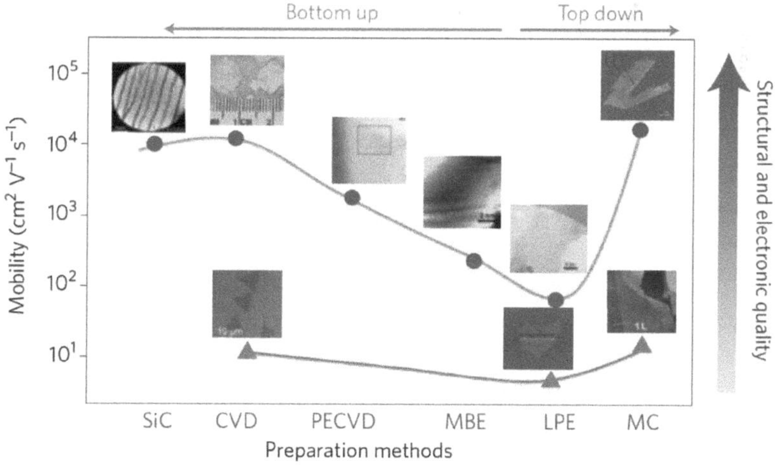

Fig. 4 Mobility values achieved for different graphene growth processes. Growth methods include epitaxial growth on SiC, CVD, plasma-enhanced chemical vapor deposition (PECVD), molecular beam epitaxy (MBE), liquid-phase exfoliation (LPE), and micromechanical cleavage (MC). The top curve is for graphene, and the bottom curve is for MoS_2 (Fiori et al., 2014). *Reproduced with permission from the Nature Publishing Group.*

chemical vapor deposition (CVD) of graphene on metal substrates, such as copper (Cu) or nickel (Ni), and the epitaxial growth of graphene on silicon carbide (SiC) (Li et al., 2009; Petrone et al., 2015; Tedesco et al., 2009). These two methods produced graphene with mobility values comparable to exfoliated material.

2.3.2 Transition Metal Dichalcogenides

Renewed research interest in TMDs was sparked in 2010 by reports of a direct bandgap transition for monolayer MoS_2 (Mak et al., 2010; Splendiani et al., 2010) and by the report of impressive electrical performance from a monolayer of MoS_2 in 2011 (Fuhrer and Hone, 2013; Radisavljevic and Kis, 2013a; Radisavljevic et al., 2011a). While the exact mobility value extracted from this MoS_2 transistor has been disputed, direct bandgap semiconducting layered materials show great potential for high-performance transistors. Among the most prominent TMDs studied are the Mo-based and W-based dichalcogenides (MoS_2, $MoSe_2$, $MoTe_2$, WS_2, WSe_2, and WTe_2). Until recently, much of the research on these dichalcogenides has been accomplished using exfoliated material from large crystalline sources. Following the example set by graphene, large-area synthesis methods are being explored for TMDs, especially for MoS_2, the most studied semiconducting 2D material. Fig. 4 shows the mobility values measured for devices made from MoS_2 deposited by CVD, liquid-phase exfoliation (LPE), and mechanical exfoliation. The mobility ranged from ~4 to 12 $cm^2/V \cdot s$ and was less dependent upon growth method than graphene. Fig. 5 shows a typical transfer curve measured from a CVD MoS_2 FET in a back-gate configuration. For this device, a mobility of 7.36 $cm^2/V \cdot s$ was extracted.

With bandgaps similar to Si and many III–V semiconductors, TMDs possess desirable electrical properties for use in low-power logic circuits. To achieve adequate I_{on}/I_{off} ratios desired for logic circuits, transistor channels typically need bandgaps larger than 0.4 eV (Schwierz, 2015). Monolayer Mo- and W-based TMDs have direct bandgaps that range from 1.5 to 1.9 eV (see Table 1 for a full listing). Transistors fabricated using MoS_2 and WSe_2 have been reported with field-effect mobility values ranging from 1 to 200 $cm^2/V \cdot s$ with I_{on}/I_{off} ratios typically reaching 10^5 to 10^7. While not as high when compared to Si and other semiconductors, they appear to be suitable for niche applications. With strain limits approaching 30% before irreversible damage occurs to the crystal lattice, TMDs have shown promise for use in flexible electronics (Akinwande et al., 2014), which compare favorably to Si lattices that are irreversibly damaged beyond 10% straining.

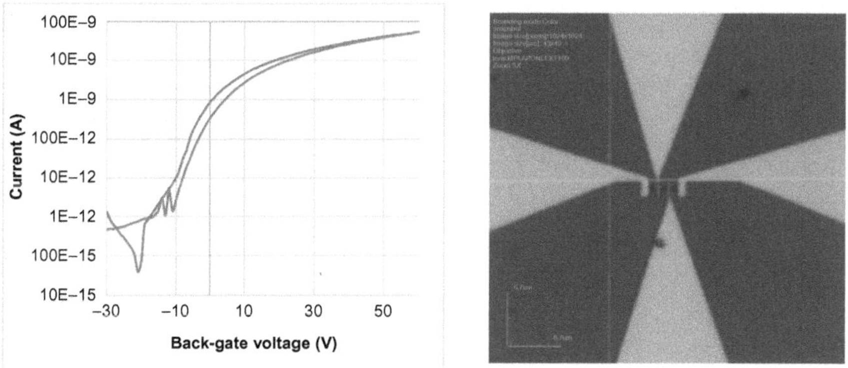

Fig. 5 Device performance of a typical CVD-grown MoS_2 FET at a drain–source bias of 100 mV and optical microscope image of a typical device. *Data and image provided courtesy of the US Army Research Laboratory.*

Additionally, compared to commonly used semiconductors for flexible device applications, such as amorphous silicon (a-Si) and organic semiconductors, TMDs possess field-effect mobility values that are at least an order of magnitude higher, providing for superior switching performance when modulated (Akinwande et al., 2014). The development of TMD transistors for these applications continues to make strides as performance obstacles are solved.

One major performance issue in TMD-based FETs is the contact resistance. In MoS_2, it was shown that the MoS_2–metal interface is strongly affected by Fermi level pinning close to the conduction band (Das et al., 2013). Based on metal work function alignment with the conduction and valence band of MoS_2 (Fig. 6A), one would expect scandium (Sc) and titanium (Ti) contacts to inject electrons efficiently and Ni and platinum (Pt) contacts to inject holes (Fig. 6A). Due to the Fermi level pinning near the conduction band, however, MoS_2 devices usually show n-type behavior (Fig. 6C), while p-type conduction has been difficult to achieve. Recently, high work function molybdenum trioxide (MoO_x) contacts were used to demonstrate improved hole injection into MoS_2 and tungsten diselenide (WSe_2) (Chuang et al., 2014). The results can be seen in Fig. 7. The MoO_x contacts were compared to palladium (Pd), another high work function metal, for both MoS_2 and WSe_2 devices. In MoS_2, the Pd contacts show no p-type behavior, and in WSe_2, the MoO_x contacted devices show over an order of magnitude higher current.

Promising performance has been demonstrated in monolayer WSe_2 for both holes and electrons. As previously mentioned, hole injection was seen

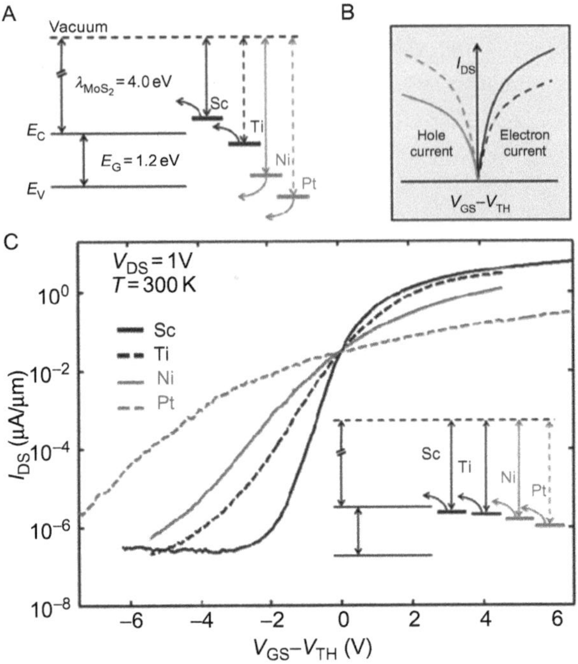

Fig. 6 (A) Work function alignment of Sc, Ti, Ni, and Pt with the conduction and valence band of MoS_2. (B) Expected transfer characteristics based on the work function alignment in (A). (C) Actual transfer characteristics seen as a result of the Fermi level pinning. The inset shows the adjusted work function alignment, with all metals pinned close to the conduction band (Das et al., 2013). *Reproduced with permission from the American Chemical Society.*

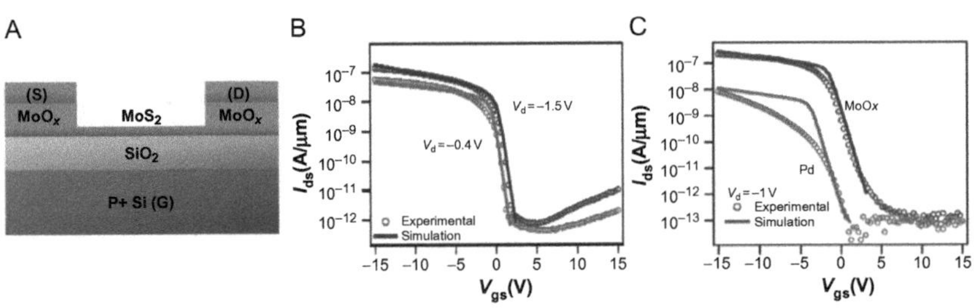

Fig. 7 (A) Cross-sectional drawing of a back-gated MoS_2 device implementing MoO_x contacts. (B) Transfer characteristics of a MoS_2 transistor with MoO_x contacts, showing clear p-type behavior. (C) Transfer characteristics of a WSe_2 transistor demonstrating the improved hole conduction using MoO_x instead of Pd (Chuang et al., 2014). *Reproduced with permission from the American Chemical Society.*

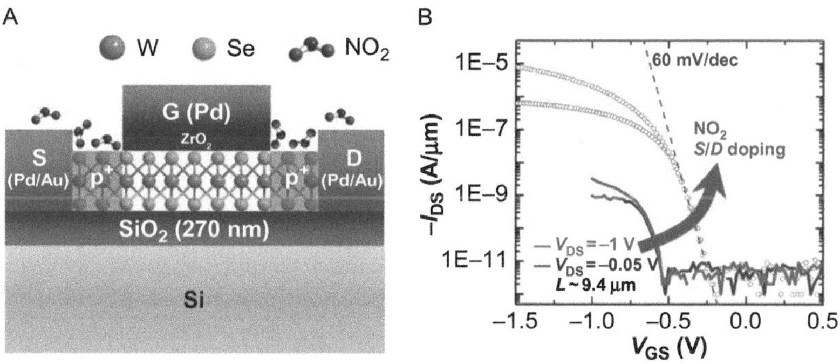

Fig. 8 (A) Cross-sectional drawing of a top-gated WSe_2 device implementing NO_2 doping of the contact region. (B) Transfer characteristics of a WSe_2 transistor with and without the NO_2 doping, showing enhanced p-type conduction with the doping (Fang et al., 2012). *Reproduced with permission from the American Chemical Society.*

with Pd contacts; however, the contact resistance remained very large. Chemisorption of nitrogen dioxide (NO_2) at the drain/source contact areas was used to degenerately dope the WSe_2 surface and caused a remarkable decrease in contact resistance. As seen in Fig. 8, a hole mobility of 250 $cm^2/V \cdot s$, $I_{on}/I_{off} > 10^6$, and a subthreshold swing of 60 mV/decade was achieved at room temperature using that technique (Fang et al., 2012). Efficient electron injection to WSe_2 was demonstrated with indium (In) contacts. A back-gated FET with In contacts had an electron mobility of 142 $cm^2/V \cdot s$ and an $I_{on}/I_{off} > 10^6$ (Liu et al., 2013). With electron and hole carrier mobilities greater than 100 $cm^2/V \cdot s$ and I_{on}/I_{off} greater than 10^6, monolayer WSe_2 shows great potential for digital logic applications.

One recent trend in TMD device fabrication development is to use graphene to improve the contact resistance for MoS_2 devices. When graphene is inserted between Ti/gold (Au) contacts and MoS_2, an improvement over threefold in contact resistance was measured (Du et al., 2014). In this configuration, the Fermi level of the MoS_2 and the graphene contact can be adjusted by the back-gate bias. A similar approach was used to decrease the contact resistance of Ni contacts on MoS_2 (Leong et al., 2015). In this case, though, the graphene is first etched in order to create many zigzag edges on the graphene surface, which enables improved conduction through the contact (Leong et al., 2014).

Gong et al. (2014) simulated the impact that the physical separation between the metal and the MoS_2 has on Fermi level pinning. To decrease the Fermi level pinning at the metal/MoS_2 interface, a buffer layer is proposed. Density functional theory calculations were performed to investigate graphene as a buffer layer between MoS_2 and different metals. The graphene was found to successfully isolate the MoS_2, but a consistent change to the Schottky barrier height (SBH) was not observed for all metals. For silver (Ag) and Ti, there was a decrease in the SBH; for Au and Pt, there was an increase. The best-case scenario was found for ruthenium (Ru), where an ohmic-type contact was formed (Chanana and Mahapatra, 2016).

Another interesting method for decreasing contact resistance in MoS_2 devices is the phase conversion of the contact, as demonstrated by Kappera et al. (2014). MoS_2 naturally occurs in a semiconducting 2H phase, but a metastable metallic 1T phase can also exist. To achieve the 2H to 1T phase transformation, selected areas were exposed using a poly(methyl methacrylate) (PMMA) mask and then soaked in *n*-butyl lithium for 2 h. At zero gate bias, the phase transition decreased the contact resistance by a factor of 5 to 200–300 Ω μm.

2.3.3 Phosphorene

Phosphorene, a 2D counterpart to the bulk material black phosphorus (BP), was exfoliated and characterized by multiple groups in early 2014 (Koenig et al., 2014; Li et al., 2014; Liu et al., 2014; Xia et al., 2014). With Ti/Au contacts, Liu et al. (2014) observed p-type conduction that is anisotropic (see Fig. 9) and a maximum room temperature hole mobility of 286 $cm^2/V \cdot s$ in a 5-nm-thick flake. Koenig et al. (2014) measured a maximum room temperature mobility of 300 $cm^2/V \cdot s$, while mobilities up to 600 and 1000 $cm^2/V \cdot s$ were reported at room temperature by Xia et al. (2014) and Li et al. (2014), respectively. The measured mobility values depended

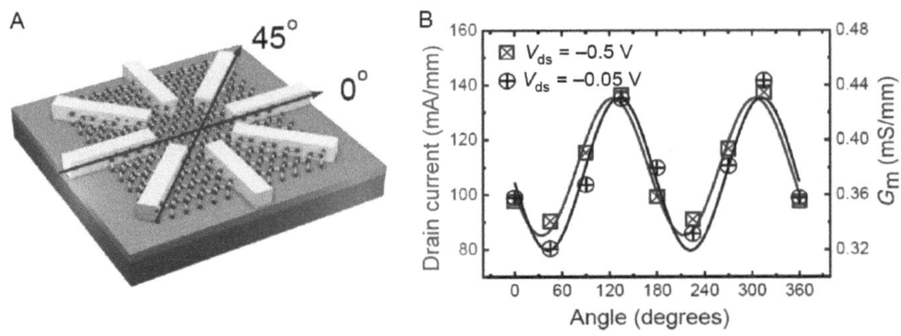

Fig. 9 (A) Top view drawing of the contact orientation used to probe the directional dependence of current flow in phosphorene. (B) Measured drain current and conductance as a function of the angle between the contacts, showing clear anisotropic and periodic behavior (Liu et al., 2014). *Reproduced with permission from the American Chemical Society.*

on the thickness of the flake and direction of current transport along the crystal. While the transport measured showed strong p-type behavior, weaker electron conduction was observed at positive gate biases. Ti contacts were used to demonstrate improved electron injection and reduced asymmetry between the electron and hole conduction (Das et al., 2014a).

As discussed in detail by Ling et al. (2015), a unique property that phosphorene has, unlike graphene or common TMDs like MoS_2, is anisotropy. To best exploit phosphorene's anisotropic behavior, they propose novel devices based on "(i) plasmonic devices with intrinsic anisotropy in their resonance properties and (ii) high-efficiency thermoelectrics using the orthogonality in the heat and electron transport directions."

Unfortunately, exfoliated phosphorene degrades in ambient conditions. Studies suggest that phosphorene possesses a hygroscopic nature, where absorbed moisture causes volumetric expansion of the material as well as chemical conversion to phosphorus oxides or oxyacids (Kim et al., 2015a). Atomic force microscopy (AFM) images of the same phosphorene flake showing deterioration due to ambient exposure are provided in Fig. 10. Effects of the degradation are manifested in the electrical properties as well, and Fig. 11 shows the electrical properties over time for various passivation schemes. A capping layer of atomic layer deposition (ALD) sapphire (Al_2O_3) followed by a hydrophobic fluoropolymer spin-cast successfully passivated the phosphorene layers, enabling air-stability over months of observation (Kim et al., 2015a).

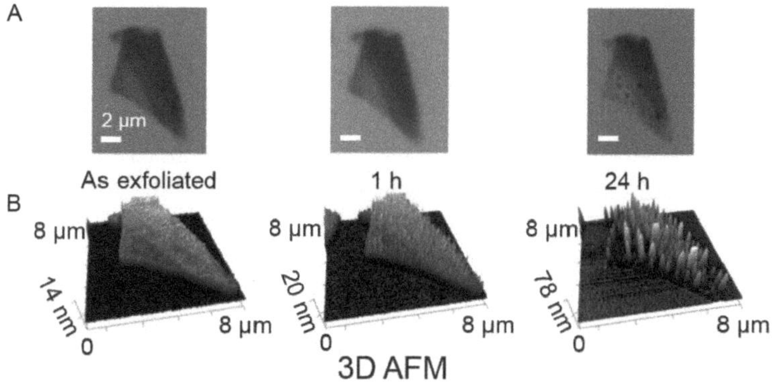

Fig. 10 (A) Optical microscope images of the same flake of BP immediately after exfoliation, 1 h later, and a day later. Degradation can be seen in the thinner areas of the flake. (B) AFM images of the same flake in (A), taken at the same time intervals, showing increased roughness as atmospheric exposure time increases (Kim et al., 2015a). *Reproduced with permission from the Nature Publishing Group.*

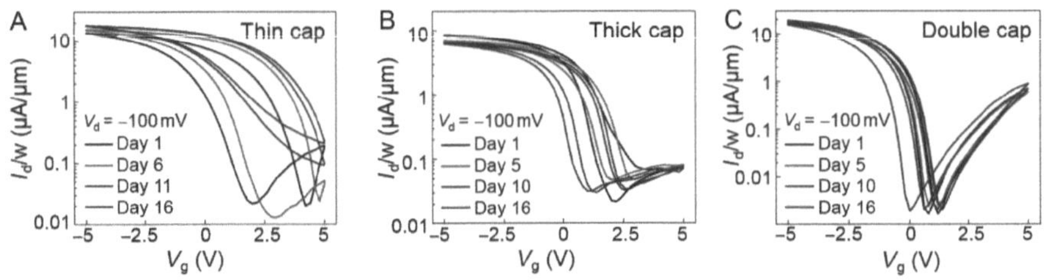

Fig. 11 Stability over time of the transfer characteristics for a phosphorene flake covered with various capping layers. (A) 5 nm of ALD Al_2O_3. (B) 25 nm of Al_2O_3. (C) 25-nm Al_2O_3 coating followed by a hydrophobic fluoropolymer spin-cast (Kim et al., 2015a). *Reproduced with permission from the Nature Publishing Group.*

2.3.4 Silicene

The advent of graphene as a promising candidate for ballistic transport and high frequency operations has sparked questions to the existence of graphene-like counterparts in other group IV elements, such as Si and germanium (Ge). While free-standing, planar, single atomic layers of these group IV elements are unlikely to form due to the prominence of sp^3 bonding in these materials, stable, slightly buckled, honeycomb atomic arrangements are possible (Cahangirov et al., 2009). Monolayer silicene, the silicon analog to graphene, has been experimentally verified to form on metals, particularly on Ag(110) and Ag(111) surfaces (Kara et al., 2012; Lalmi et al., 2010; Le Lay et al., 2009). It has a zero bandgap, possesses a Dirac-cone band structure very similar to graphene and is compatible with current Si technology. These features make silicene an intriguing material for many electronic applications.

Because of its instability in air due to rapid oxidation, silicene has been difficult to produce let alone studied to determine its electrical properties. However, a unique synthesis-transfer-fabrication process noted as silicene encapsulated delamination with native electrodes (SEDNE), reported by Tao et al. (2015), has been able to overcome this challenge and create the first silicene-based FETs. As shown in Fig. 12A, the process makes use of the silver thin film substrate as native electrodes and an Al_2O_3 capping layer as the gate dielectric to form the FET while minimizing air exposure. Electrical testing reveals that Ag forms an ohmic contact with silicene and that the silicene exhibits ambipolar behavior like graphene (Fig. 12B) and I_{on}/I_{off} values of ~10 (Fig. 12C). Electron and hole mobility values calculated for a series of FET devices ranged from 33 to 129 $cm^2/V \cdot s$. Furthermore, the transfer characteristics show a low residual carrier density compared to graphene (more than an order of magnitude lower) and hints to the small

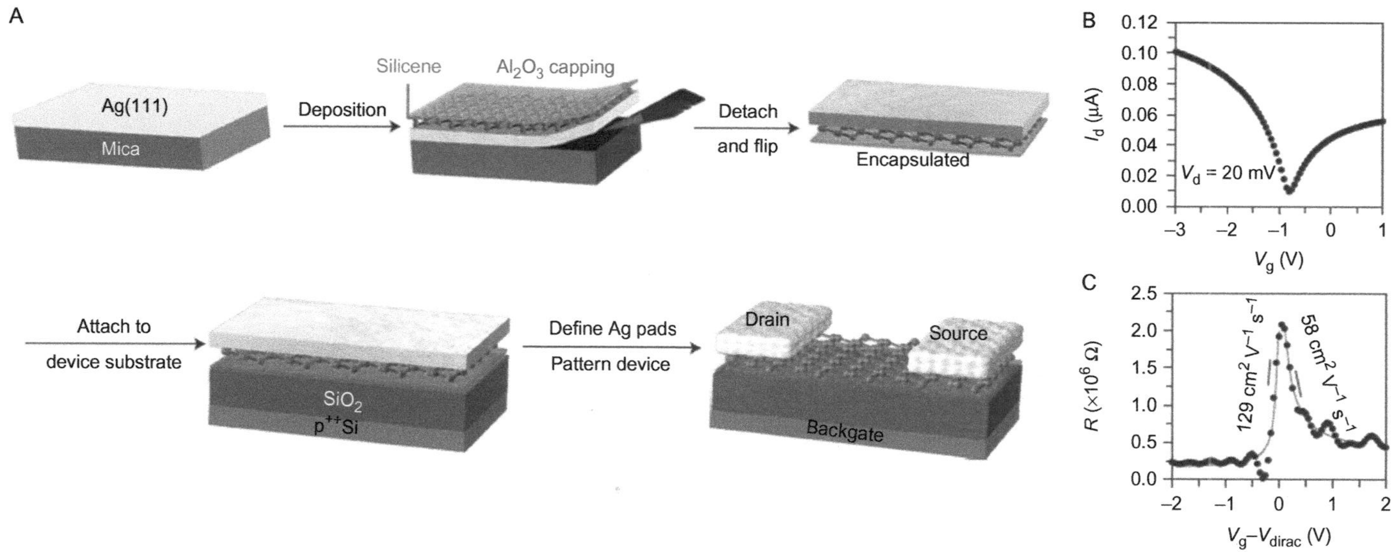

Fig. 12 Silicene synthesis–transfer–fabrication process and room temperature electric characterization of silicene transistor devices. (A) Schematic of the SEDNE process. (B) I_d vs V_g curve of a silicene device displays ambipolar electron–hole symmetry expected from silicene. (C) R vs ($V_g - V_{dirac}$) of a silicene device with calculated electron and hole mobility values (Tao et al., 2015). *Reproduced with permission from the Nature Publishing Group.*

bandgap opening of ~210 meV in the material. Additional studies are needed to confirm the existence of a bandgap opening in silicene.

2.4 Experimental Considerations

Device configurations, assumptions, and testing conditions can all heavily impact the measured electrical properties (eg, carrier mobility) of 2D material-based FETs. It is worthwhile to address a few items that should be considered when creating devices, performing measurements, or reviewing results in the literature.

A common source of error when extracting field-effect mobility values from transfer curves is using the incorrect gate capacitance value. Underestimating the capacitance leads to an overestimated mobility and vice versa. Fuhrer and Hone (2013) addressed this issue, especially for the case of dual-gated devices (see Fig. 13) when leaving the top-gate electrode floating while sweeping the back gate. Using the back-gate capacitance in this configuration to estimate the mobility, an overestimation of mobility by $\sim C_{tg}/C_{bg}$ will occur. They stress that one should always ground the gate not in use for well-defined capacitances.

Radisavljevic and Kis (2013b) found that encapsulation with a 30 nm hafnium oxide (HfO_2) top dielectric also increased the capacitive coupling of their back-gated device by 2.4, reinforcing the idea that accurate capacitance values are necessary when extracting field-effect mobility in devices with a complex dielectric geometry. To ensure correct capacitance values, Hall measurements or capacitance–voltage (CV) techniques should be used for corroboration.

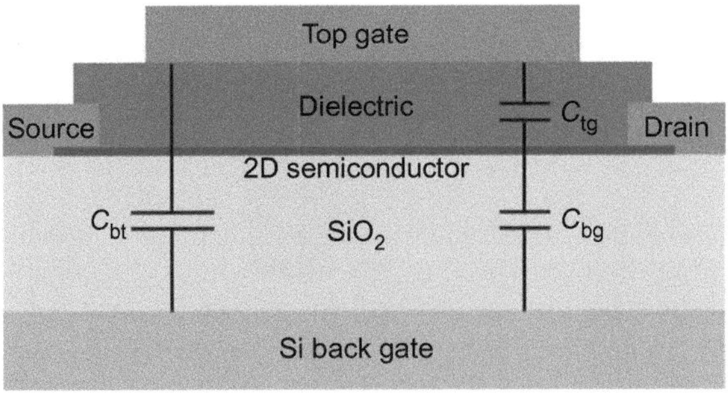

Fig. 13 Cross-sectional drawing of the dual-gated geometry with capacitances labeled. C_{tg} represents the capacitance of the top gate with the channel, C_{bg} represents the capacitance of the back gate with the channel, and C_{bt} represents the capacitive coupling between the top and bottom gate. *Adapted from Fuhrer and Hone (2013).*

Another potential source of error when using capacitances to calculate field-effect mobility is the negligence of the quantum capacitance, C_q. The total gate capacitance in a FET is the series combination of the geometrical capacitance, modeled as a parallel plate capacitor, and the quantum capacitance. The total capacitance is dominated by the smaller capacitance, which is usually the geometrical capacitance. However, as demonstrated by Ma and Jena (2015), for 2D material transistors with very thin gate dielectrics, C_{ox} is no longer much smaller than C_q so the combination of the two capacitances in series should be used, $C_{ox}C_q/(C_{ox}+C_q)$.

The testing environment has a large impact on the measured electrical properties of 2D materials. Especially if the devices are not encapsulated, one needs to test in vacuum and perform annealing to achieve the best electrical performance. Chan et al. (2012) demonstrated the importance of annealing by testing CVD graphene in a vacuum probe station and comparing the mobility of devices that were just fabricated, annealed at 300°C in ultrahigh vacuum, and annealed at 80°C in the vacuum probe station after the 300°C anneal (Fig. 14). The mobility of the as-fabricated devices averaged 1000–2000 $cm^2/V \cdot s$. After annealing at 300°C for 3 h, the average mobility increased to roughly 3000 $cm^2/V \cdot s$. The devices

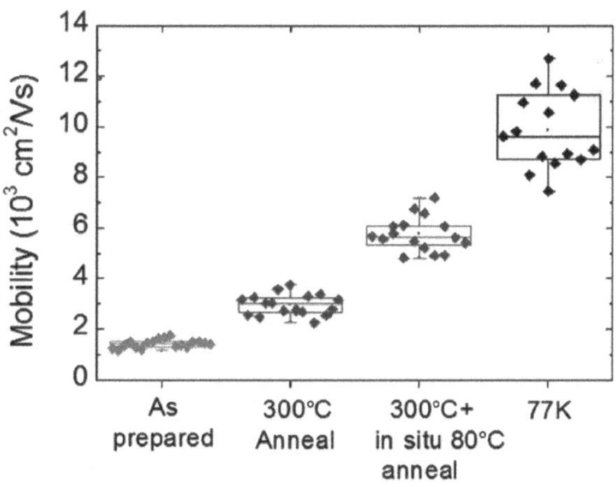

Fig. 14 Effect of annealing on the mobility measured in CVD graphene. The "as prepared" samples were tested in vacuum with no anneal. The "300°C Anneal" samples were annealed outside the probe station at 300°C and then moved into the probe station and tested in vacuum. The "300°C+ in-situ 80°C anneal" samples were annealed outside the probe station at 300°C, moved into the probe station, and then annealed at 80°C in vacuum for 15 h before testing (Chan et al., 2012). *Reproduced with permission from the American Chemical Society.*

were then measured after annealing in a vacuum probe station at 80°C for 15 h, which led to an average mobility around 6000 $cm^2/V \cdot s$. The mobility improvement was attributed to the removal of contaminants introduced during the transfer process and desorption of ambient environmental adsorbates.

Ovchinnikov et al. (2014) investigated the dependence of measured tungsten disulfide (WS_2) mobility with annealing time. WS_2 devices were annealed in vacuum at 115°C for up to 145 h, and the results are plotted in Fig. 15. 4PP mobility steadily increased until plateauing after 80 h around 50 $cm^2/V \cdot s$. The 2PP mobility on the other hand continued to increase for the entire 145 h, indicating that the contact resistance was still decreasing. This result highlights how sensitive the 2D material devices can be to the testing conditions, and why the conditions must be considered when one is reviewing, comparing, or testing device performance.

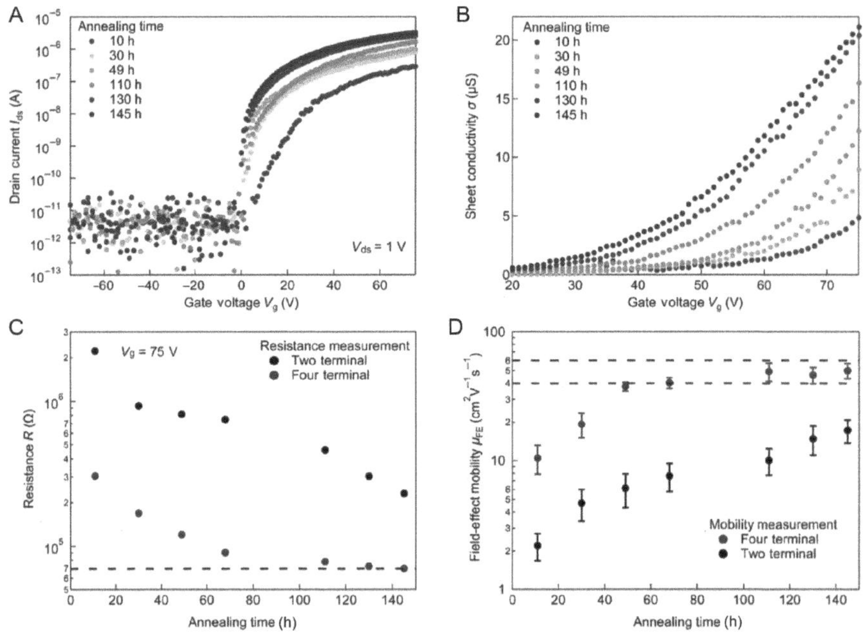

Fig. 15 Dependence of the electrical properties of a WS_2 transistor on the length of the annealing time. Annealing was performed in the vacuum probe station at 115°C (A) I_{ds} vs V_{gs} dependence on annealing time. (B) Measured conductance for gate voltages of 20–75 V. (C) 2PP and 4PP resistance measurements at a gate bias of 75 V. (D) Maximum measured 2PP and 4PP mobility (Ovchinnikov et al., 2014). *Reproduced with permission from the American Chemical Society.*

3. 2D MATERIALS IN LOGIC CIRCUITS

With the numerous advancements in the development of transistors and electronic devices based on 2D materials, and the reliability of fabricating these structures increasing, it has become possible to produce multi-transistor circuits based on these van der Waals (vdW) materials. The feasibility of producing useful electronic circuit building blocks has been demonstrated by a number of research groups since 2009 (Amani et al., 2015; Das et al., 2014b; Fang et al., 2012; Jeon et al., 2015; Li et al., 2010; Lin et al., 2014; Liu et al., 2014; Radisavljevic et al., 2011b; Rizzi et al., 2012; Tosun et al., 2014; Traversi et al., 2009; Wang et al., 2012b; Yu et al., 2014). Compared to Si-based electronic circuits and manufacturing techniques, electronic circuits based on 2D materials might not have superior performance, manufacturability, or cost, but there are certainly applications and environments where electronic technologies based on 2D materials have advantages over the current state-of-the-art electronics produced using Si or III–V transistors (Amani et al., 2015; Guerriero et al., 2013; Schall et al., 2013; Tosun et al., 2014; Wang et al., 2012b; Yu et al., 2014, 2015; Zhu et al., 2015). FETs based on 2D TMDs have been touted by a number of research groups for their ability to reduce short channel effects and power consumption compared to similarly sized transistors based on Si (Radisavljevic et al., 2011b, 2012; Wang et al., 2012b; Yu et al., 2015).

Flexible and transparent electronic circuits would be able to take advantage of the robustness and transparency of ultrathin 2D material channels (Amani et al., 2015; Radisavljevic et al., 2011b, 2012; Wang et al., 2012b; Yu et al., 2014, 2015). Examples of 2D circuits on flexible substrates that have been realized include inverters based on MoS_2 (Amani et al., 2015; Cheng et al., 2014) and BP (Zhu et al., 2015) fabricated on polyimide. Amorphous Si-based and organic-based transistor technologies are typically used in these applications, but they generally lack the superior electronic properties of Si. Both organic semiconductors and a-Si typically have electronic mobility values of less than 1 $cm^2/V \cdot s$. As discussed previously, transistors produced from TMDs have electronic mobility values ranging in the 10s to 100s of $cm^2/V \cdot s$, which allows for devices to be produced with orders-of-magnitude superior performance (Wang et al., 2012b; Yu et al., 2014).

This section details the advances being made in the development of multi-transistor ICs based on 2D materials. While many groups have demonstrated the ability to fabricate a functional transistor with a 2D channel material, substantially few have been able to produce more than one active circuit element on the same substrate. Fewer still have been able to produce several transistors on a single substrate based on 2D materials. The difficulty of repeatable 2D material transfer and growth techniques along with targeted lithography alignment processes are current road blocks to consistent, repeatable circuit fabrication. The ability to reliably produce transistors with similar electrical characteristics, if not the same characteristics, is an important step toward the creation of viable electronic circuits.

3.1 Inverters and the Beginning of 2D Integrated Logic Circuits

A key development for logic circuits based on 2D materials was the demonstration of the inverter, or NOT gate. Common inverter configurations include the complementary inverter and the depletion-load inverter design. Both configurations are composed of two transistors in series. Schematic diagrams of these two inverter designs are shown in Fig. 16. In the case of complementary inverters, the transistors possess opposite channel polarities, with one being an n-channel MOSFET (NMOS) transistor and the other being a p-channel MOSFET (PMOS) transistor. The gates of the two transistors are both connected to the input voltage source. In the case of depletion-load inverters, two NMOS transistors are used, with the input voltage applied to the gate of the transistor connected to ground and the gate of the transistor connected to the system voltage tied to the output.

The first reported demonstration of an inverter using integrated transistors with a 2D material was presented by Traversi et al. (2009). The complementary inverter circuit was achieved using graphene as the channel

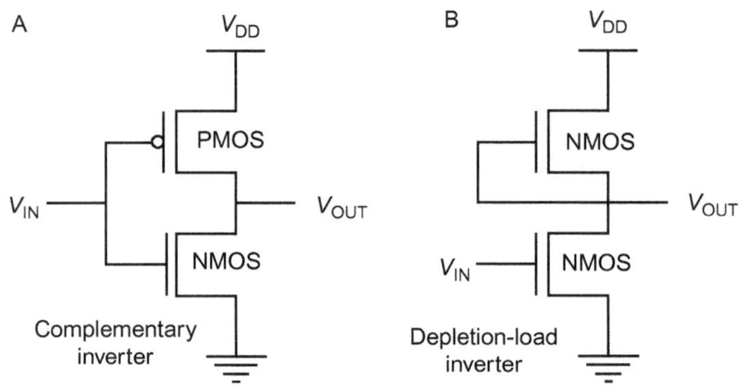

Fig. 16 Depictions of common inverter configurations are shown: (A) transistor diagram of a complementary inverter and (B) transistor diagram of a NMOS-based depletion-load inverter.

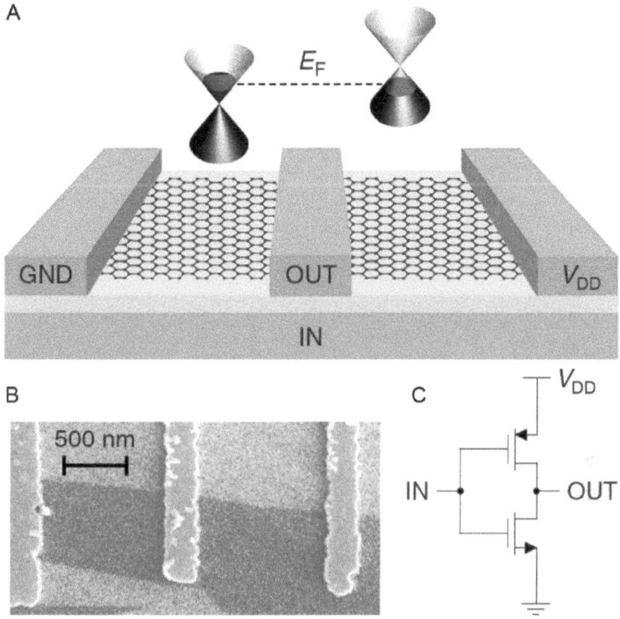

Fig. 17 Integrated complementary graphene inverters. (A) A schematic of a fabricated inverter is shown with three electrodes patterned on the same graphene monolayer flake, which defines two transistors. The part of the flake between the two leftmost electrodes is electrically annealed to obtain an n-type transistor. The other part of the flake between the two rightmost electrodes is a pristine p-type FET. The flake is electrically insulated from the input (highly doped Si labeled with "IN") by a layer of SiO_2. (B) SEM image of the fabricated inverter is show in a channel length of 1 μm. (C) The circuit layout is shown with a power supply of $V_{DD}=3.3$ V (Traversi et al., 2009). *Reproduced with permission from the American Institute of Physics.*

material for the two transistors depicted in Fig. 17. The PMOS transistor was air-doped, while the NMOS transistor was annealed to shift its Dirac point. The removal of oxygen adsorbed on the graphene channel surface due to annealing tends to skew the devices p-type. The Boolean inversion was obtained by operating the transistors between their Dirac points (Traversi et al., 2009). Since the demonstration of this inverter circuit, a number of other groups have since been able to produce similar inverter circuits using graphene transistors (Guerriero et al., 2012; Kim et al., 2015b; Li et al., 2010, 2011; Schall et al., 2013). The majority of the inverters based on graphene channels make use of charge neutrality point splitting, where the symmetry in Dirac points of the two graphene transistors is purposely broken electrostatically. This is typically done by either applying an appropriately large bias across the transistors in series or making use of dissimilar gate dielectrics (varying the thickness and/or material composition) to create a difference between Dirac points of the transistors.

As a result of graphene's lack of a bandgap and poor on/off current ratio, graphene transistors tend to have high power consumption compared to most other FET technologies. TMDs have been considered as an alternative because they possess a bandgap, with MoS_2 and WSe_2 being the most studied among the TMDs and, as such, a number of research groups have developed transistors and circuits based on these material systems. The first demonstration of a TMD-based inverter was reported in 2011 by Radisavljevic et al. (2011b) using two n-type MoS_2 transistors in a depletion-load inverter configuration. Due to the fact that MoS_2 transistors are naturally doped n-type and n-type transistors are relatively easy to produce, the majority of demonstrations using MoS_2 transistors to produce inverter circuits make use of the NMOS-based depletion-load configuration (Amani et al., 2015; Cheng et al., 2014; Radisavljevic et al., 2011b, 2012; Yu et al., 2014). A method for electrostatically doping MoS_2 to create enhancement-mode and depletion-mode MoS_2 transistors to create a pseudo-complementary inverter has been shown by Wang et al. (2012b) where dissimilar metal gates were used to separate the threshold voltages of the two transistors. Low work function Al gates were used to bend the bands of the transistors further into electron accumulation while high work function Pd gates were used to bend the bands into electron depletion, illustrated in Fig. 18. Inverters based on a MoS_2 channel, but using graphene contacts have been demonstrated by Yu et al. (2014) showing integration of multiple 2D materials for the fabrication of an IC. As discussed previously, the use of graphene as a contact has opened up the possibilities of electrostatically tuning the contact to reduce the difference between the contact's effective work function and the semiconductor channel's electron affinity, which decreases the contact resistance of the devices.

Other TMDs have been used to create functional inverter circuits. Complementary inverter circuits based on WSe_2 have been reported with voltage gains greater than 30 (Yu et al., 2015). The WSe_2 channels are doped n-type or p-type through the use of either contact metal work function pinning (Das and Roelofs, 2014; Das et al., 2014b) or a combination of the contact metal work function doping and chemical doping (Tosun et al., 2014; Yu et al., 2015). Molybdenum telluride ($MoTe_2$)-based transistors have also been used to produce a functional integrated inverter, exploiting the ambipolar properties of the $MoTe_2$ transistors (Lin et al., 2014). Inverters based on few-layer BP has also been reported (Zhu et al., 2015). Due to

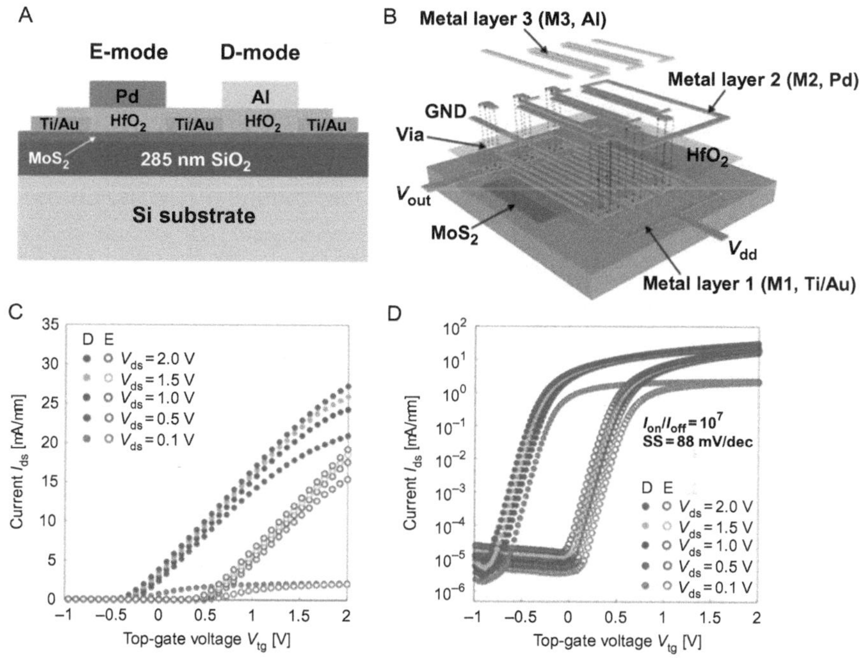

Fig. 18 (A) A schematic representation of an enhancement-mode (E-mode) and a depletion-mode (D-mode) device is shown. (B) A schematic illustration of an integrated 5-stage ring oscillator circuit on MoS_2 thin films, which is constructed by integrated 12 MoS_2 FETs is shown. Three distinct metal layers of the MoS_2 integrated circuit are represented by M1, M2, and M3. M1 is directly in contact with the bilayer MoS_2 thin films while M2 and M3 are the Pd and Al gate layers, respectively. Via holes are etched through the HfO_2 dielectric layer to allow connections from M2 and M3 to M1. (C) The transfer characteristics of D-mode and E-mode bilayer MoS_2 FETs are shown with D-mode FETs using Al metal gates and E-mode FETs using Pd metal gates. In these bilayer MoS_2 FETs, the on-state current density exceeds 23 μA/μm at $V_{ds}=1$ V, and the transconductance is above 12 μS/μm. The difference between the work functions of Al and Pd (~1.04 V in vacuum) gates results in a 0.76 V shift in the threshold voltage. The discrepancy between the work function difference in vacuum and in HfO_2 can be attributed to the dipoles at the metal/HfO_2 interface, resulting from charge transfer across this boundary. (D) The transfer characteristics in logarithmic scale of D-mode and E-mode bilayer MoS_2 FETs are shown with an I_{on}/I_{off} ratio exceeding 10^7 for V_{ds} above 0.5 V and is about 10^6 at $V_{ds}=0.1$ V. The subthreshold slope (SS) is 88 mV/dec. Device dimensions are $L_g=1$ μm and $L_{ds}=1$ μm. The substrate is grounded (Wang et al., 2012b). *Reproduced with permission from the American Chemical Society.*

the ambipolar properties of phosphorene, a similar charge neutrality point splitting technique was used to electrostatically dope the transistor n-type and p-type to create a complementary pair.

Complementary inverters based on a single 2D material have proven challenging as conventional chemical implant doping techniques are not possible due to the atomically thin nature of the channel materials. Complementary inverters with transistors that take advantage of the natural doping of two different material systems have been shown to produce complementary metal–oxide semiconductor (CMOS) inverters that do not rely upon electrostatic doping. With MoS_2 as the n-type transistor, integrated inverter circuits have been produced using a variety of material systems for the p-type transistor including $Bi_2Sr_2Co_2O_8$ (Yu et al., 2013), few-layer BP (Zhu et al., 2015), and WSe_2 (Jeon et al., 2015). Micrographs and illustrations of a complementary MoS_2/WSe_2 inverter are shown in Fig. 19. Recent examples of integrated inverter circuits based on 2D materials are provided in Table 4 along with the reported maximum voltage gains and the V_{DD} used to achieve the reported gain.

3.2 More Complex Logic Circuits

Beyond inverters, a number of other logic gate circuits have been demonstrated using integrated sets of transistors based on 2D materials. AND, OR, NAND, and NOR logic gates have been reported using graphene, MoS_2, and WSe_2 transistors (Amani et al., 2015; Jeon et al., 2015; Li et al., 2011; Radisavljevic et al., 2011b; Wang et al., 2012b; Yu et al., 2014). These simple logic operations are important building blocks for logic circuits that implement more advanced components such as registers, multiplexers, memory elements, and shifters. Additionally, more complex logic components in the form of memory elements based on 2D FETs have been shown by Wang et al. (2012b), which make use of four (4) MoS_2-based transistors. Fig. 20 provides logic output examples of a NAND gate, NOR gate, and a memory element based on MoS_2 transistors on a flexible polyimide substrate along with the transfer characteristics of the memory element (Amani et al., 2015). Beyond DC operation, a number of the inverter demonstrations described earlier made use of a modulated voltage at the input, demonstrating the ability to operate some of these 2D ICs at frequency (Amani et al., 2015; Cheng et al., 2014; Guerriero et al., 2012; Kim et al., 2015b; Li et al., 2010; Radisavljevic et al., 2012; Traversi et al., 2009; Wang et al., 2012b; Zhu et al., 2015). An example set of waveforms from Traversi et al.

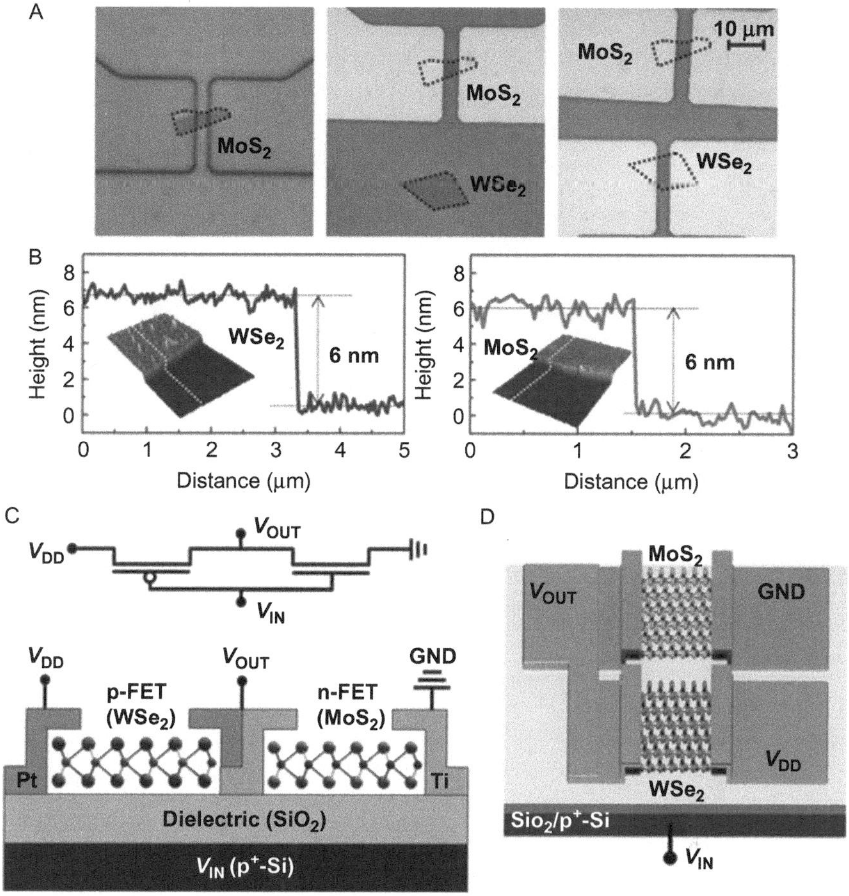

Fig. 19 (A) Optical microscope images display the device fabrication processes, which were implemented on a 285-nm-thick SiO_2/p^+-Si wafer using the direct imprint method. The left image indicates a MoS_2 nanosheet along with ultraviolet-patterned photoresist prior to Au/Ti S/D deposition, while the central image shows a fabricated n-MoS_2 FET and a direct imprint-transferred WSe_2. The right image displays completed n- and p-FETs. (B) Line profiles and topographic images of (*left*) WSe_2 and (*right*) MoS_2 nanosheets as obtained from AFM. (C and D) A schematic circuit diagram, cross section, and a 3D illustration of the hetero-CMOS inverters on a 285-nm-thick SiO_2/p^+-Si universal gate are shown (Jeon et al., 2015). *Reproduced with permission from the American Chemical Society.*

(2009) is shown in Fig. 21 showing the output voltage for input voltage frequencies ranging from 100 Hz to 10 kHz. A graphene-based inverter used as a complementary push–pull amplifier was able to achieve a unity voltage gain frequency of 360 kHz for an open load with a $V_{DD}=2.5$ V (Guerriero et al., 2012). Recent efforts in MoS_2-based inverter modulation

Table 4 Compiled Table of Two-Transistor Inverter Examples That Use 2D Material Channels in the Integrated Transistors

No.	Inverter Type	Channel Material	V_{DD}	Voltage Gain	Refs.
1	Complementary	Graphene	3.3 V	0.044	Traversi et al. (2009)
2	Complementary	Graphene	2 V	7	Li et al. (2010)
3	Complementary	Graphene	1 V	2.5	Li et al. (2011)
4	NMOS depletion-load	MoS_2	2 V	>4	Radisavljevic et al. (2011b)
5	Complementary	Graphene	2.5 V	3.7	Guerriero et al. (2012)
6	NMOS depletion-load	MoS_2	0.1 V	>4	Radisavljevic et al. (2012)
7	Complementary	MoS_2	2 V	~5	Wang et al. (2012b)
8	Complementary	$MoS_2/Bi_2Sr_2Co_2O_8$	-2 V	~1.7	Yu et al. (2013)
9	Complementary	Graphene	-3 V	6.4	Schall et al. (2013)
10	Complementary	WSe_2	3 V	>12	Tosun et al. (2014)
11	Complementary	MoS_2/black phosphorus	1 V	~1.4	Liu et al. (2014)
12	Complementary	$MoTe_2$	2 V	~1.5	Lin et al. (2014)
13	NMOS depletion-load	MoS_2	3 V	>12	Yu et al. (2014)
14	Complementary	WSe_2	5 V	>12	Das and Roelofs (2014)
15	Complementary	WSe_2	5 V	>25	Das et al. (2014b)
16	NMOS depletion-load	MoS_2	5 V	2	Cheng et al. (2014)
17	Complementary	Black phosphorus	-2 V	~1.68	Zhu et al. (2015)
18	NMOS depletion-load	MoS_2	15 V	>16	Amani et al. (2015)
19	Complementary	Graphene	4 V	2.8	Kim et al. (2015b)
20	Complementary	WSe_2	3 V	~38	Yu et al. (2015)
21	Complementary	MoS_2/WSe_2	5 V	~27	Jeon et al. (2015)

Various mechanisms were used to "dope" the 2D channel materials.
The maximum reported voltage gain is listed with each entry along with the supply voltage, V_{DD}, used to achieve the reported voltage gain.
Note that the table is listed chronologically based on the date of publication.

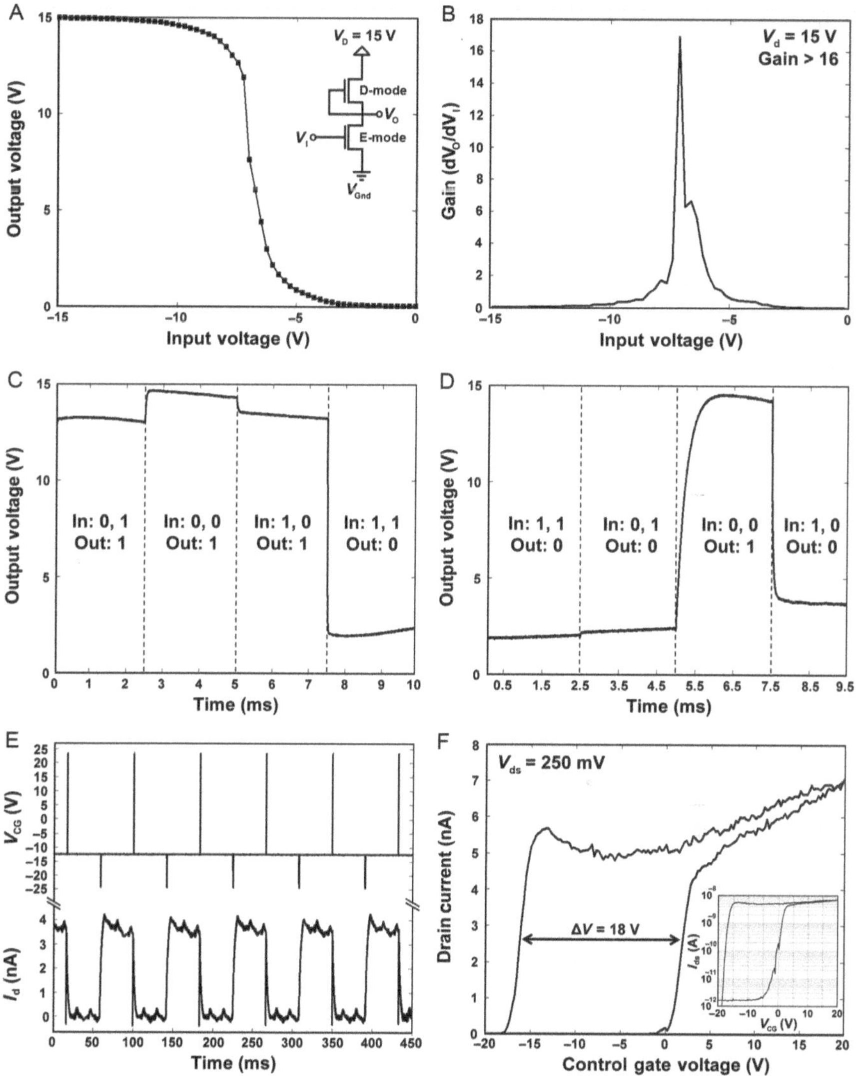

Fig. 20 (A) The output characteristics and schematic for the inverter circuit is shown. (B) The gain of an inverter circuit as a function of input voltage taken at a supply voltage of 20 V is presented. (C and D) The output voltage for a NAND gate and NOR gate was taken at a supply voltage of 20 V for the four possible input voltage combinations (0,0), (0,1), (1,0), and (1,1). (E) Performance of a MoS_2 nonvolatile memory cell with a graphene floating gate, showing the drain current at varying program states with a bipolar pulse train centered at −10 V (T_{pulse} = 100 ms, V_{ds} = 0.25 V), is provided. (F) The transfer characteristics for the nonvolatile memory cell showing a large hysteresis window of 18 V, which is caused by the accumulation of charge on the graphene floating gate, are shown. The insert shows the log plot of the same data (Amani et al., 2015). *Reproduced with permission from the Institute of Physics.*

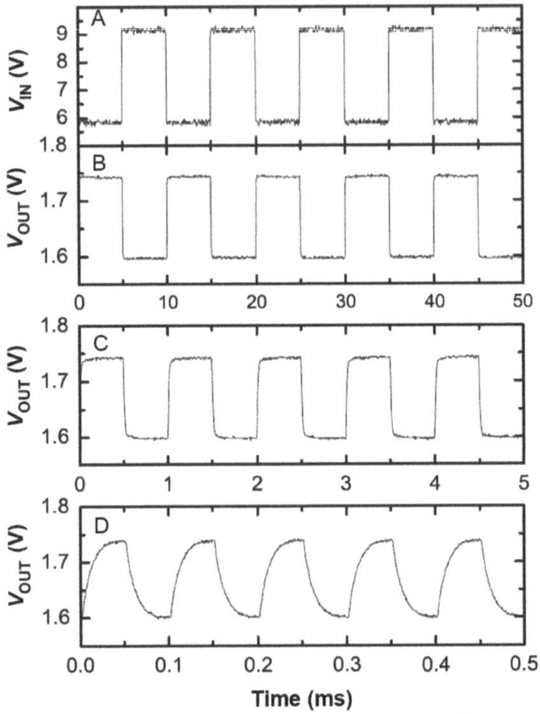

Fig. 21 (A) Digital waveforms measured on the fabricated inverter, with the input voltage possessing the following: Offset voltage is 7.5 V, voltage swing $V_{DD}=3.3$ V, and frequency $f=100$ Hz. (B) Output voltage is shown at $T=3$ K with input signal frequency of $f=100$ Hz. (C) Output voltage is shown at $T=3$ K with input signal frequency of $f=1$ kHz. (D) Output voltage is shown at $T=3$ K with input signal frequency of $f=10$ kHz (Traversi et al., 2009). *Reproduced with permission from the American Institute of Physics.*

have achieved unity gain operational frequencies as high as 1 GHz (Cheng et al., 2014). As most logic blocks of more complex circuits operate with a clock frequency, the feasibility of modulating these 2D circuits has been an important milestone for their use in electronic applications. Additionally, beyond logic operations, 2D-based circuits and devices have been demonstrated and show promise in a variety of RF applications, crucial for a number of important fields such as communications and radar.

4. 2D MATERIALS IN RF APPLICATIONS

Alternating current (AC) and RF circuits are extremely important for a number of applications including communications and radar detection.

New form factors, such as flexible and/or transparent circuits, have provided possible avenues for circuit technologies based on 2D materials to show benefits over current Si, a-Si, and organic semiconductor technologies. With relatively higher carrier mobility compared to a-Si and organic semiconductors and superior robustness and transparency compared to Si devices, initial studies have been performed using integrated AC and RF circuits based on 2D materials. With continued progress in this area, 2D materials may just fill the need for flexible electronics or devices in the THz gap (300 GHz to 3 THz).

4.1 RF Requirements: Device Physics

As with all RF applications, the main metrics used to qualify 2D active region RF transistors for information transmission and sensing applications are f_{max}, the maximum frequency limit for RF power gain given matched input and output impedances, and f_t, the current gain cutoff frequency. These frequency limits can be related back to the physical elements of the RF device, which includes the transconductance (G_m), a measure of the drain current (I_{ds}) gain with gate-source voltage variation, and the output conductance (G_{ds}), a measure of the drain current change with variation in the drain–source voltage. In terms of these device output and transfer characteristics, a small G_{ds}, large G_m, and good drain current saturation (related to the presence of an energy bandgap of the channel material) are necessary to achieve large f_{max} and f_t values (Schwierz, 2013). Additional considerations for RF performance include moderate mobility (light effective mass and low scattering) and low contact resistance. Schwierz et al. (2015) has recommended a wish list of properties for 2D material-based transistors and within it, the characteristics pertaining to RF FETs include a 300-meV bandgap, mobility in the 10,000 $cm^2/V \cdot s$ range, and a contact resistance around 0.03 Ω mm. For RF devices, AC voltage gain, an attribute required in most analog circuits, can also be used as a figure of merit (Han et al., 2011).

Unfortunately, for graphene, obtaining these device characteristics has been problematic. In a typical graphene-based field-effect transistor (GFET), the drain current does not saturate because the carrier velocity does not reach saturation as happens in classic MOSFETs. This is due to charge carrier traps, tunneling, and/or the fact that the saturation velocity is just too high to achieve in graphene (Wang et al., 2012a). Secondly, the transconductance of GFETs is zero for some regions of the transfer characteristics due to the lack of a bandgap and transition of the Fermi level through the Dirac point as the channel potential bias changes and current transitions from an electron current to a hole current (Schwierz, 2013). Observations from the literature indicate that more investigations are needed to understand how to design GFET devices with better G_m control. For example, there have been GFET device demonstrations exhibiting very large G_m (7900 μF/V·s) due to the simultaneous presence of a large interface capacitance and large carrier mobility (Hess et al., 2011; Xu et al., 2011). However, others have observed small transconductance values even though ballistic transport had occurred (Chari et al., 2015). Given the desired device requirements and performance complications, achieving a large f_{max} is a challenge for graphene FETs. However, predictions based on semiclassical ballistic transport simulations indicate that $f_{max} > 400$ GHz is possible for a 50-nm gate length graphene MOSFET (Koswatta et al., 2011).

Theoretically, TMDs are good candidates for RF devices because their charge carrier mobility values are as good as Si. MoS_2 FETs exhibit large G_m, due to a low density of states (DOS) (Yoon et al., 2011). When considering speed, monolayer WS_2 is perhaps the best TMD material because of its low effective mass and a phonon-limited electron mobility over 1000 $cm^2/V\cdot s$ (Zhang et al., 2014). Yet despite the promising properties, there are many barriers to fully realizing TMDs as the active channel material in many RF applications.

High quality vdW interfaces between the channel material and the dielectric can greatly improve carrier mobilities by eliminating charge carrier trapping or scattering at the interfaces. Good interfaces would also help to minimize hysteresis in device current–voltage characteristics. Lee has demonstrated a 10-layer MoS_2 metal–semiconductor field-effect transistors (MESFETs) with intrinsic-like mobilities of 200–500 $cm^2/V\cdot s$ at low voltages by using a NiO_x Schottky gate electrode that makes a vdW interface to the MoS_2 (Lee et al., 2015). Furthermore, using more than one 2D monolayer reduces the effect of interface traps because the trapped charge can be screened by the layers close to the interface, reducing the effect on current flow further away from the interface. Therefore, for high speed there may be an optimal number of layers depending on the 2D material and the Debye screening length for charge scattering at the interface. Interface traps can lead to hysteresis as well (Shah et al., 2014).

In a 2D FET, when projecting performance metrics, the gate capacitance dominates the transit time. This capacitance is modeled as a series combination of the quantum capacitance and oxide capacitance. The oxide capacitance is constant while the quantum capacitance is a function of the voltage drop at each point in the channel. Therefore, the gate capacitance depends strongly on V_{ds} and V_{gs}. This is opposite to CMOS where the overlap capacitance C_{GD} is independent of bias voltage. Therefore, the peak f_t is obtained only for a very narrow range of I_{ds} (Rodriguez et al., 2012).

Besides large f_{max} and f_t, analog applications would also benefit from having a minimum noise figure that is as low as possible. Extracting a noise-related figure of merit based on minimum noise temperature and gate length at 8 GHz, Schwierz indicates that the noise performance of graphene transistors significantly lags behind indium phosphide (InP) and gallium arsenide (GaAs) high-electron-mobility transistors but is close to that of RF Si MOSFETs (Schwierz et al., 2015). Noise figure analysis has been performed for the design of a low noise graphene amplifier, and the results indicate a strong dependence on operating frequency and gate and drain bias (Yeh et al., 2014). Using a compact model, the noise figure for a graphene transistor was determined to be 3.9 dB at 800 GHz (Frégonèse et al., 2012).

4.2 RF FET Devices

In this section, we discuss the electrical results from fabricated devices—first considering single layer 2D FETs made from graphene and BP, then followed by the TMD MoS_2. Next, we review the performance of flexible transistors made from various 2D materials. For radar, communications, electronic warfare and munitions applications, high RF power is desirable.

Epitaxially grown graphene FETs have exhibited the best RF power performance to date. Using a Si-face 4H-SiC (0001) substrate (etched in hydrogen at 1600°C to remove surface scratches prior to growth), a self-aligned 100-nm gate length device exhibited a de-embedded $f_{max} = 105$ GHz and $f_t = 93$ GHz. This device also had a 6-nm-thick Al_2O_3 dielectric formed by oxidizing e-beam deposited aluminum (Al) (Feng et al., 2014) and 100-nm access regions. In a previous report, a 100-nm gate length FET, made from epitaxially grown graphene on C-face SiC (000-1) and consisting of a 15-nm-thick ALD Al_2O_3 dielectric, exhibited a f_{max} of 70 GHz and a f_t of 300 GHz (Guo et al., 2013). Using a c-plane Al_2O_3 substrate, CVD-epitaxially grown graphene FETs with a gate length of 100 nm exhibited a de-embedded $f_t = 32$ GHz and an f_{max} of 21.5 GHz. The gate dielectric consists of 2 nm of e-beam evaporated Al that is oxidized for 10 min (Qing-Bin et al., 2014). This provides a more cost effective route for fabricating GFETs because the graphene does not have to be transferred from the growth substrate to another substrate. On the other hand, a mechanical transfer process was used to assemble a graphene-on-boron nitride back-gated FET with a 600-nm gate length. This device exhibited a de-embedded $f_t = 44$ GHz and $f_{max} = 34$ GHz (Meric et al., 2011).

Compared to graphene, the presence of an energy bandgap in BP makes it a more promising candidate for power RF applications. Recently, phosphorene transistors have been fabricated reporting a f_t of 12 GHz and an f_{max} of 20 GHz. This 300-nm channel length FET with bottom/top passivation of SiO_2 and HfO_2, respectively, was fabricated from an exfoliated BP layer 6–10 nm thick (Wang et al., 2014). Already, these initial results are on par with results from CVD-epitaxial GFETs; with more research, they should match the best performing GFETs.

With regard to TMDs, research to develop large-area RF MoS_2 FETs, funded by the US Army Research Laboratory, resulted in devices exhibiting current densities of 200 μA/μm, a maximum transconductance of 38 μS/μm, and a contact resistance corrected low field mobility of 55 $cm^2/V \cdot s$ (Sanne et al., 2015). This FET with a gate length of 250 nm was fabricated from MoS_2 grown by CVD on a SiO_2/Si wafer and reached f_t and f_{max} values of 6.7 and 5.3 GHz, respectively. This device exhibited an extrinsic voltage gain of 6 dB at 100 MHz with voltage amplification until 3 GHz (Sanne et al., 2015). Previously, an exfoliated MoS_2 RF FET series exhibited $f_t = 2$ GHz for single layer, 4 GHz for bilayer, and 6 GHz for trilayer structures. The trilayer structure also exhibited an f_{max} of 8.2 GHz (Krasnozhon et al., 2014). Another exfoliated MoS_2 FET exhibited a cutoff frequency f_t of 42 GHz and f_{max} of 50 GHz sized with a 6-nm gate length (Cheng et al., 2014). Thus far, the best results have been realized using exfoliated material. While promising, true mass production capability will rely on CVD-grown MoS_2 of which extensive research is currently underway.

Early flexible electronics results also indicate promise, although postbending stress results are limited. A flexible graphene transistor exhibited $f_t = 32$ GHz and $f_{max} = 20$ GHz, which reduced to 22 and 13 GHz after 2.5% strain (Yeh et al., 2014). More recently, a flexible 260-nm gate graphene MOSFET with f_t of 198 GHz, f_{max} of 28 GHz, and a strain limit of 2% was fabricated using transferred CVD graphene (Petrone et al., 2015). A record strain limit of 8.2% was observed for a transferred CVD graphene FET on a 10-nm Al_2O_3 dielectric layer used in a gate-first fabrication process (Lee et al., 2013). In addition to graphene,

MoS_2 has been explored as the active material in flexible device. A flexible MoS_2 MOSFET transferred to a polyimide substrate with a 68-nm gate exhibited $f_t = 13.5$ GHz and $f_{max} = 10.5$ GHz and survived over 1000 bending cycles (Cheng et al., 2014).

4.3 RF Circuit Demonstrations

Many RF circuits have been demonstrated using 2D materials; however, most have been made using graphene as the active channel material. Integrated cascaded inverters based on graphene transistors first showed the possibilities of creating amplifier circuits based on 2D materials (Rizzi et al., 2012). Recent advances in material integration have even allowed for the possibility of 2D material integration into Si-based fabrication processes. An IBM study demonstrated a 3-stage RF receiver, fabricated from graphene, capable of transmitting data on a wireless carrier frequency of 4.3 GHz (Han et al., 2014). The three stages performed as a receiver providing amplification, filtering, and down conversion mixing. Their results showed the feasibility of heterogeneous 3D integration of graphene-based devices onto a Si CMOS IC using the graphene transfer process and avoiding process steps higher than 400°C (Han et al., 2014).

Other elements for complete RF integration have also been explored, including ink-printed graphene for antenna and transmission lines (Huang et al., 2015). Furthermore, graphene transistor based active balun architectures have been fabricated and demonstrated good symmetry. Baluns are important for high frequency circuit designs. They employ differential signal topologies for immunity to common mode noise couplings and cross talk through the substrate and supply rails and alleviate the negative impact of parasitic inductances on gain, output power, and stability (Zimmer and Fregonese, 2015).

Ring oscillators play an important role in digital technology, primarily acting as a clock or time reference for signal synchronization. Thus, there is great interest creating these oscillators from 2D materials because it will allow complete digital systems to be integrated using only flexible semiconductors. Ring oscillators based on integrated 2D transistors were demonstrated by a number of groups (Guerriero et al., 2013; Schall et al., 2013; Wang et al., 2012b). The first published demonstration of an operational 5-stage ring oscillator based on MoS_2 transistors had a maximum oscillating frequency of 1.6 MHz (Wang et al., 2012b). A graphene-based 3-stage ring oscillator published soon afterward showed the capability of possessing a maximum oscillating frequency in the gigahertz range with a reported value of 1.28 GHz (Guerriero et al., 2013).

Graphene technology frequency mixers have been developed based on two techniques. The first technique makes use of the nonlinearity introduced by graphene's dual charge transport. These ambipolar GFET mixers have demonstrated conversion losses of −55 dB at 4 GHz (Liao et al., 2012). Using a second technique, channel resistance modulation controlled by the gate and drain bias leads to a single GFET mixer with a conversion loss of −26 dB operation at 5 GHz (Lyu et al., 2015). Subharmonic (achieved by operating around the Dirac point) resistive mixers operating up to 30 GHz with a conversion loss of 19 dB over the frequency range of 24–31 GHz have been demonstrated (Habibpour et al., 2013). Additionally, a flexible graphene-based resistive mixer exhibiting −9.6 dB conversion losses at 3.8 GHz showed maximum degradation within 30% following bending (Yeh et al., 2014).

In addition to graphene, other 2D materials have been used for RF frequency mixing applications. A BP ambipolar frequency doubler has recently been fabricated that was demonstrated to double an input 64-kHz signal (Zhu et al., 2015). Optimization, including proper gate design, is needed to push the operating frequency higher. Also, a recent demonstration of a MoS_2 mixer exhibited a −15 dB conversion loss at ~1 MHz (Sanne et al., 2015).

4.4 RF Circuit Design Considerations

Compact large signal RF models of active devices that can properly model switching and linearity characteristics are needed to design and optimize analog circuits. The major difference between 2D device compact models and conventional RF device large signal models is the inclusion of quantum capacitance and effects of the 2D DOS. Models are being developed in Verilog A for easy implementation in harmonic balance simulations (Aguirre-Morales et al., 2015; Landauer et al., 2014). Graphene FET models that account for the charge neutrality point, where the FET has maximum resistance, have been developed. This model includes an interlayer capacitance to account for the bandgap's effect on the charge neutrality point for an arbitrary number of graphene layers (Umoh et al., 2014). Though monolayer graphene does not exhibit a bandgap, there have been a number of techniques investigated that result in a bandgap such as doping, substrate effects, patterning, and hydrogenation. References given in Park and Choi (2015) provide specific examples. Additionally, rather than basing the large signal model equations on the drift-diffusion model, an alternative transport model based on the concept of virtual source charge/velocity has been implemented to more accurately model the short channel and nonsaturating performance of graphene devices (Rakheja et al., 2014).

For RF power applications, accurate impedance matching circuits are needed at the input and output of the GFET because the transistors have difficulty amplifying power in a 50-Ω environment. Source and load pull measurements are necessary to determine the needed impedance values of the device. A measurement attempt has been reported for epitaxial graphene FETs fabricated on SiC (Si-face thermal decomposition) with an 8-nm Al_2O_3 gate dielectric layer (Fregonese et al., 2015). The results indicate that although good f_t and f_{max} values are possible, it is hard to achieve power gain using a GFET in a circuit configuration. This is because the high impedance at the gate makes impedance matching difficult. A gain of up to −15 dB in a 50-Ω environment was measured, and with source and load tuners, the gain increased to 0.9 dB. A simulated power gain with theoretically optimum matching indicated a power gain of 6 dB at 900 MHz and 5 dB at 2.4 GHz (including effect of pads) (Fregonese et al., 2015). It is noted that the impedance mismatch of graphene devices is not as bad as carbon nanotube-based devices.

For RF applications, linear performance with minimum distortion over the signal and bias of interest is important. Promising linearity has been indicated from analysis using gate lengths over 250 nm. However, to investigate the potential of short gate length devices, nonlinear small signal circuit models have been used. In one analysis, an 18-nm gate length device exhibiting ballistic transport and a bias away from the Dirac point were considered. The results, based on analysis of the input third order intercept power, reveal poor linearity at peak cutoff frequency. The GFET linearity is otherwise comparable to that of MOSFETs and carbon nanotube-based FETs with intermodulation distortion dominated by drain injected carriers due to the lack of a bandgap (Alam et al., 2015).

5. NOVEL DEVICES AND OPERATING MECHANISMS

The channel length scaling in MOSFETs has led the path in achieving higher speeds and higher densities of microprocessors. Yet, existing complimentary MOSFET technologies will soon face a major challenge in reducing the supply voltage and leakage current for high-performance devices. These challenges are directly associated with the increasingly important problem of power consumption in MOSFET technologies. The thermionic nature of charge transport in MOSFETs also suggests a theoretical limit for how fast one can switch the device between the on and off states. As a result, device concepts beyond conventional MOSFET technologies are the only solution to get beyond these limits. Several efforts to solve the power and speed problems have garnered remarkable attention over the past few years, including approaches based on layered materials and the unique properties that they have to offer. Tunneling, valleytronics, and memristor device structures are a few of the approaches being explored. Here, we review some aspects of these topics in relationship to the growing research interests in 2D materials.

5.1 Tunneling Devices in 2D Materials

Power dissipation is the single most important problem facing high-performance transistor technology (Franklin, 2015). Conventional MOSFETs require roughly 60 mV of gate voltage for every order of magnitude change in current. This stems from the nature of charge injection that is thermally driven over potential barriers in the device and imposes a theoretical limit on the switching power dissipation. In contrast, devices based on quantum tunneling promise fast operational speeds (in the terahertz regime) and low power consumption beyond the theoretical limits of MOSFETs. One essential requirement for advancements in this area of research is the control of tunneling properties, such as barrier and band structure characteristics of the constituent device material.

The ultrathin nature of 2D materials with good gate control, direct bandgap in monolayer form, and the possibility of complex heterostructure assembly of these materials are attractive and promising properties permitting their exploration in tunneling device architectures. Several recent efforts devoted to the design and optimization of device architectures using multilayered single- and hetero-type 2D materials have shown some success (Roy et al., 2015; Sarkar et al., 2015). These experiments present major gains toward alleviating some challenges associated with power dissipation. For instance, swing slopes as low as 3.9 mV per decade at driving voltages as low as 0.1 V were demonstrated using bilayer MoS_2 tunneling FETs (see Fig. 22A and B). Additionally, resonant tunneling devices made of MoS_2/WSe_2 stack structures with significant negative differential resistance are shown to be an alternative approach to achieving lower power consumption (see Fig. 22C and D). Both reports focus on the design of vertical tunneling devices that allow for gate modulation of the tunneling current. The choice of heterostructure material properties and band alignment in these device architectures can lead to resonance transport and negative differential resistances that can be highly useful in the development of low-power devices. Nevertheless, the large effective mass of charge carriers and large bandgaps in TMDs are major challenges to achieving large current

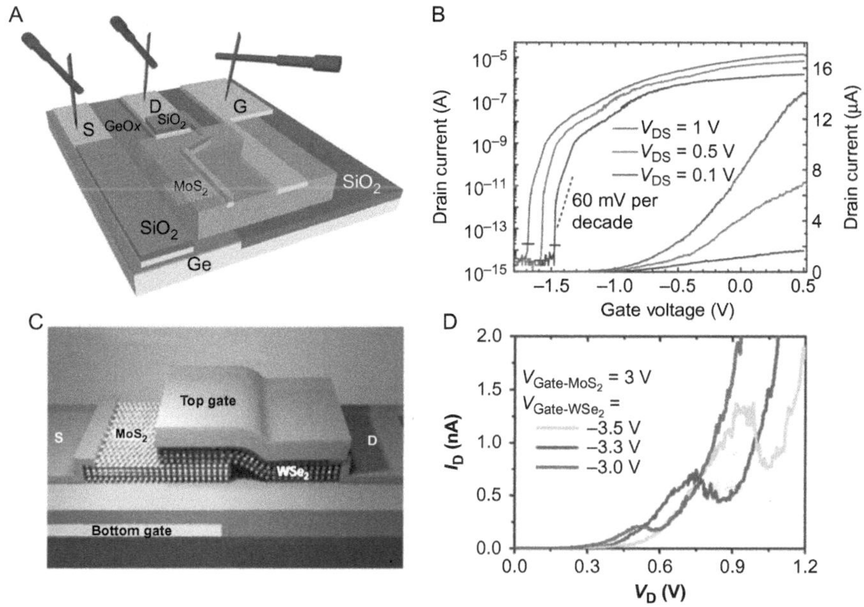

Fig. 22 Tunneling field-effect devices based on TMDs and their heterostructures: (A) device architecture for a field-effect tunneling transistor made of MoS_2 and germanium heterostructures. (B) The gate control over the tunneling current in MoS_2/germanium tunneling devices showing extremely low swing slopes for the switching behavior. (C) The design of gate controlled field-effect tunneling device made of MoS_2 and WS_2 atomic layers. (D) The bias and gate voltage dependence of current–voltage characteristics of MoS_2/WS_2 tunneling devices showing the resonance peaks and the negative differential resistance in this device architecture (Roy et al., 2015; Sarkar et al., 2015). *Reproduced with permission from the Nature Publishing Group and the American Chemical Society.*

densities. However, the rich diversity of 2D material properties and device designs show great promise in improving current densities.

5.2 Valleytronics in 2D Materials

Conventional electronics are based on manipulating the flow of charge carriers through the control of potential barriers in the material. But the innate limitations of power and speed of operation imposed by the thermionic nature of such switches have inspired the exploration of novel concepts in the development of next-generation electronics. Two such concepts are spintronics and valleytronics. The field of spintronics is based on controlling the charge carrier spin as a method to store, manipulate, and retrieve bits of information. Valleytronics, coined as an analog to spintronics, involves controlling the interaction of charge carriers with the local minimum/maximum (or valleys) found in the conduction/valence band to channel current flow in a particular way (Nebel, 2013). Valley degeneracy in the momentum space of the conduction and valence band presents an additional degree of freedom for charge carrier manipulation and is theoretically possible in many 2D materials. The feasibility of this area of electronics is still under examination. As it is with spin states in spintronics, controlling the population of valley states is essential to the development of valley-based electronics. This has proven to be a difficult task. Therefore, exploring and developing the means for producing and manipulating the valley polarization are an essential aspect of research toward the realization of valleytronics.

In their monolayer form, MoS_2 and other TMDs have great advantages that would solve some of the limitations found with commonly used materials for valleytronics research. Key features enabling a deliberate population of valley states in TMDs are (1) the breakdown of inversion symmetry at the conduction and valence band edges at the K point of the hexagonal Brillouin zone, (2) the generally strong spin–orbit interactions, and (3) the direct bandgap in their single layer form (Behnia, 2012). Recent efforts show that right- and left-hand circularly polarized light can efficiently excite electrons into distinguishable valley states (Cao et al., 2012; Kioseoglou et al., 2012; Mak et al., 2012; Zeng et al., 2012). This dynamic approach allows for significant valley polarization with efficiencies as high as 85%. Using this approach, one can maintain the charge carriers in one valley state for more than 1 ns. This is sufficient stability for effective manipulation and realization of some valleytronic devices. Although demonstration and operation of valleytronic devices remains to be realized, the accomplishments in control of valley states is a major step toward achieving this goal.

5.3 2D Materials Memristors

In a memristor, the resistance is not constant and depends on the history of current passed through the device. Such a circuit element was first proposed in the 1970s and has many advantages as compared to commonplace solid-state memory devices (Chua, 1971). Memristors use less power for operation and can store more information per unit area (Bessonov et al., 2015). In addition, they can be used to enable modern computing and memory device technologies such as neuromorphic computing architectures and nanoelectronic memories (Bessonov et al., 2015; Sangwan et al., 2015). However, realization of the theoretically proposed concept of memristors has proven to be a challenging task. The most successful demonstration of memristors has been based on titanium oxide (TiO_2) metal–insulator–metal (MIM) designs (Strukov et al., 2008). The general switching behavior has been explained through the coupling of ionic and electronic diffusions. It has been proposed that the diffusion of oxygen from low density defect regions to defect rich regions constitutes the switching behavior observed. Nevertheless, the more recent examinations of the original models provided for understanding the TiO_2 base memristor current–voltage behavior has been extensively scrutinized and shows fundamental inconsistencies (Vongehr and Meng, 2015). The overall conclusion is that the observed behavior in TiO_2 MIM devices may not be the exact behavior described in the first theoretical descriptions of memristors.

Nonetheless, several studies in this area have been conducted in recent years that show memristor behavior in layered materials, such as MoS_2 (Bessonov et al., 2015; Sangwan et al., 2015). For instance, Sangwan et al. (2015) show that if MoS_2 devices are intentionally made to contain a grain boundary region, the current–voltage (*I–V*) curve exhibits hysteresis under both forward and reverse bias (Fig. 23A and B). Similarly Bessonov et al. (2015) report that devices containing MoO_x/MoS_2 interfaces can also behave like memristive devices with large hysteretic *I–V* characteristics with clear on-set and reset transitions (Fig. 23C and D). The mechanism of the behavior in both cases has been attributed to a nonuniform density of defects or stoichiometry. Variations in point defect densities close to grain boundaries and oxygen atoms in the thermally oxidize MoO_x/MoS_2 heterojunction create a stoichiometric gradient. Similar to the TiO_2-based memristor, such devices show a voltage–current behavior that is reliant on the history of charge and ion diffusion. Additionally, the distinct electronic phases in several TMDs, such as tantalum selenide ($TaSe_2$), can show great dependency on the voltage history of the device (Yoshida et al., 2015). Nonvolatile phase switching between a quenched glassy state and a crystalline state in $TaSe_2$ was observed upon application of an electric field. This phase transition was observed as the material was thinned down to a few layers and thought to arise due to a slowdown of the kinetics, which allow the formation of metastable phases. Although many questions remain regarding the mechanisms and feasibility of 2D material-based memristors, the existing research shows great promise for the future of this type of device.

5.4 vdW Heterostructures

The immense attention of researchers to the 2D material systems is a testament to the unique and promising properties of these materials. In essence, the most important distinctiveness of these materials stems from the weak

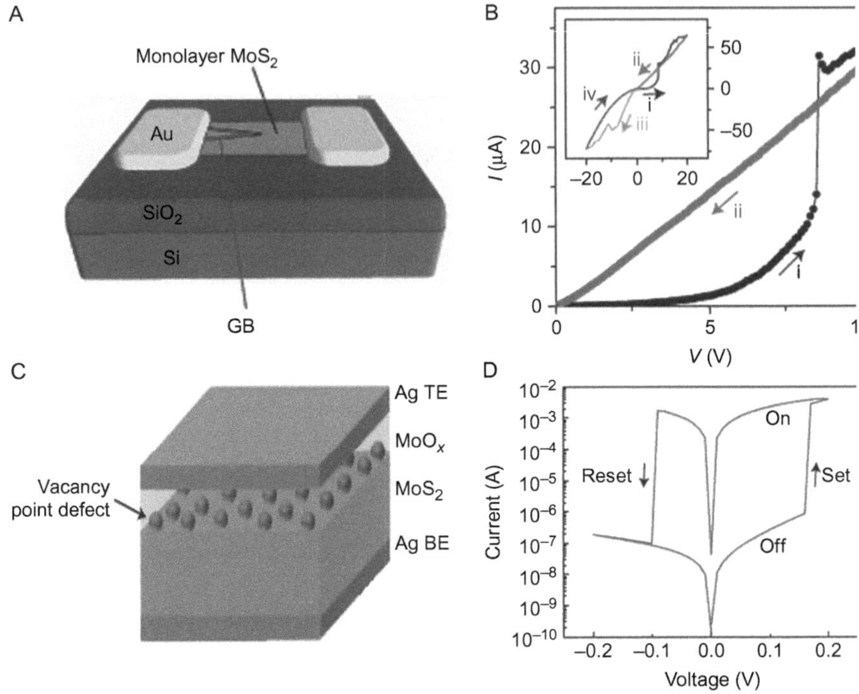

Fig. 23 Memristive behavior in TMD base devices: (A) the architecture of the memristor device made of MoS_2 layers containing a grain boundary. (B) Showing the forward and revered bias voltage dependency of current and the hysteric behavior resembling a memristor behavior. (C) The device geometry of memristor made of MoO_x/MoS_2 heterojunctions. (D) The current–voltage characteristics of MoO_x/MoS_2 devices showing the clear onset and reset characteristics of the memristor device (Bessonov et al., 2015; Sangwan et al., 2015). *Reproduced with permission from the Nature Publishing Group.*

vdW forces between their layers that enable layer isolation down to the monolayer. Despite their physical weakness, vdW forces between the layers play a significant role in the layer properties. For example, the band structures of most TMDs exhibit major thickness dependency. More notably, a transition from an indirect to a direct bandgap is observed as they are cleaved from bilayer to monolayer. These findings suggest new physics and properties can be achieved by vdW assembly of layered materials or heterostructures. The first example of such an approach has been the application of few layered h-BN as an ideal substrate for graphene transistors (Dean et al., 2010). It has been shown that substrate-based scattering is dominant and impedes charge carrier mobility in 2D materials (Chen et al., 2008). Therefore, tailoring the interfacial properties between layered materials has been used as a viable solution to these challenges. Additionally, the richness of properties in the 2D material library along with the endless ways and designs that can be used to assemble the vdW heterostructures guarantees many years of research that will put to test the feasibility, strengths, and weaknesses of these approaches. In fact, Geim and Grigorieva (2013) compared 2D materials to individual "Lego blocks," where one is able to construct a custom device by simply stacking different Lego blocks (ie, hetero-layered structures) (see Fig. 24). We have already seen the impact of the substrate on the properties of monolayer material. It will be fascinating to observe the interaction between different 2D materials and their impact on one another. We potentially have the ability to affect the combination of materials, stacking order/alignment, spacing, and orientation of the layers.

One obvious goal that can be explored using vdW heterostructures is building p–n junctions. The p–n junction is a fundamental element of electronics and replicating its properties can have great importance in the development of 2D technologies. Many research groups have focused on such goals and some success has been achieved by demonstration of basic p–n operation of common TMD heterostructures (Lee et al., 2014a). The

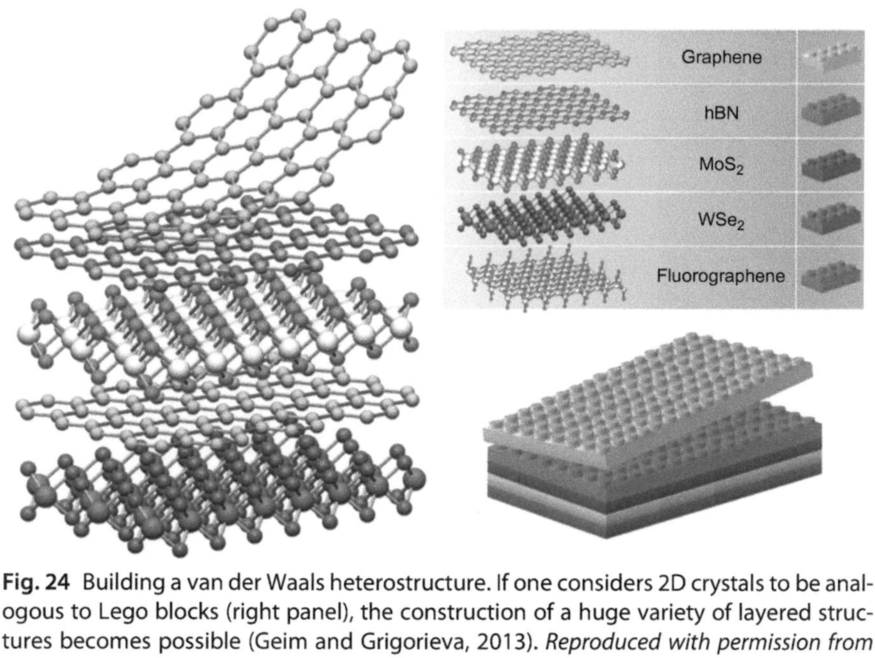

Fig. 24 Building a van der Waals heterostructure. If one considers 2D crystals to be analogous to Lego blocks (right panel), the construction of a huge variety of layered structures becomes possible (Geim and Grigorieva, 2013). *Reproduced with permission from the Nature Publishing Group.*

current approach preparing heterostructures is to transfer these layers one by one on top of each other. Naturally, the arbitrary assembly of heterostructures results in complexities that require greater attention going forward. For example, the mismatch in rotational angle of the assembled layers results in so-called Moiré patterns that can modify electronic states and result in major change in the optical and electrical properties of these heterostructures (Kang et al., 2013).

As a result of the difficulty in making 2D vdW heterostructures, much of the research has been guided and supported by theories developed to predict their important properties. One example for such research is related to the absorption properties of TMD heterostructures to develop photodetector and photovoltaic devices. The limited absorption properties of single layers and their transparency to light are an obstacle to the viability of this research. Theoretical modeling of TMD heterostructures suggests that the changes in DOS associated with van Hove singularities result in significantly enhanced absorption in these materials if the right combination of TMD layers are ideally stacked (Bernardi et al., 2013; Britnell et al., 2013). Yet, the challenge with the preparation of ideal vdW stacks and control of interfaces motivates the development of direct synthesis approaches that guarantee pristine and perfect ordering between the layers. Nevertheless, the extraordinary properties in these complex crystalline structures will continue to be the driving force in further exploration of vdW heterostructures.

6. CONCLUSIONS AND PERSPECTIVES

In this chapter, we focused on the electronic properties of 2D materials and discussed the issues making these materials the active channel in transistors, novel electrical devices, and logic and RF circuits. First, a brief discussion about the advantages of 2D materials for electronic applications was given. The continuing scaling and miniaturization of microelectronic transistors gives promise to the use of atomic layered materials.

The basic electrical characterization of the 2D materials is typically done through the fabrication of FETs using the 2D material of choice as the active channel material. Electrical testing focused on gaining information about important metrics, such as contact resistance, on/off current ratio, and charge carrier mobility. While there exists a large range of 2D materials to consider, there is particular interest in semimetal and semiconductor materials. To that end, we discussed in-depth the electrical performance results of graphene and silicene (semimetals), phosphorene (small bandgap semiconductor), and select TMDs (mid-bandgap semiconductors). Although testing of these 2D materials in theory should be simple, actual experimental considerations need to be recognized when testing and analyzing the results. Device testing configurations, the existence of a quantum capacitance, and the atmospheric exposure of the 2D-based devices all play a role to confound the determination of the device metrics.

Next, we touched upon how these 2D materials can be used not just as individual standalone devices, but how they can be formed into actual circuits. While some materials lend themselves more readily to one particular application than another, it is important to demonstrate just what can be done with 2D materials and how far they can go. In particular, RF applications for these materials are of interest, and research has shown that they are promising for these applications as well. Considering these unconventional materials have unique properties when compared to traditional semiconductor materials, we imagined their role in novel device concepts and operations.

While we were only able to discuss a few applications for 2D materials here, but it should be noted that the number of potential electronics applications for 2D materials is quite complex since they encompass insulators, semiconductors, metals, and superconductors. We referenced several different interest areas, including low power/logic/RF electronics, valleytronics, and memristors; however, there exists a wide range of applications for 2D materials, including photonics and optoelectronics (photodetectors, light-emitting diodes [LEDs], and lasers), sensors (chemical, gas, biological, magnetic, and strain), energy storage and conversion (supercapacitors, solar cells, thermoelectric devices, and fuel cells), composites, and even biomedical applications (Bhimanapati et al., 2015; Bonaccorso et al., 2015; Butler et al., 2013; Das et al., 2015; Ferrari et al., 2015; Han et al., 2015; Rao et al., 2015).

As the field becomes more mature, we should gain a better understanding of applications where 2D materials can make an impact. While it is unlikely that 2D materials will replace Si CMOS technology in the upcoming years, they certainly have the potential to complement it in niche applications where 2D materials are found to be superior to the current technology. Significant progress is being made in energy applications, transparent displays, and foldable/flexible electronics where 2D materials can have an impact in the near future. For each of these areas, 2D materials already possess the potential to outperform the current technology even though they have yet to reach their optimal quality material or device performance. We are already seeing the potential for graphene to break into the market for cell phone components (Hill, 2014; Peleg, 2015; Son et al., 2015; Spasenovic, 2014; Withers et al., 2015).

However, if 2D materials are going to develop into commercially viable products, reliable, cost-efficient processes will need to be developed. With this in mind, the main challenges to overcome for 2D materials to reach market include the development of (1) large-area growth, specifically roll-to-roll production, (2) reliable and repeatable fabrication processes, (3) low resistance contacts, and (4) characterization techniques specific to 2D materials. Obviously, one key to success will be the ability to synthesis 2D materials on the wafer scale. By doing so, it will be possible to mass produce 2D devices in a fabrication facility and minimize the cost of the product. To date, wafer scale deposition techniques have been developed for graphene, MoS_2, and WS_2 (Bae et al., 2010; Kang et al., 2015; Lee et al., 2014b). The second challenge is the development of reliable and repeatable fabrication processes. Fabrication processes include transfer (in situ growth would be ideal), doping, etching/patterning, and deposition of dielectrics. With 2D materials being a novelty, new processes need to be developed to minimize contamination due to transfer, selectively etch various 2D materials, prevent the formation of defects, circumvent oxidation and stability issues, and alternatively dope the material. Since 2D materials can be one atom or molecule thick, contact resistance has proven to be a major issue. Developing processes to minimize the impact of the contacts and contend with doping and Fermi level pinning issues concerning 2D materials will be pertinent. Finally, it is necessary to develop new characterization techniques to analyze the materials. Due to the thinness of 2D materials, common techniques such as scanning electron microscopy (SEM) and transmission electron microscopy (TEM) have been shown to modify the 2D material (Hashimoto et al., 2004). There are also detection issues for other techniques due to the minimal amount of material available for analysis. Hopefully with time and patience, these issues will be resolved.

In the meantime, the research community will continue to push forward with the exploration of new 2D materials. As each new material is discovered, we will gain a better understanding of the differences between the monolayer, few layer, and bulk forms of the material. Through research, we will be able to create a master chart of 2D materials including their properties and potential applications and determine how they may be used in conjunction with one another to form a functional device. Furthermore, as we look to stacking these layers and creating heterostructures, our master chart will expand to account for the impact of the stack on an individual layer and new composite devices with tailored properties will be realized. We have only scratched the surface for 2D materials, and only time will tell where the field progresses.

The discussion in this chapter is intended to give an overview of the electronic properties in 2D materials and describe where electronics research is progressing in this new and exciting field. For those looking for a more in-depth explanation into the electronic performance of a specific class of 2D materials or a particular electronic application, we refer the reader to the notable reviews by Akinwande et al.(2014), Fiori et al.(2014), Franklin (2015), Hsu et al. (2013), Li et al. (2016), and Schwierz et al. (2015).

ACKNOWLEDGMENTS

The authors would like to recognize the support of the US Army Research Laboratory (ARL), under the direction of Dr. Thomas Russell, for funding research into the growth, characterization, device fabrication and electrical testing, and modeling of graphene and 2D stacked atomic layers through the Director's Strategic Initiative (DSI) program. They acknowledge the support and active encouragement of Dr. Philip Perconti, Dr. Paul Amirtharaj, and Dr. Brett Piekarski of the ARL Sensors and Electron Devices Directorate. Special thanks is extended to Dr. A. Glen Birdwell of ARL and Matin Amani of the University of California, Berkeley for the motivation, enthusiasm, and hard work given to our research efforts into 2D materials. We sincerely thank Eugene Zakar and Carol Johnson of ARL for their support of this project.

REFERENCES

Aguirre-Morales, J.-D., Fregonese, S., Mukherjee, C., Maneux, C., Zimmer, T., 2015. An accurate physics-based compact model for dual-gate bilayer graphene FETs. IEEE Trans. Electron Devices 62, 4333–4339.

Akinwande, D., Petrone, N., Hone, J., 2014. Two-dimensional flexible nanoelectronics. Nat. Commun. 5, 5678.

Alam, A.U., Holland, K.D., Wong, M., Ahmed, S., Kienle, D., Vaidyanathan, M., 2015. RF linearity performance potential of short-channel graphene field-effect transistors. IEEE Trans. Microw. Theory Tech. 63, 3874–3887.

Amani, M., Burke, R.A., Proie, R.M., Dubey, M., 2015. Flexible integrated circuits and multifunctional electronics based on single atomic layers of MoS_2 and graphene. Nanotechnology 26, 115202.

Bae, S., Kim, H., Lee, Y., Xu, X., Park, J.-S., Zheng, Y., Balakrishnan, J., Lei, T., Ri Kim, H., Song, Y.I., Kim, Y.-J., Kim, K.S., Özyilmaz, B., Ahn, J.-H., Hong, B.H., Iijima, S., 2010. Roll-to-roll production of 30-inch graphene films for transparent electrodes. Nat. Nanotechnol. 5, 574–578.

Behnia, K., 2012. Condensed-matter physics: polarized light boosts valleytronics. Nat. Nanotechnol. 7, 488–489.

Bernardi, M., Palummo, M., Grossman, J.C., 2013. Extraordinary sunlight absorption and one nanometer thick photovoltaics using two-dimensional monolayer materials. Nano Lett. 13, 3664–3670.

Bessonov, A.A., Kirikova, M.N., Petukhov, D.I., Allen, M., Ryhänen, T., Bailey, M.J.A., 2015. Layered memristive and memcapacitive switches for printable electronics. Nat. Mater. 14, 199–204.

Bhimanapati, G.R., Lin, Z., Meunier, V., Jung, Y., Cha, J., Das, S., Xiao, D., Son, Y., Strano, M.S., Cooper, V.R., Liang, L., Louie, S.G., Ringe, E., Zhou, W., Kim, S.S., Naik, R.R., Sumpter, B.G., Terrones, H., Xia, F., Wang, Y., Zhu, J., Akinwande, D., Alem, N., Schuller, J.A., Schaak, R.E., Terrones, M., Robinson, J.A., 2015. Recent advances in two-dimensional materials beyond graphene. ACS Nano 9, 11509–11539.

Bolotin, K.I., Sikes, K.J., Jiang, Z., Klima, M., Fudenberg, G., Hone, J., Kim, P., Stormer, H.L., 2008. Ultrahigh electron mobility in suspended graphene. Solid State Commun. 146, 351–355.

Bonaccorso, F., Colombo, L., Yu, G., Stoller, M., Tozzini, V., Ferrari, A.C., Ruoff, R.S., Pellegrini, V., 2015. Graphene, related two-dimensional crystals, and hybrid systems for energy conversion and storage. Science 347, 1246501.

Britnell, L., Ribeiro, R.M., Eckmann, A., Jalil, R., Belle, B.D., Mishchenko, A., Kim, Y.-J., Gorbachev, R.V., Georgiou, T., Morozov, S.V., Grigorenko, A.N., Geim, A.K., Casiraghi, C., Neto, A.H.C., Novoselov, K.S., 2013. Strong light-matter interactions in heterostructures of atomically thin films. Science 340, 1311–1314.

Butler, S.Z., Hollen, S.M., Cao, L., Cui, Y., Gupta, J.A., Gutiérrez, H.R., Heinz, T.F., Hong, S.S., Huang, J., Ismach, A.F., Johnston-Halperin, E., Kuno, M., Plashnitsa, V.V., Robinson, R.D., Ruoff, R.S., Salahuddin, S., Shan, J., Shi, L., Spencer, M.G., Terrones, M., Windl, W., Goldberger, J.E., 2013. Progress, challenges, and opportunities in two-dimensional materials beyond graphene. ACS Nano 7, 2898–2926.

Cahangirov, S., Topsakal, M., Aktürk, E., Şahin, H., Ciraci, S., 2009. Two- and one-dimensional honeycomb structures of Silicon and Germanium. Phys. Rev. Lett. 102, 236804.

Cao, T., Wang, G., Han, W., Ye, H., Zhu, C., Shi, J., Niu, Q., Tan, P., Wang, E., Liu, B., Feng, J., 2012. Valley-selective circular dichroism of monolayer molybdenum disulphide. Nat. Commun. 3, 887.

Chan, J., Venugopal, A., Pirkle, A., McDonnell, S., Hinojos, D., Magnuson, C.W., Ruoff, R.S., Colombo, L., Wallace, R.M., Vogel, E.M., 2012. Reducing extrinsic performance-limiting factors in graphene grown by chemical vapor deposition. ACS Nano 6, 3224–3229.

Chanana, A., Mahapatra, S., 2016. Prospects of zero Schottky barrier height in a graphene-inserted MoS_2-metal interface. J. Appl. Phys. 119, 014303.

Chari, T., Meric, I., Dean, C., Shepard, K., 2015. Properties of self-aligned short-channel graphene field-effect transistors based on boron-nitride-dielectric encapsulation and edge contacts. IEEE Trans. Electron Devices 62, 4322–4326.

Chen, J.-H., Jang, C., Xiao, S., Ishigami, M., Fuhrer, M.S., 2008. Intrinsic and extrinsic performance limits of graphene devices on SiO_2. Nat. Nanotechnol. 3, 206–209.

Cheng, R., Jiang, S., Chen, Y., Liu, Y., Weiss, N., Cheng, H.-C., Wu, H., Huang, Y., Duan, X., 2014. Few-layer molybdenum disulfide transistors and circuits for high-speed flexible electronics. Nat. Commun. 5, 5143.

Chua, L.O., 1971. Memristor—the missing circuit element. IEEE Trans. Circuit Theory 18, 507–519.

Chuang, S., Battaglia, C., Azcatl, A., McDonnell, S., Kang, J.S., Yin, X., Tosun, M., Kapadia, R., Fang, H., Wallace, R.M., Javey, A., 2014. MoS_2 p-type transistors and diodes enabled by high work function MoO_x contacts. Nano Lett. 14, 1337–1342.

Das, S., Roelofs, A., 2014. Electrostatically doped WSe_2 CMOS inverter. In: Device Research Conference (DRC), 2014 72nd Annual. Presented at the Device Research Conference (DRC), 2014 72nd Annual, IEEE, Santa Barbara, CA, pp. 185–186.

Das, S., Chen, H.-Y., Penumatcha, A.V., Appenzeller, J., 2013. High performance multi-layer MoS_2 transistors with scandium contacts. Nano Lett. 13, 100–105.
Das, S., Demarteau, M., Roelofs, A., 2014a. Ambipolar phosphorene field effect transistor. ACS Nano 8, 11730–11738.
Das, S., Dubey, M., Roelofs, A., 2014b. High gain, low noise, fully complementary logic inverter based on bi-layer WSe_2 field effect transistors. Appl. Phys. Lett. 105, 083511.
Das, S., Robinson, J.A., Dubey, M., Terrones, H., Terrones, M., 2015. Beyond graphene: progress in novel two-dimensional materials and van der Waals solids. Annu. Rev. Mater. Res. 45, 1–27.
Dean, C.R., Young, A.F., Meric, I., Lee, C., Wang, L., Sorgenfrei, S., Watanabe, K., Taniguchi, T., Kim, P., Shepard, K.L., Hone, J., 2010. Boron nitride substrates for high-quality graphene electronics. Nat. Nanotechnol. 5, 722–726.
Du, Y., Yang, L., Zhang, J., Liu, H., Majumdar, K., Kirsch, P.D., Ye, P.D., 2014. MoS_2 field-effect transistors with graphene/metal heterocontacts. IEEE Electron Device Lett. 35, 599–601.
Fang, H., Chuang, S., Chang, T.C., Takei, K., Takahashi, T., Javey, A., 2012. High-performance single layered WSe_2 p-FETs with chemically doped contacts. Nano Lett. 12, 3788–3792.
Feng, Z.H., Yu, C., Li, J., Liu, Q.B., He, Z.Z., Song, X.B., Wang, J.J., Cai, S.J., 2014. An ultra clean self-aligned process for high maximum oscillation frequency graphene transistors. Carbon 75, 249–254.
Ferrari, A.C., Bonaccorso, F., Fal'ko, V., Novoselov, K.S., Roche, S., Bøggild, P., Borini, S., Koppens, F.H.L., Palermo, V., Pugno, N., Garrido, J.A., Sordan, R., Bianco, A., Ballerini, L., Prato, M., Lidorikis, E., Kivioja, J., Marinelli, C., Ryhänen, T., Morpurgo, A., Coleman, J.N., Nicolosi, V., Colombo, L., Fert, A., Garcia-Hernandez, M., Bachtold, A., Schneider, G.F., Guinea, F., Dekker, C., Barbone, M., Sun, Z., Galiotis, C., Grigorenko, A.N., Konstantatos, G., Kis, A., Katsnelson, M., Vandersypen, L., Loiseau, A., Morandi, V., Neumaier, D., Treossi, E., Pellegrini, V., Polini, M., Tredicucci, A., Williams, G.M., Hong, B.H., Ahn, J.-H., Kim, J.M., Zirath, H., van Wees, B.J., van der Zant, H., Occhipinti, L., Matteo, A.D., Kinloch, I.A., Seyller, T., Quesnel, E., Feng, X., Teo, K., Rupesinghe, N., Hakonen, P., Neil, S.R.T., Tannock, Q., Löfwander, T., Kinaret, J., 2015. Science and technology roadmap for graphene, related two-dimensional crystals, and hybrid systems. Nanoscale 7, 4598–4810.
Fiori, G., Bonaccorso, F., Iannaccone, G., Palacios, T., Neumaier, D., Seabaugh, A., Banerjee, S.K., Colombo, L., 2014. Electronics based on two-dimensional materials. Nat. Nanotechnol. 9, 768–779.
Franklin, A.D., 2015. Nanomaterials in transistors: from high-performance to thin-film applications. Science 349, aab2750.
Frégonèse, S., Meng, N., Nguyen, H.-N., Majek, C., Maneux, C., Happy, H., Zimmer, T., 2012. Electrical compact modelling of graphene transistors. Solid State Electron. 73, 27–31.
Fregonese, S., De Matos, M., Mele, D., Maneux, C., Happy, H., Zimmer, T., 2015. Source-pull and load-pull characterization of graphene FET. IEEE J. Electron Devices Soc. 3, 49–53.
Fuhrer, M.S., Hone, J., 2013. Measurement of mobility in dual-gated MoS_2 transistors. Nat. Nanotechnol. 8, 146–147.
Geim, A.K., Grigorieva, I.V., 2013. Van der Waals heterostructures. Nature 499, 419–425.
Gold, A., 1987. Electronic transport properties of a two-dimensional electron gas in a silicon quantum-well structure at low temperature. Phys. Rev. B 35, 723–733.
Gong, C., Colombo, L., Wallace, R.M., Cho, K., 2014. The unusual mechanism of partial fermi level pinning at Metal–MoS_2 interfaces. Nano Lett. 14, 1714–1720.

Guerriero, E., Polloni, L., Rizzi, L.G., Bianchi, M., Mondello, G., Sordan, R., 2012. Graphene audio voltage amplifier. Small 8, 357–361.

Guerriero, E., Polloni, L., Bianchi, M., Behnam, A., Carrion, E., Rizzi, L.G., Pop, E., Sordan, R., 2013. Gigahertz integrated graphene ring oscillators. ACS Nano 7, 5588–5594.

Guo, Z., Dong, R., Chakraborty, P.S., Lourenco, N., Palmer, J., Hu, Y., Ruan, M., Hankinson, J., Kunc, J., Cressler, J.D., Berger, C., de Heer, W.A., 2013. Record maximum oscillation frequency in C-face epitaxial graphene transistors. Nano Lett. 13, 942–947.

Habibpour, O., Vukusic, J., Stake, J., 2013. A 30-GHz integrated subharmonic mixer based on a multichannel graphene FET. IEEE Trans. Microw. Theory Tech. 61, 841–847.

Han, S.-J., Jenkins, K.A., Valdes Garcia, A., Franklin, A.D., Bol, A.A., Haensch, W., 2011. High-frequency graphene voltage amplifier. Nano Lett. 11, 3690–3693.

Han, S.-J., Garcia, A.V., Oida, S., Jenkins, K.A., Haensch, W., 2014. Graphene radio frequency receiver integrated circuit. Nat. Commun. 5, 3086.

Han, S.A., Bhatia, R., Kim, S.-W., 2015. Synthesis, properties and potential applications of two-dimensional transition metal dichalcogenides. Nano Converg. 2, 1–14.

Hashimoto, A., Suenaga, K., Gloter, A., Urita, K., Iijima, S., 2004. Direct evidence for atomic defects in graphene layers. Nature 430, 870–873.

Hess, L.H., Hauf, M.V., Seifert, M., Speck, F., Seyller, T., Stutzmann, M., Sharp, I.D., Garrido, J.A., 2011. High-transconductance graphene solution-gated field effect transistors. Appl. Phys. Lett. 99, 033503.

Hill, S., 2014. Meet the world's first 2D material: 5 ways Graphene could transform your phone. [WWW Document] http://www.digitaltrends.com/mobile/graphene-explained/(accessed 2.17.16).

Hsu, A., Wang, H., Shin, Y.C., Mailly, B., Zhang, X., Yu, L., Shi, Y., Lee, Y.H., Dubey, M., Kim, K.K., Kong, J., Palacios, T., 2013. Large-area 2-D electronics: materials, technology, and devices. Proc. IEEE 101, 1638–1652.

Huang, X., Leng, T., Zhu, M., Zhang, X., Chen, J., Chang, K., Aqeeli, M., Geim, A.K., Novoselov, K.S., Hu, Z., 2015. Highly flexible and conductive printed graphene for wireless wearable communications applications. Sci. Rep. 5, 18298.

Ieong, M., Doris, B., Kedzierski, J., Rim, K., Yang, M., 2004. Silicon device scaling to the sub-10-nm regime. Science 306, 2057–2060.

Jeon, P.J., Kim, J.S., Lim, J.Y., Cho, Y., Pezeshki, A., Lee, H.S., Yu, S., Min, S.-W., Im, S., 2015. Low power consumption complementary inverters with n-MoS_2 and p-WSe_2 dichalcogenide nanosheets on glass for logic and light-emitting diode circuits. ACS Appl. Mater. Interfaces 7, 22333–22340.

Kam, K.-K., 1982. Electrical properties of WSe_2, WS_2, $MoSe_2$, MoS_2, and their use as photoanodes in a semiconductor liquid junction solar cell. (Retrospective Theses and Dissertations Paper 8356).

Kang, J., Li, J., Li, S.-S., Xia, J.-B., Wang, L.-W., 2013. Electronic structural Moiré pattern effects on MoS_2/$MoSe_2$ 2D heterostructures. Nano Lett. 13, 5485–5490.

Kang, K., Xie, S., Huang, L., Han, Y., Huang, P.Y., Mak, K.F., Kim, C.-J., Muller, D., Park, J., 2015. High-mobility three-atom-thick semiconducting films with wafer-scale homogeneity. Nature 520, 656–660.

Kappera, R., Voiry, D., Yalcin, S.E., Branch, B., Gupta, G., Mohite, A.D., Chhowalla, M., 2014. Phase-engineered low-resistance contacts for ultrathin MoS_2 transistors. Nat. Mater. 13, 1128–1134.

Kara, A., Enriquez, H., Seitsonen, A.P., Lew Yan Voon, L.C., Vizzini, S., Aufray, B., Oughaddou, H., 2012. A review on silicene—new candidate for electronics. Surf. Sci. Rep. 67, 1–18.

Kim, J.-S., Liu, Y., Zhu, W., Kim, S., Wu, D., Tao, L., Dodabalapur, A., Lai, K., Akinwande, D., 2015a. Toward air-stable multilayer phosphorene thin-films and transistors. Sci. Rep. 5, 8989.

Kim, W., Li, C., Chekurov, N., Arpiainen, S., Akinwande, D., Lipsanen, H., Riikonen, J., 2015b. All-graphene three-terminal-junction field-effect devices as rectifiers and inverters. ACS Nano 9, 5666–5674.

Kioseoglou, G., Hanbicki, A.T., Currie, M., Friedman, A.L., Gunlycke, D., Jonker, B.T., 2012. Valley polarization and intervalley scattering in monolayer MoS_2. Appl. Phys. Lett. 101, 221907.

Koenig, S.P., Doganov, R.A., Schmidt, H., Neto, A.H.C., Özyilmaz, B., 2014. Electric field effect in ultrathin black phosphorus. Appl. Phys. Lett. 104, 103106.

Koswatta, S.O., Valdes-Garcia, A., Steiner, M.B., Lin, Y.-M., Avouris, P., 2011. Ultimate RF performance potential of carbon electronics. IEEE Trans. Microw. Theory Tech. 59, 2739–2750.

Krasnozhon, D., Lembke, D., Nyffeler, C., Leblebici, Y., Kis, A., 2014. MoS_2 transistors operating at gigahertz frequencies. Nano Lett. 14, 5905–5911.

Lalmi, B., Oughaddou, H., Enriquez, H., Kara, A., Vizzini, S., Ealet, B., Aufray, B., 2010. Epitaxial growth of a silicene sheet. Appl. Phys. Lett. 97, 223109.

Landauer, G.M., Jimenez, D., Gonzalez, J.L., 2014. An accurate and Verilog-A compatible compact model for graphene field-effect transistors. IEEE Trans. Nanotechnol. 13, 895–904.

Lee, J., Ha, T.-J., Li, H., Parrish, K.N., Holt, M., Dodabalapur, A., Ruoff, R.S., Akinwande, D., 2013. 25 GHz embedded-gate graphene transistors with high-K dielectrics on extremely flexible plastic sheets. ACS Nano 7, 7744–7750.

Lee, C.-H., Lee, G.-H., van der Zande, A.M., Chen, W., Li, Y., Han, M., Cui, X., Arefe, G., Nuckolls, C., Heinz, T.F., Guo, J., Hone, J., Kim, P., 2014a. Atomically thin p–n junctions with van der Waals heterointerfaces. Nat. Nanotechnol. 9, 676–681.

Lee, J.-H., Lee, E.K., Joo, W.-J., Jang, Y., Kim, B.-S., Lim, J.Y., Choi, S.-H., Ahn, S.J., Ahn, J.R., Park, M.-H., Yang, C.-W., Choi, B.L., Hwang, S.-W., Whang, D., 2014b. Wafer-scale growth of single-crystal monolayer graphene on reusable hydrogen-terminated germanium. Science 344, 286–289.

Lee, H.S., Baik, S.S., Lee, K., Min, S.-W., Jeon, P.J., Kim, J.S., Choi, K., Choi, H.J., Kim, J.H., Im, S., 2015. Metal semiconductor field-effect transistor with MoS_2/conducting NiOx van der Waals Schottky interface for intrinsic high mobility and photoswitching speed. ACS Nano 9, 8312–8320.

Le Lay, G., Aufray, B., Léandri, C., Oughaddou, H., Biberian, J.-P., De Padova, P., Dávila, M.E., Ealet, B., Kara, A., 2009. Physics and chemistry of silicene nano-ribbons. Appl. Surf. Sci. 256, 524–529.

Leong, W.S., Gong, H., Thong, J.T.L., 2014. Low-contact-resistance graphene devices with nickel-etched-graphene contacts. ACS Nano 8, 994–1001.

Leong, W.S., Luo, X., Li, Y., Khoo, K.H., Quek, S.Y., Thong, J.T.L., 2015. Low resistance metal contacts to MoS_2 devices with nickel-etched-graphene electrodes. ACS Nano 9, 869–877.

Li, X., Cai, W., An, J., Kim, S., Nah, J., Yang, D., Piner, R., Velamakanni, A., Jung, I., Tutuc, E., Banerjee, S.K., Colombo, L., Ruoff, R.S., 2009. Large-area synthesis of high-quality and uniform graphene films on copper foils. Science 324, 1312–1314.

Li, S.-L., Miyazaki, H., Kumatani, A., Kanda, A., Tsukagoshi, K., 2010. Low operating bias and matched input–output characteristics in graphene logic inverters. Nano Lett. 10, 2357–2362.

Li, S.-L., Miyazaki, H., Lee, M.V., Liu, C., Kanda, A., Tsukagoshi, K., 2011. Complementary-like graphene logic gates controlled by electrostatic doping. Small 7, 1552–1556.

Li, L., Yu, Y., Ye, G.J., Ge, Q., Ou, X., Wu, H., Feng, D., Chen, X.H., Zhang, Y., 2014. Black phosphorus field-effect transistors. Nat. Nanotechnol. 9, 372–377.

Li, S.-L., Tsukagoshi, K., Orgiu, E., Samorì, P., 2016. Charge transport and mobility engineering in two-dimensional transition metal chalcogenide semiconductors. Chem. Soc. Rev. 45, 118–151.

Liao, L., Bai, J., Cheng, R., Zhou, H., Liu, L., Liu, Y., Huang, Y., Duan, X., 2012. Scalable fabrication of self-aligned graphene transistors and circuits on glass. Nano Lett. 12, 2653–2657.

Lin, Y.-F., Xu, Y., Wang, S.-T., Li, S.-L., Yamamoto, M., Aparecido-Ferreira, A., Li, W., Sun, H., Nakaharai, S., Jian, W.-B., Ueno, K., Tsukagoshi, K., 2014. Ambipolar $MoTe_2$ transistors and their applications in logic circuits. Adv. Mater. 26, 3263–3269.

Ling, X., Wang, H., Huang, S., Xia, F., Dresselhaus, M.S., 2015. The renaissance of black phosphorus. Proc. Natl. Acad. Sci. 112, 4523–4530.

Liu, W., Kang, J., Sarkar, D., Khatami, Y., Jena, D., Banerjee, K., 2013. Role of metal contacts in designing high-performance monolayer n-type WSe_2 field effect transistors. Nano Lett. 13, 1983–1990.

Liu, H., Neal, A.T., Zhu, Z., Luo, Z., Xu, X., Tománek, D., Ye, P.D., 2014. Phosphorene: an unexplored 2D semiconductor with a high hole mobility. ACS Nano 8, 4033–4041.

Lyu, H., Lu, Q., Huang, Y., Ma, T., Zhang, J., Wu, X., Yu, Z., Ren, W., Cheng, H.-M., Wu, H., Qian, H., 2015. Graphene distributed amplifiers: generating desirable gain for graphene field-effect transistors. Sci. Rep. 5, 17649.

Ma, N., Jena, D., 2015. Carrier statistics and quantum capacitance effects on mobility extraction in two-dimensional crystal semiconductor field-effect transistors. 2D Materials 2, 015003.

Majumdar, K., Hobbs, C., Kirsch, P.D., 2014. Benchmarking transition metal dichalcogenide MOSFET in the ultimate physical scaling limit. IEEE Electron Device Lett. 35, 402–404.

Mak, K.F., Lee, C., Hone, J., Shan, J., Heinz, T.F., 2010. Atomically thin MoS_2: a new direct-gap semiconductor. Phys. Rev. Lett. 105, 136805.

Mak, K.F., He, K., Shan, J., Heinz, T.F., 2012. Control of valley polarization in monolayer MoS_2 by optical helicity. Nat. Nanotechnol. 7, 494–498.

Mayorov, A.S., Gorbachev, R.V., Morozov, S.V., Britnell, L., Jalil, R., Ponomarenko, L.A., Blake, P., Novoselov, K.S., Watanabe, K., Taniguchi, T., Geim, A.K., 2011. Micrometer-scale ballistic transport in encapsulated graphene at room temperature. Nano Lett. 11, 2396–2399.

Meric, I., Dean, C.R., Han, S.-J., Wang, L., Jenkins, K.A., Hone, J., Shepard, K.L., 2011. High-frequency performance of graphene field effect transistors with saturating IV-characteristics. In: Electron Devices Meeting (IEDM), 2011 IEEE International. Presented at the Electron Devices Meeting (IEDM), 2011 IEEE International, pp. 2.1.1–2.1.4.

Miró, P., Audiffred, M., Heine, T., 2014. An atlas of two-dimensional materials. Chem. Soc. Rev. 43, 6537–6554.

Nebel, C.E., 2013. Valleytronics: electrons dance in diamond. Nat. Mater. 12, 690–691.

Novoselov, K.S., Geim, A.K., Morozov, S.V., Jiang, D., Zhang, Y., Dubonos, S.V., Grigorieva, I.V., Firsov, A.A., 2004. Electric field effect in atomically thin carbon films. Science 306, 666–669.

Ovchinnikov, D., Allain, A., Huang, Y.-S., Dumcenco, D., Kis, A., 2014. Electrical transport properties of single-layer WS_2. ACS Nano 8, 8174–8181.

Park, J.-S., Choi, H.J., 2015. Band-gap opening in graphene: a reverse-engineering approach. Phys. Rev. B 92, 045402.

Peleg, R., 2015. How graphene may revolutionize the mobile industry. [WWW Document]. http://www.androidauthority.com/graphene-may-revolutionize-mobile-industry-596375/ (accessed 2.17.16).

Petrone, N., Meric, I., Chari, T., Shepard, K.L., Hone, J., 2015. Graphene field-effect transistors for radio-frequency flexible electronics. IEEE J. Electron Devices Soc. 3, 44–48.

Qing-Bin, L., Cui, Y., Jia, L., Xu-Bo, S., Ze-Zhao, H., Wei-Li, L., Guo-Dong, G., Yuan-Gang, W., Zhi-Hong, F., 2014. Radio-frequency performance of epitaxial graphene field-effect transistors on sapphire substrates. Chin. Phys. Lett. 31, 078104.

Radisavljevic, B., Kis, A., 2013a. Reply to "Measurement of mobility in dual-gated MoS_2 transistors". Nat. Nanotechnol. 8, 147–148.

Radisavljevic, B., Kis, A., 2013b. Mobility engineering and a metal–insulator transition in monolayer MoS_2. Nat. Mater. 12, 815–820.

Radisavljevic, B., Radenovic, A., Brivio, J., Giacometti, V., Kis, A., 2011a. Single-layer MoS_2 transistors. Nat. Nanotechnol. 6, 147–150.

Radisavljevic, B., Whitwick, M.B., Kis, A., 2011b. Integrated circuits and logic operations based on single-layer MoS_2. ACS Nano 5, 9934–9938.

Radisavljevic, B., Whitwick, M.B., Kis, A., 2012. Small-signal amplifier based on single-layer MoS_2. Appl. Phys. Lett. 101, 043103.

Rakheja, S., Wu, Y., Wang, H., Palacios, T., Avouris, P., Antoniadis, D.A., 2014. An ambipolar virtual-source-based charge-current compact model for nanoscale graphene transistors. IEEE Trans. Nanotechnol. 13, 1005–1013.

Rao, C.N.R., Gopalakrishnan, K., Maitra, U., 2015. Comparative study of potential applications of graphene, MoS_2, and other two-dimensional materials in energy devices, sensors, and related areas. ACS Appl. Mater. Interfaces 7, 7809–7832.

Rizzi, L.G., Bianchi, M., Behnam, A., Carrion, E., Guerriero, E., Polloni, L., Pop, E., Sordan, R., 2012. Cascading wafer-scale integrated graphene complementary inverters under ambient conditions. Nano Lett. 12, 3948–3953.

Rodriguez, S., Vaziri, S., Ostling, M., Rusu, A., Alarcon, E., Lemme, M.C., 2012. RF performance projections of graphene FETs vs. silicon MOSFETs. ECS Solid State Lett. 1, Q39–Q41.

Roy, T., Tosun, M., Cao, X., Fang, H., Lien, D.-H., Zhao, P., Chen, Y.-Z., Chueh, Y.-L., Guo, J., Javey, A., 2015. Dual-gated MoS_2/WSe_2 van der Waals tunnel diodes and transistors. ACS Nano 9, 2071–2079.

Samuel Reich, E., 2014. Phosphorene excites materials scientists. Nat. News 506, 19.

Sangwan, V.K., Jariwala, D., Kim, I.S., Chen, K.-S., Marks, T.J., Lauhon, L.J., Hersam, M.C., 2015. Gate-tunable memristive phenomena mediated by grain boundaries in single-layer MoS_2. Nat. Nanotechnol. 10, 403–406.

Sanne, A., Ghosh, R., Rai, A., Yogeesh, M.N., Shin, S.H., Sharma, A., Jarvis, K., Mathew, L., Rao, R., Akinwande, D., Banerjee, S., 2015. Radio frequency transistors and circuits based on CVD MoS_2. Nano Lett. 15, 5039–5045.

Sarkar, D., Xie, X., Liu, W., Cao, W., Kang, J., Gong, Y., Kraemer, S., Ajayan, P.M., Banerjee, K., 2015. A subthermionic tunnel field-effect transistor with an atomically thin channel. Nature 526, 91–95.

Schall, D., Otto, M., Neumaier, D., Kurz, H., 2013. Integrated ring oscillators based on high-performance graphene inverters. Sci. Rep. 3, 2592.

Schroder, D.K., 1990. Semiconductor Material and Device Characterization. Wiley, New York.

Schwierz, F., 2013. Graphene transistors: status, prospects, and problems. Proc. IEEE 101, 1567–1584.

Schwierz, F., 2015. Graphene and Beyond: Two-Dimensional Materials for Transistor Applications. In: Proc. SPIE 9467, Micro- and Nanotechnology Sensors, Systems, and Applications VII, 94670W.

Schwierz, F., Pezoldt, J., Granzner, R., 2015. Two-dimensional materials and their prospects in transistor electronics. Nanoscale 7, 8261–8283.

Shah, P.B., Amani, M., Chin, M.L., O'Regan, T.P., Crowne, F.J., Dubey, M., 2014. Analysis of temperature dependent hysteresis in MoS_2 field effect transistors for high frequency applications. Solid State Electron. 91, 87–90.

Son, I.H., Hwan Park, J., Kwon, S., Park, S., Rümmeli, M.H., Bachmatiuk, A., Song, H.J., Ku, J., Choi, J.W., Choi, J., Doo, S.-G., Chang, H., 2015. Silicon carbide-free graphene growth on silicon for lithium-ion battery with high volumetric energy density. Nat. Commun. 6, 7393.

Spasenovic, M., 2014. Samsung files patent for graphene-based touch screen. [WWW Document]. http://www.graphenetracker.com/samsung-patents-graphene-based-touch-screen/ (accessed 2.16.16).

Splendiani, A., Sun, L., Zhang, Y., Li, T., Kim, J., Chim, C.-Y., Galli, G., Wang, F., 2010. Emerging photoluminescence in monolayer MoS_2. Nano Lett. 10, 1271–1275.

Strukov, D.B., Snider, G.S., Stewart, D.R., Williams, R.S., 2008. The missing memristor found. Nature 453, 80–83.

Tao, L., Cinquanta, E., Chiappe, D., Grazianetti, C., Fanciulli, M., Dubey, M., Molle, A., Akinwande, D., 2015. Silicene field-effect transistors operating at room temperature. Nat. Nanotechnol. 10, 227–231.

Tedesco, J.L., VanMil, B.L., Myers-Ward, R.L., McCrate, J.M., Kitt, S.A., Campbell, P.M., Jernigan, G.G., Culbertson Jr., J.C., Eddy, C.R., Gaskill, D.K., 2009. Hall effect mobility of epitaxial graphene grown on silicon carbide. Appl. Phys. Lett. 95, 122102.

Tosun, M., Chuang, S., Fang, H., Sachid, A.B., Hettick, M., Lin, Y., Zeng, Y., Javey, A., 2014. High-gain inverters based on WSe_2 complementary field-effect transistors. ACS Nano 8, 4948–4953.

Traversi, F., Russo, V., Sordan, R., 2009. Integrated complementary graphene inverter. Appl. Phys. Lett. 94, 223312.

Uchida, K., Takagi, S., 2003. Carrier scattering induced by thickness fluctuation of silicon-on-insulator film in ultrathin-body metal–oxide–semiconductor field-effect transistors. Appl. Phys. Lett. 82, 2916–2918.

Umoh, I.J., Kazmierski, T.J., Al-Hashimi, B.M., 2014. Multilayer graphene FET compact circuit-level model with temperature effects. IEEE Trans. Nanotechnol. 13, 805–813.

Vongehr, S., Meng, X., 2015. The missing memristor has not been found. Sci. Rep. 5, 11657.

Wang, J., Lundstrom, M., 2002. Does source-to-drain tunneling limit the ultimate scaling of MOSFETs? In: Electron Devices Meeting, 2002. IEDM '02. International. Presented at the Electron Devices Meeting, 2002. IEDM '02. International, pp. 707–710.

Wang, H., Hsu, A.L., Palacios, T., 2012a. Graphene electronics for RF applications. IEEE Microw. Mag. 13, 114–125.

Wang, H., Yu, L., Lee, Y.-H., Shi, Y., Hsu, A., Chin, M.L., Li, L.-J., Dubey, M., Kong, J., Palacios, T., 2012b. Integrated circuits based on bilayer MoS_2 transistors. Nano Lett. 12, 4674–4680.

Wang, H., Wang, X., Xia, F., Wang, L., Jiang, H., Xia, Q., Chin, M.L., Dubey, M., Han, S., 2014. Black phosphorus radio-frequency transistors. Nano Lett. 14, 6424–6429.

Withers, F., Del Pozo-Zamudio, O., Mishchenko, A., Rooney, A.P., Gholinia, A., Watanabe, K., Taniguchi, T., Haigh, S.J., Geim, A.K., Tartakovskii, A.I., Novoselov, K.S., 2015. Light-emitting diodes by band-structure engineering in van der Waals heterostructures. Nat. Mater. 14, 301–306.

Xia, F., Wang, H., Jia, Y., 2014. Rediscovering black phosphorus as an anisotropic layered material for optoelectronics and electronics. Nat. Commun. 5, 4458.

Xu, H., Zhang, Z., Xu, H., Wang, Z., Wang, S., Peng, L.-M., 2011. Top-gated graphene field-effect transistors with high normalized transconductance and designable dirac point voltage. ACS Nano 5, 5031–5037.

Yan, R.-H., Ourmazd, A., Lee, K.F., 1992. Scaling the Si MOSFET: from bulk to SOI to bulk. IEEE Trans. Electron Devices 39, 1704–1710.
Yeh, C.-H., Lain, Y.-W., Chiu, Y.-C., Liao, C.-H., Moyano, D.R., Hsu, S.S.H., Chiu, P.-W., 2014. Gigahertz flexible graphene transistors for microwave integrated circuits. ACS Nano 8, 7663–7670.
Yoon, Y., Ganapathi, K., Salahuddin, S., 2011. How good can monolayer MoS_2 transistors be? Nano Lett. 11, 3768–3773.
Yoshida, M., Suzuki, R., Zhang, Y., Nakano, M., Iwasa, Y., 2015. Memristive phase switching in two-dimensional 1T-TaS_2 crystals. Sci. Adv. 1, e1500606.
Yu, W.J., Li, Z., Zhou, H., Chen, Y., Wang, Y., Huang, Y., Duan, X., 2013. Vertically stacked multi-heterostructures of layered materials for logic transistors and complementary inverters. Nat. Mater. 12, 246–252.
Yu, L., Lee, Y.-H., Ling, X., Santos, E.J.G., Shin, Y.C., Lin, Y., Dubey, M., Kaxiras, E., Kong, J., Wang, H., Palacios, T., 2014. Graphene/MoS_2 hybrid technology for large-scale two-dimensional electronics. Nano Lett. 14, 3055–3063.
Yu, L., Zubair, A., Santos, E.J.G., Zhang, X., Lin, Y., Zhang, Y., Palacios, T., 2015. High-performance WSe_2 complementary metal oxide semiconductor technology and integrated circuits. Nano Lett. 15, 4928–4934.
Zeng, H., Dai, J., Yao, W., Xiao, D., Cui, X., 2012. Valley polarization in MoS_2 monolayers by optical pumping. Nat. Nanotechnol. 7, 490–493.
Zhang, W., Huang, Z., Zhang, W., Li, Y., 2014. Two-dimensional semiconductors with possible high room temperature mobility. Nano Res. 7, 1731–1737.
Zhu, W., Yogeesh, M.N., Yang, S., Aldave, S.H., Kim, J.-S., Sonde, S., Tao, L., Lu, N., Akinwande, D., 2015. Flexible black phosphorus ambipolar transistors, circuits and AM demodulator. Nano Lett. 15, 1883–1890.
Zimmer, T., Fregonese, S., 2015. Graphene transistor-based active balun architectures. IEEE Trans. Electron Devices 62, 3079–3083.

CHAPTER SEVEN

Black Phosphorus-Based Nanodevices

J.O. Island*,[1], A. Castellanos-Gomez†,[1]

*Kavli Institute of Nanoscience, Delft University of Technology, Delft, The Netherlands
†Instituto Madrileño de Estudios Avanzados en Nanociencia (IMDEA-Nanociencia), Madrid, Spain
[1]Corresponding authors: e-mail address: j.o.island@tudelft.nl; andres.castellanos@imdea.org

Contents

1. INTRODUCTION

The isolation of graphene in 2004 (Novoselov et al., 2004) has unleashed the isolation of many other two-dimensional materials by mechanical exfoliation of bulk-layered crystals whose layers are held together by weak van der Waals forces (Novoselov et al., 2005). Among two-dimensional crystals obtained by exfoliation of layered 3D materials, graphene is still the most studied by far due to its outstanding mechanical and electronic properties. Other 2D crystals, however, have recently gained considerable interest, since their properties are complementary to those of graphene. For instance, the lack of a band gap in graphene, which yields small current on/off ratios in

Semiconductors and Semimetals, Volume 95
ISSN 0080-8784
http://dx.doi.org/10.1016/bs.semsem.2016.03.002

graphene-based field-effect transistors (FETs), has motivated research on other 2D semiconductor crystals such as transition metal dichalcogenides, which possess a large intrinsic band gap. However, this material family has demonstrated relatively slow photoresponse and, because of the large band gap of Mo- and W-based compounds, they are suited for applications in only part of the visible range of the electromagnetic spectrum. A material with a direct and small band gap together with fast photoresponse is needed to extend the detection range accessible with 2D materials.

In 2013, Zhang and coworkers presented a talk in the American Physical Society March meeting entitled "Electronic Properties of Few-Layer Black Phosphorus," demonstrating that graphene was not the only elemental two-dimensional material that could be isolated by mechanical exfoliation of a bulk-layered crystal (Li et al., 2013). In fact, atomically thin layers (and now one-atom thick layers) of phosphorus can be prepared by mechanical exfoliation of bulk black phosphorus, a layered allotrope of phosphorus. Black phosphorus single crystals were synthesized for the first time in 1914 by Bridgman during his investigations of phase transformations of white phosphorus under high pressure (Bridgman, 1914). In bulk, black phosphorus is a semiconductor with a direct band gap of 0.35 eV and charge carrier mobilities in the order of 10,000 $cm^2/V\ s$ (Rudenko and Katsnelson, 2014). Thin flakes of black phosphorus have been recently used in FETs showing mobility values up to 1000 $cm^2/V\ s$. Moreover, due to its small and direct band gap black phosphorus is a bright perspective for broadband photodetection where transition metal dichalcogenides are limited because of their large band gap.

In this chapter we review the recent works on atomically thin black phosphorus and we analyze the reasons behind the surge of experimental and theoretical works on this novel two-dimensional material. We focus on its thickness-dependent band gap value and its rather unusual in-plane anisotropy that has drawn the attention of the scientific community toward this two-dimensional material. Here, we also review the main challenges that black phosphorus will face in the near future.

2. ISOLATION OF ULTRATHIN BLACK PHOSPHORUS

The crystal structure of black phosphorus resembles that of graphite: atoms are strongly bonded in-plane forming layers, while the layers weakly interact through van der Waals forces. While in graphite each carbon atom bonds with three neighboring atoms through sp^2-hybridized orbitals, in black phosphorus the phosphorus atoms have five valence shell electrons available for bonding with a valence shell configuration $3s^23p^3$. Therefore, each phosphorus atom bonds to three neighboring phosphorus atoms through sp^3-hybridized orbitals, making the phosphorus atoms form a puckered honeycomb lattice (orthorhombic, with space group *Cmca*) and each phosphorus atom also has a lone pair, which makes phosphorus very reactive to air. Fig. 1 shows a schematic diagram of the black phosphorus crystal structure. As can be seen in Fig. 1, this atomic arrangement yields two inequivalent directions: the so-called zigzag direction (parallel to the atomic ridges) and the so-called armchair direction (perpendicular to the ridges).

This layered structure of bulk black phosphorus permits to employ mechanical exfoliation to extract thin black phosphorus layers from a bulk macroscopic crystal. Atomically thin flakes can be extracted by using an adhesive tape (Nitto Denko SPV 224 is the one typically used although other conventional tapes such as Magic tape from Scotch 3M® can be also used). However, the mechanical exfoliation protocol used to obtain graphene or transition metal dichalcogenides typically yields a very low density of atomically thin black phosphorus flakes. Therefore, a modified mechanical exfoliation technique should be employed: an intermediate viscoelastic surface to exfoliate the flakes substantially increases the yield and reduces the contamination of the fabricated flakes. Exfoliation and transfer of thin flakes are typically accomplished in the following manner. First, a commercially available bulk black phosphorus crystal (99.998%, Smart Elements) is cleaved several times using Nitto SPV 224 tape. The tape containing the thin black phosphorus crystals is then pressed onto a poly-dimethylsiloxane (PDMS)

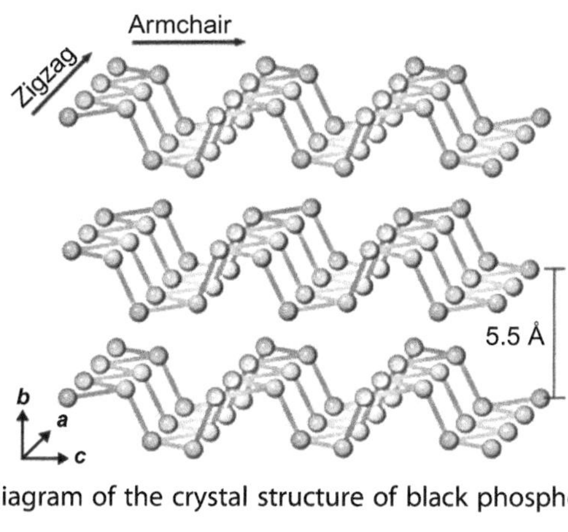

Fig. 1 Schematic diagram of the crystal structure of black phosphorus.

substrate and it is peeled off rapidly. Transmission optical microscopy is used to find thin black phosphorus flakes. The thin flakes on the surface of the PDMS substrate can be transferred to other substrates by simply putting the PDMS substrate in gentle contact with the new acceptor substrate and peeling it off slowly (it takes about 5–10 min to peel off the stamp completely from the surface).

Fig. 2A shows an example of a transmission mode optical image of a black phosphorus flake deposited by mechanical exfoliation onto a PDMS substrate. Fig. 2B shows an optical microscopy image of the same flake after having been transferred onto a silicon chip with 285 nm of thermally grown SiO_2 (note that part of the flake broke during the transfer). Fig. 2C shows

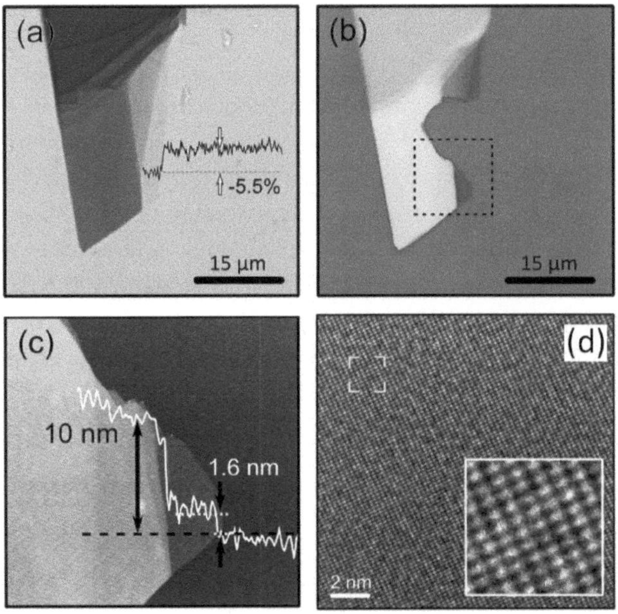

Fig. 2 Example of isolation of black phosphorus flakes by mechanical exfoliation. (A) Transmission mode optical microscopy image of a few-layer black phosphorus flake exfoliated onto a PDMS substrate. (B) Optical microscopy image of the same flake after transferring it onto a SiO_2/Si substrate. Note that part of the flake was broken during the transfer. (C) Atomic force microscopy topography image of the region highlighted with a *square* in (B). A topographic line profile, acquired along the horizontal *dashed black line*, has been included in the image. (D) High-resolution transmission electron microscopy image of the multilayered region of the flake (~13–21 layers). *Reprinted from Castellanos-Gomez, A., Vicarelli, L., Prada, E., Island, J.O., Narasimha-Acharya, K.L., Blanter, S.I., Groenendijk, D.J., Buscema, M., Steele, G.A., Alvarez, J.V., et al., 2014. Isolation and characterization of few-layer black phosphorus. 2D Mater. 1 (2), 025001 with permission.*

an atomic force microscopy (AFM) topographic image of the same black phosphorus flake.

Most of the experiments reported on black phosphorus so far are based on mechanically exfoliated samples, which provide high-quality samples but is not a scalable technology. Liquid-phase exfoliation, on the other hand, has been recently demonstrated as a cheap and scalable alternative to isolate ultrathin black phosphorus layers, but the quality of the resulting material is insufficient for certain electronic or optical applications as thin films are based on the stacking of microscopic black phosphorus crystallites (Hanlon et al., 2015; Jiang et al., 2015; Kang et al., 2015; Luo et al., 2015a; Sresht et al., 2015; Yasaei et al., 2015). Another alternative method to produce black phosphorus thin films is based on pulsed laser deposition which is a technology that can be scaled up (Yang et al., 2015a). Fig. 3A and B shows transmission electron microscopy (TEM) images of the resulting film. While the scaled-up growth is advantageous, the fabricated layers lack crystallinity and their electrical properties still pale in comparison to those of pristine black phosphorus material. Therefore in order to transfer black phosphorus from research laboratories to the first real-life applications, one should solve the challenge of developing a large-area synthesis method to fabricate highly crystalline black phosphorus layers. There are also preliminary results demonstrating large-area growth of nanocrystalline black phosphorus on flexible substrates by employing a multiple anvil-diamond cell to subject a red phosphorus thin film to high pressure (Li et al., 2015a). Fig. 3C and D shows TEM images of the resulting film using this method. Nanocrystalline regions can be seen in Fig. 3D.

3. THICKNESS-DEPENDENT BAND GAP

Black phosphorus, unlike graphene, is a semiconductor with a sizeable gap (Asahina and Morita, 1984; Keyes, 1953; Morita, 1986). Fig. 4A shows the calculated band structure for monolayer, bilayer, and trilayer black phosphorus by ab initio calculations with the *GW* approximation (Rudenko and Katsnelson, 2014). The band gap is direct and located at the Γ point of the Brillouin zone, independent of the number of layers. Semiconducting transition metal dichalcogenides, on the other hand, present a gap at the K point and it is only direct for single-layers (Mak et al., 2010; Splendiani et al., 2010). Interestingly, the band gap of black phosphorus strongly depends on the number of layers.

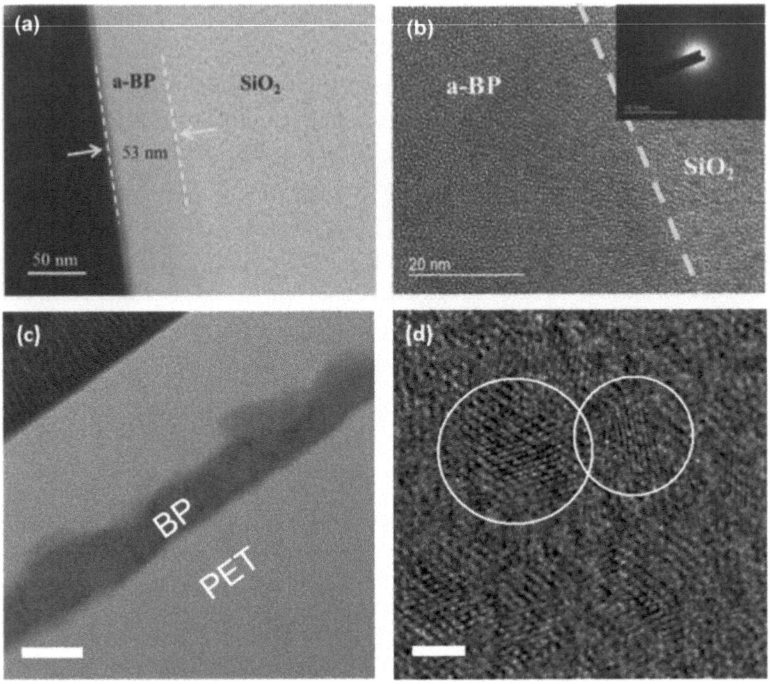

Fig. 3 Resulting amorphous films from pulse laser deposition and high-pressure multi-anvil cell growth methods. (A) Cross-sectional TEM image of a thin film produced by pulsed laser deposition. (B) High-resolution TEM image with selected area electron diffraction pattern (inset) indicating the amorphous nature of the deposited films from pulsed laser deposition. (C) Cross-sectional TEM image of a thin film produced by a high-pressure multi-anvil cell. The scale bar is 50 nm. (D) High-resolution TEM image of a BP film produced using the multi-anvil cell method showing nanocrystalline portions indicated with *white circles*. The scale bar is 5 nm. *Panel (B): Adapted from Yang, Z., Hao, J., Yuan, S., Lin, S., Yau, H.M., Dai, J., Lau, S.P., 2015. Field-effect transistors based on amorphous black phosphorus ultrathin films by pulsed laser deposition. Adv. Mater. 27 (25), 3748–3754 with permission. Panel (D): Adapted from Li, X., Deng, B., Wang, X., Chen, S., Vaisman, M., Karato, S., Pan, G., Larry Lee, M., Cha, J., Wang, H., et al., 2015. Synthesis of thin-film black phosphorus on a flexible substrate. 2D Mater. 2 (3), 031002 with permission.*

Fig. 4B summarizes the thickness dependence of the black phosphorus band gap, calculated with different ab initio methods (Castellanos-Gomez et al., 2014a; Qiao et al., 2014; Tran et al., 2014), and measured through photoluminescence (Liu et al., 2014; Wang et al., 2015a; Yang et al., 2015b), infrared spectroscopy (Xia et al., 2014a), and scanning tunneling spectroscopy (Liang et al., 2014; Zhang et al., 2009) techniques. The magnitude of the theoretical band gap strongly depends on the approximation

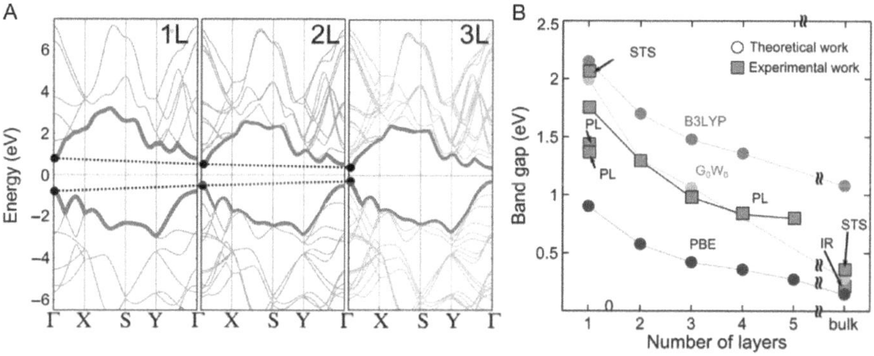

Fig. 4 (A) Calculated band structure of single-, bi-, and trilayer black phosphorus. (B) Thickness dependence of the black phosphorus band gap, calculated with different approaches. *Data extracted from Castellanos-Gomez, A., Vicarelli, L., Prada, E., Island, J.O., Narasimha-Acharya, K.L., Blanter, S.I., Groenendijk, D.J., Buscema, M., Steele, G.A., Alvarez, J.V., et al., 2014. Isolation and characterization of few-layer black phosphorus. 2D Mater. 1 (2), 025001; Liang, L., Wang, J., Lin, W., Sumpter, B.G., Meunier, V., Pan, M., 2014. Electronic bandgap and edge reconstruction in phosphorene materials. Nano Lett. 14 (11), 6400–6406; Liu, H., Neal, A.T., Zhu, Z., Luo, Z., Xu, X., Tománek, D., Ye, P.D., 2014. Phosphorene: an unexplored 2D semiconductor with a high hole mobility. ACS Nano 8 (4), 4033–4041; Xia, F., Wang, H., Jia, Y., 2014. Rediscovering black phosphorus as an anisotropic layered material for optoelectronics and electronics. Nat. Commun. 5, 4458; Yang, J., Xu, R., Pei, J., Myint, Y.W., Wang, F., Wang, Z., Zhang, S., Yu, Z., Lu, Y., 2015. Optical tuning of exciton and trion emissions in monolayer phosphorene. Light Sci. Appl. 4 (7), e312; Zhang, C.D., Lian, J.C., Yi, W., Jiang, Y.H., Liu, L.W., Hu, H., Xiao, W.D., Du, S.X., Sun, L.L., Gao, H.J., 2009. Surface structures of black phosphorus investigated with scanning tunneling microscopy. J. Phys. Chem. C 113 (43), 18823–18826. Figure reprinted from Castellanos-Gomez, A., 2015. Black phosphorus: narrow gap wide applications. J. Phys. Chem. Lett. 6 (21), 4280–4291 with permission.*

employed to calculate the band structure, but in all the cases, there is a marked thickness dependence: from a large gap (close to 2 eV) for single-layer that monotonically decreases to a narrow band gap value (about 0.3 eV) for bulk black phosphorus.

Therefore, black phosphorus' direct and narrow band gap is a significant advantage (in comparison with other 2D materials) for infrared light detection, free-space and fiber optics-based communication. Fig. 5 summarizes the band gaps of most commonly studied 2D, semiconducting materials as well as conventional semiconductors. The current technology of mid-infrared detection is based on mercury cadmium telluride and InSb photodetectors or quantum well infrared photodetectors which require expensive materials that have to be epitaxially grown and are not easy to integrate with

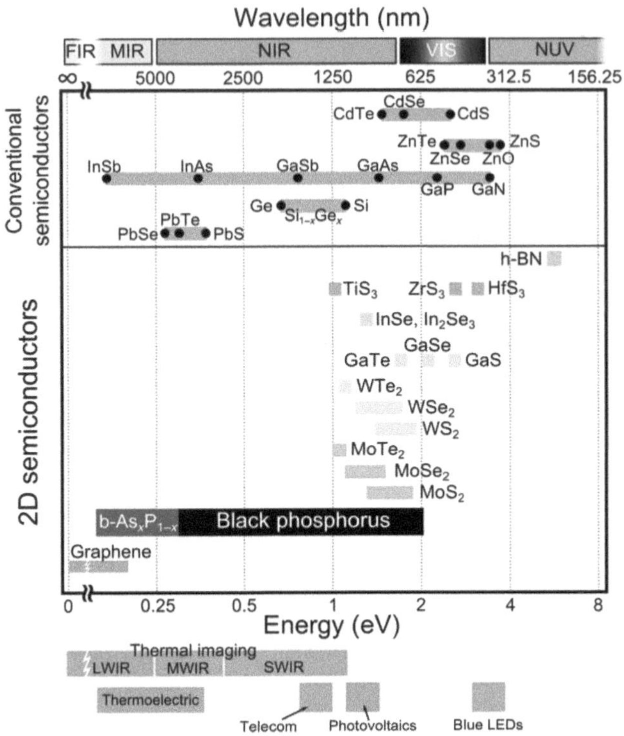

Fig. 5 Comparison of the band gap values for different 2D semiconductor materials. The band gap values for conventional semiconductors have been also included for comparison. The horizontal bars spanning a range of band gap values indicate that the band gap can be tuned over that range by changing the number of layers, straining, or alloying. In conventional semiconductors, the bar indicates that the band gap can be continuously tuned by alloying the semiconductors (eg, $Si_{1-x}Ge_x$ or $In_{1-x}Ga_xAs$). The range of band gap values required for certain applications has been highlighted at the bottom part of the figure to illustrate the potential applications of the different semiconductors. *Reprinted from Castellanos-Gomez, A., 2015. Black phosphorus: narrow gap wide applications. J. Phys. Chem. Lett. 6 (21), 4280–4291 with permission.*

silicon technology. Moreover, thermoelectric power generation utilizing semiconductor materials with band gaps in the order of 0.2–0.3 eV are out of reach of the two-dimensional semiconductors isolated to date. The Seebeck coefficient (S) is approximately proportional to $1/n$ (thus it is higher for semiconductors than for metals), but for power generation high electrical conductivity is needed. Therefore, the best thermoelectric materials for thermopower generation are thus narrow band gap semiconductors with a good combination of finite band gap and high conductivity. The band gap value range spanned by black phosphorus nanolayers makes this 2D material especially suitable for thermal imaging, thermoelectric, telecom, and photovoltaic applications and also makes it a prospective replacement to conventional narrow gap semiconductors in applications requiring thin, flexible, and quasi-transparent material where using those 3D materials might be challenging.

4. IN-PLANE ANISOTROPIC ELECTRICAL, MECHANICAL, AND OPTICAL PROPERTIES

As discussed earlier, the crystal structure of black phosphorus (sketched in Fig. 1) is the origin of its in-plane anisotropic properties (Chaves et al., 2015; Fei and Yang, 2014; Ge et al., 2015; He et al., 2015; Hong et al., 2014; Jiang, 2015a,b; Jiang and Park, 2014; Li et al., 2014a,b; Liu et al., 2014; Lu et al., 2015; Luo et al., 2015b; Ong et al., 2014; Popović et al., 2015; Qiao et al., 2014; Qin et al., 2014a,b, 2015; Ribeiro et al., 2015; Rodin et al., 2014; Schuster et al., 2015; Tran et al., 2014; Wang et al., 2015a; Wei and Peng, 2014; Wu et al., 2015; Xia et al., 2014a; Yuan et al., 2015; Zhang et al., 2014). On the other hand, most 2D materials isolated and studied to date (graphene, boron nitride, or Mo- and W-based transition metal dichalcogenides) do not present noticeable in-plane anisotropy.

Regarding its electrical anisotropy, field-effect devices with electrodes arranged at different angles (see inset in Fig. 6A) have been used to probe the anisotropy of the charge carrier mobility and the electrical conductance, finding a higher mobility and conductance along the armchair direction (up to 50% higher than along the zigzag direction at room temperature) (Li et al., 2014a; Liu et al., 2014; Lu et al., 2015; Xia et al., 2014a). Fig. 6A shows a polar plot with the angular dependence of the DC conductivity measured in a black phosphorus device with electrodes patterned with an angular spacing of 30 degree (see the inset) (Xia et al., 2014a).

Fig. 6B shows the relative optical extinction measured on the black phosphorus device in Fig. 6A for different polarization angles (relative to the horizontal axis) (Xia et al., 2014a). This marked linear dichroism of black phosphorus, optical absorption that depends on the relative orientation between the materials lattice and an incident linearly polarized light, has also strong implications in its Raman spectra, plasmonic and screening effects, photoresponse, and photoluminescence emission yield (Cakir et al., 2015; Hong et al., 2014; Low et al., 2014a; Lu et al., 2015; Ribeiro et al., 2015; Tran et al., 2014; Wang et al., 2015a; Yuan et al., 2015). For example,

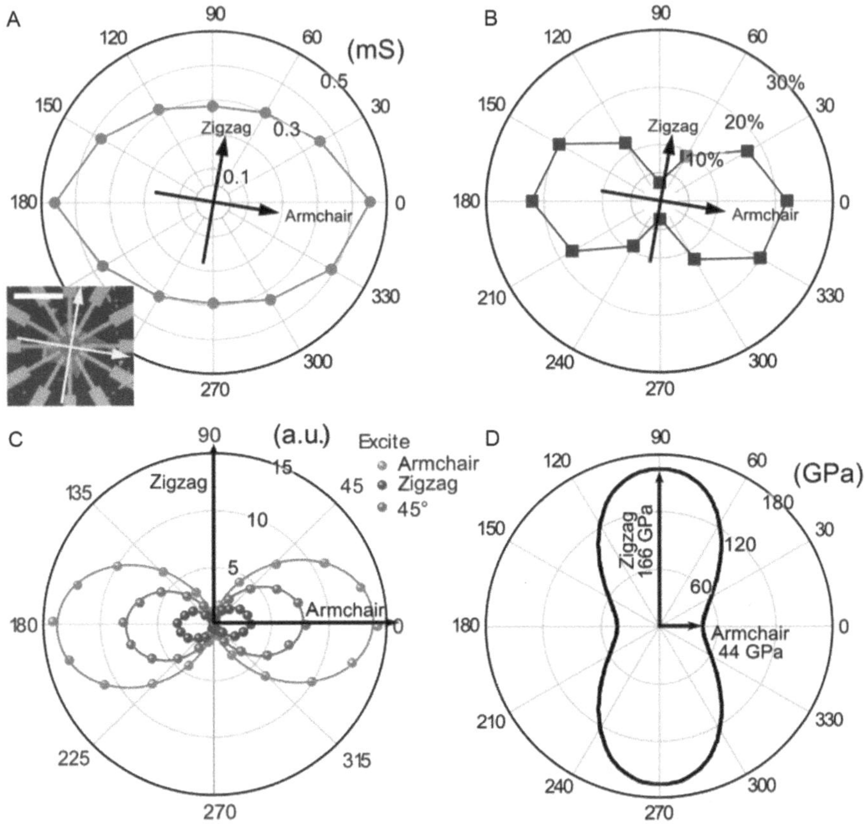

Fig. 6 In-plane anisotropic properties of black phosphorus. (A) Angular dependence of the DC electrical conductance of a black phosphorus transistor. (B) Angular dependence of the relative optical extinction measured on the same flake displayed in (A). (C) Angular-dependent photoluminescence yield of a black phosphorus flake with laser excitation ($\lambda = 532$ nm) polarized along the zigzag direction, armchair direction, and a 45 degree direction. (D) Calculated angular dependence of the black phosphorus Young's modulus. The calculation predicts that the zigzag direction is about four times stiffer than the armchair direction. *The data displayed in (A) and (B) have been extracted from Xia, F., Wang, H., Jia, Y., 2014. Rediscovering black phosphorus as an anisotropic layered material for optoelectronics and electronics. Nat. Commun. 5, 4458. Panel (C): Data extracted from Wang, X., Jones, A.M., Seyler, K.L., Tran, V., Jia, Y., Zhao, H., Wang, H., Yang, L., Xu, X., Xia, F., 2015. Highly anisotropic and robust excitons in monolayer black phosphorus. Nat. Nanotechnol. 10 (6), 517–521. Panel (D): Data extracted from Wei, Q., Peng, X., 2014. Superior mechanical flexibility of phosphorene and few-layer black phosphorus. Appl. Phys. Lett. 104 (25), 251915. This figure has been reprinted from Castellanos-Gomez, A., 2015. Black phosphorus: narrow gap wide applications. J. Phys. Chem. Lett. 6 (21), 4280–4291 with permission.*

Fig. 6C shows a polar plot with the polarization angle of the photoluminescence yield of a black phosphorus flake excited with a laser linearly polarized along the zigzag direction, armchair direction, and a direction at 45 degree between the zigzag and armchair directions. For all the different excitation laser polarization angles, the photoluminescence yield always shows a high degree of polarization along the armchair direction of the flake (Wang et al., 2015a). Similar to the optical and electrical properties, the mechanical properties also display a strong in-plane anisotropy. Fig. 6D shows the calculated angular dependence of the Young's modulus of black phosphorus which is predicted to be up to four times higher along the zigzag direction ($E_{zz} = 166$ GPa) than along the armchair direction ($E_{ac} = 44$ GPa) (Appalakondaiah et al., 2012; Wang and Feng, 2015; Wei and Peng, 2014).

This characteristic in-plane anisotropy of black phosphorus can be exploited to study quasi-1D physics in a system with 2D geometry, which is easier to integrate with other devices and interact with because of its large surface area. Moreover, the in-plane anisotropy can also provide novel functionality to devices based on black phosphorus (see Section 5).

5. NANODEVICES BASED ON BLACK PHOSPHORUS

All the nanofabrication experience gathered during the years after the isolation of graphene has made it possible to fabricate and characterize black phosphorus nanodevices very rapidly (Buscema et al., 2014b; Castellanos-Gomez et al., 2014a; Koenig et al., 2014; Li et al., 2014a; Liu et al., 2014; Xia et al., 2014a). Fig. 7 shows several examples of black phosphorus nanodevices with different functionalities.

5.1 Field-Effect Transistors

FETs were some of the first devices fabricated and measured using exfoliated black phosphorus flakes on Si/SiO_2 substrates (Li et al., 2014a). Fig. 7A shows a schematic of a typical device with chromium/gold electrodes and typical transfer curves for a black phosphorus FET. Being one of the most attractive properties of black phosphorus, FETs in which black phosphorus is used as the channel material display high hole mobilities (of 10–1000 $cm^2/V\,s$) in addition to large current switching ratios of 100–10,000 depending on flake thickness and details of the fabrication and measurements (Buscema et al., 2014b; Castellanos-Gomez et al., 2014a; Koenig et al., 2014; Li et al., 2014a; Liu et al., 2014; Xia et al., 2014a). This makes black phosphorus a superior material in terms of FET performance when directly compared with materials from

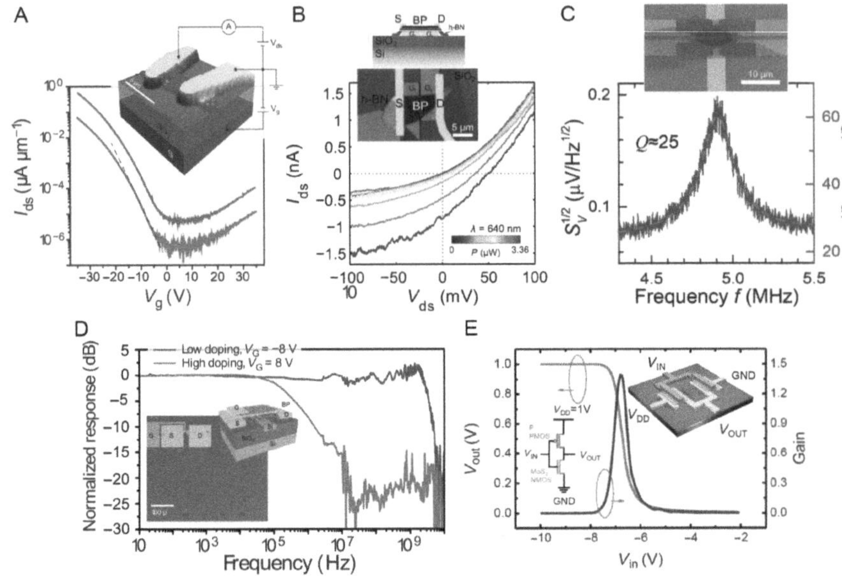

Fig. 7 Black phosphorus-based nanodevices. (A) Ambipolar field-effect transistor. (B) Electrostatically gated PN junction displaying photovoltaic effect. (C) Nanoelectromechanical resonator vibrating in the very high-frequency regime. (D) High-frequency photodetector coupled to a silicon waveguide. (E) Inverter amplifier based on MoS_2 and black phosphorus transistors, demonstrating the potential of black phosphorus for logic circuits. *Panel (A): Adapted from Li, L., Yu, Y., Ye, G.J., Ge, Q., Ou, X., Wu, H., Feng, D., Chen, X.H., Zhang, Y., 2014. Black phosphorus field-effect transistors. Nat. Nanotechnol. 9 (5), 372–377. Panel (B): Adapted from Buscema, M., Groenendijk, D.J., Steele, G.A., van der Zant, H.S.J., Castellanos-Gomez, A., 2014. Photovoltaic effect in few-layer black phosphorus PN junctions defined by local electrostatic gating Nat. Commun. 5, 4651. Panel (C): Adapted from Wang, Z., Jia, H., Zheng, X., Yang, R., Wang, Z., Ye, G., Chen, X.H., Shan, J., Feng, P., 2015. Black phosphorus nanoelectromechanical resonators vibrating at very high frequencies. Nanoscale 7, 877–884. Panel (D): Adapted from Youngblood, N., Chen, C., Koester, S.J., Li, M., 2015. Waveguide-integrated black phosphorus photodetector with high responsivity and low dark current. Nat. Photonics 9 (4), 247–252. Panel (E): Adapted from Liu, H., Neal, A.T., Zhu, Z., Luo, Z., Xu, X., Tománek, D., Ye, P.D., 2014. Phosphorene: an unexplored 2D semiconductor with a high hole mobility. ACS Nano 8 (4), 4033–4041. Reproduced from Castellanos-Gomez, A., 2015. Black phosphorus: narrow gap wide applications. J. Phys. Chem. Lett. 6 (21), 4280–4291 with permission.*

the transition metal dichalcogenides. Additionally, these devices display ambipolar transport with simple electrostatic gating giving researchers access to both electron and hole accumulation regimes. The combined high mobility and current switching ratios fill the missing gap between MoS_2 (large switching ratio) and graphene (high mobility) and make black phosphorus a prospective material for next-generation electronics.

5.2 PN Junctions

The ambipolarity of black phosphorus FETs is an asset in the fabrication of more complex nanodevices such as PN junctions. Note that MoS_2 and WS_2 (the most studied semiconducting transition metal dichalcogenides) tend to show a marked unipolar n-type behavior and chemical doping (Laskar et al., 2014), ionic gating (Zhang et al., 2012), or specific metal contacts (Chuang et al., 2014; Fontana et al., 2013) have to be used to enable p-type conduction (Ovchinnikov et al., 2014; Radisavljevic et al., 2011). Fig. 7B shows an example where the ambipolar field-effect has been exploited to fabricate a PN junction with electrostatic split-gate geometry (Buscema et al., 2014a). Alternative to the split-gate architecture, another concept of a black phosphorus-based photovoltaic device relies on artificial vertical stacking of two nanosheets with n-type and p-type doping to form a vertical PN junction with a built-in electric field due to the difference in doping between the two stacked layers (Deng et al., 2014).

5.2.1 Electrostatically Gated

Building on the ambipolar field effect of black phosphorus, Buscema et al. have fabricated PN junctions via local electrostatic gating. The devices are based on two local split gates (Buscema et al., 2014a). An hexagonal boron nitride layer, used as an atomically flat and disorder-free gate dielectric, is transferred on top of two electrodes that will be used as local gates. Then a few-layer black phosphorus flake is transferred for use as the channel material using an all-dry deterministic transfer method (Castellanos-Gomez et al., 2014b). Electrical contacts are defined in a final step by e-beam lithography. The generated electric field across PN junction with the two local gates can be used to separate photogenerated electron–hole pairs giving rise to a short-circuit current, a feature of photovoltaic effect in solar cells. Even near-infrared (NIR) photons give rise to photocurrent opening the possibility to harvest NIR photons in photovoltaic applications.

Compared to electrostatically gated solar cells realized with other layered materials such as WSe_2 (Baugher et al., 2014; Groenendijk et al., 2014; Pospischil et al., 2014; Ross et al., 2014), black phosphorus solar cells present a more extended wavelength operation range with similar external quantum efficiencies *EQE* but lower open-circuit voltages (V_{oc}). The lower open-circuit voltage is consistent with the smaller band gap of the black phosphorus compared to that of WSe_2.

5.2.2 *van der Waals Heterostructures*

Another concept of a black phosphorus-based solar cell device, alternative to the split-gate architecture, relies on artificial vertical stacking of two nanosheets with n-type and p-type doping to form a vertical PN junction with a built-in electric field due to the difference in doping between the two stacked layers (Deng et al., 2014).

Deng et al. (2014) fabricated a black phosphorus-based solar cell by exploiting the intrinsic p-type doping of exfoliated black phosphorus and the strong n-type behavior of CVD-grown MoS_2 in a vertically stacked geometry. This device relies on the strong difference in doping between the two layers (n-type and p-type) to form a vertical PN junction with a built-in electric field (Deng et al., 2014). Upon illumination, the built-in electric field at the interface between the two materials was strong enough to separate the photogenerated electron–hole pairs. Similar concepts of PN junction-based solar cells with black phosphorus have been fabricated by combining black phosphorus with conventional n-type semiconductors: Gehring et al. (2015) recently reported a device based on the vertical stacking of black phosphorus on top of an n-type GaAs substrate. Upon illumination, the electric field at the heterojunction also separates the photogenerated carriers, reaching EQE up to 10%.

5.3 Nanoelectromechanical Resonators

The reduced dimensions (and thus low mass) of black phosphorus nanosheets make them extremely advantageous for use as nanoelectromechanical resonators. Fig. 7C shows a nanoelectromechanical resonator with a natural resonance frequency in the MHz regime (Wang et al., 2015b). The quality factor of the black phosphorus mechanical resonators is in the same order of magnitude as those of other 2D materials such as graphene (Barton et al., 2011; Bunch et al., 2007) or molybdenum disulfide.

5.4 Photodetectors

Soon after the mechanical isolation of black phosphorus, several works reported fabrication and characteristics of black phosphorus photodetectors based on the simple construction of phototransistors (Buscema et al., 2014a; Engel et al., 2014; Hong et al., 2014; Low et al., 2014b; Yuan et al., 2015). The phototransistor in these examples is essentially the fabrication of FETs (like the schematic in Fig. 7A) in which the active photocurrent generation element is the channel material itself, in this case, few-layer black phosphorus flakes. When the channel is illuminated with laser light (having energy greater than that of the black phosphorus band gap), photons generate electron–hole pairs which can be swept out of the channel using an applied bias voltage. The added current (photocurrent) due to electron–hole pairs increases the total drain current of the phototransistor, and in this way, light can be detected by monitoring the drain current using this type of photodetector. This is the simplest scenario in which the photocurrent is completely generated by the photoconductive effect given by creation and separation of electron–hole pairs in the active material. It is well known now that several processes bedsides the photoconductive effect can be relevant for these types of phototransistors in which a 2D material is the active element in an FET-like geometry. For a complete discussion on the various mechanisms, we refer the interested reader to Buscema et al. (2015), Furchi et al. (2014), and Island et al. (2015a). Response times in the first reported devices were modest ($\sim$1 ms) although better than those reported for transitional metal dichalcogenides (Buscema et al., 2014a,b; Engel et al., 2014; Yuan et al., 2015). Experimental setup restrictions in the form of slower RC times could be responsible for the lower switching times reported in the first devices. By coupling a black phosphorus photodetector to a silicon photonic waveguide (see Fig. 7D), experimental setup restrictions were all evaded and response times of $<$1 ns could be achieved (Youngblood et al., 2015). To date, the number of reported photodetectors using black phosphorus has rapidly increased and is a testament to the prospects of this material in future optoelectronic applications (Buscema et al., 2014a,b, 2015; Deng et al., 2014; Engel et al., 2014; Hong et al., 2014; Low et al., 2014b; Xia et al., 2014b; Youngblood et al., 2015; Yuan et al., 2015).

5.5 Logic Circuits

Similar to other 2D semiconductors, black phosphorus can be exploited to fabricate logic circuits. Fig. 7E displays the electrical characteristics of an inverter amplifier device fabricated by coupling a p-type black phosphorus transistor and an n-type MoS_2 transistor (Liu et al., 2014), demonstrating the potential of ultrathin semiconductor materials for logic circuit applications (Du et al., 2014; Liu et al., 2014; Zhu et al., 2015).

5.6 Thermoelectric Generator

The potential of bulk black phosphorus for thermoelectric applications has been recently explored. Its Seebeck coefficient has been measured in the temperature range from 300 to 385 K, finding a value of $S = +335 \pm 10\ \mu V/K$ at room temperature (indicating a naturally occurring p-type conductivity) (see Fig. 8; Flores et al., 2015). The Seebeck coefficient increases with temperature (which can be attributed to an increase of the charge carrier density), while the electrical resistance drops up to 40%. As a consequence, the power factor at 385 K is 2.7 times higher than that at room temperature. This indicates the potential of black phosphorus in thermoelectric applications such as thermal energy scavenging, which typically require devices with high performance at temperatures near room temperature.

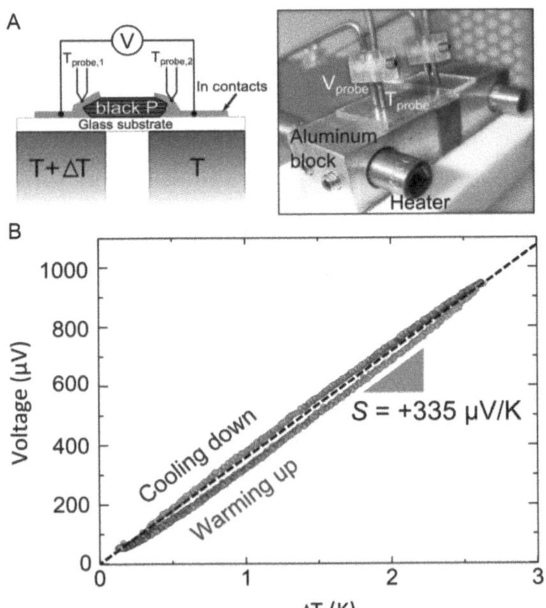

Fig. 8 Black phosphorus-based thermoelectric power generation. (A) Sketch and picture of the experimental setup employed to determine the Seebeck coefficient of bulk black phosphorus. (B) Thermoelectric generated voltage (ΔV) as a function of the temperature difference along the black phosphorus sample during a warming up and cooling down cycle. The Seebeck coefficient can be extracted from the slope $S = +335 \pm 10\ \mu V/K$. *Reproduced from Flores, E., Ares, J.R., Castellanos-Gomez, A., Barawi, M., Ferrer, I.J., Sánchez, C., 2015. Thermoelectric power of bulk black-phosphorus. Appl. Phys. Lett. 106 (2), 022102 with permission.*

6. FUTURE CHALLENGES: ISOLATION AND STABILITY

Most of the experimental works on black phosphorus in the literature employed mechanically exfoliated samples. Although this fabrication method provides high quality material for research, it is not easy to scale it up. Alternative to mechanical exfoliation, liquid-phase exfoliation can produce cheap and large quantity ultrathin black phosphorus layers, but the quality of the produced material and the size of the single-crystalline domains might not be sufficient for certain electronic or optical applications (Hanlon et al., 2015; Jiang et al., 2015; Kang et al., 2015; Luo et al., 2015a; Sresht et al., 2015; Yasaei et al., 2015). Other deposition techniques are being explored to overcome these difficulties; for example, very recent pulsed laser deposition has been used to deposit amorphous black phosphorus (Yang et al., 2015a), but the fabricated layers lack of crystallinity and their electrical properties are still far below those of pristine black phosphorus material.

A serious initial consideration of the use and prospects of black phosphorus was its relatively poor stability in ambient conditions (Castellanos-Gomez et al., 2014a; Favron et al., 2015; Island et al., 2015b; Kim et al., 2015; Koenig et al., 2014; Wood et al., 2014). While most of the 2D materials studied to date are largely inert and stable in air, black phosphorus on the other hand displays strong reactivity (Kulish et al., 2015; Ziletti et al., 2015). Several initial studies report on quick water accumulation at the surface of exfoliated flakes (Island et al., 2015b; Koenig et al., 2014; Wood et al., 2014). Fig. 9A shows AFM images of an exfoliated flake left in ambient conditions that progressively accumulate water with time (Island et al., 2015b). This water accumulation has shown to be associated with degradation of the crystal material as evidenced from electrical measurements of transistors and microwave impedance measurements over time (Island et al., 2015b; Kim et al., 2015; Wood et al., 2014). Favron et al. have deduced that the degradation most likely occurs as a result of photoinduced oxidation from oxygen absorbed in accumulated water at the surface of exfoliated flakes exposed to ambient conditions (Favron et al., 2015).

Despite these earlier reports of crystal degradation, the timescales seem long enough to fabricate fairly complex devices without considerations to ambient exposure. In cases where further protection is needed, several groups have reported the use of an encapsulating layer created at the early

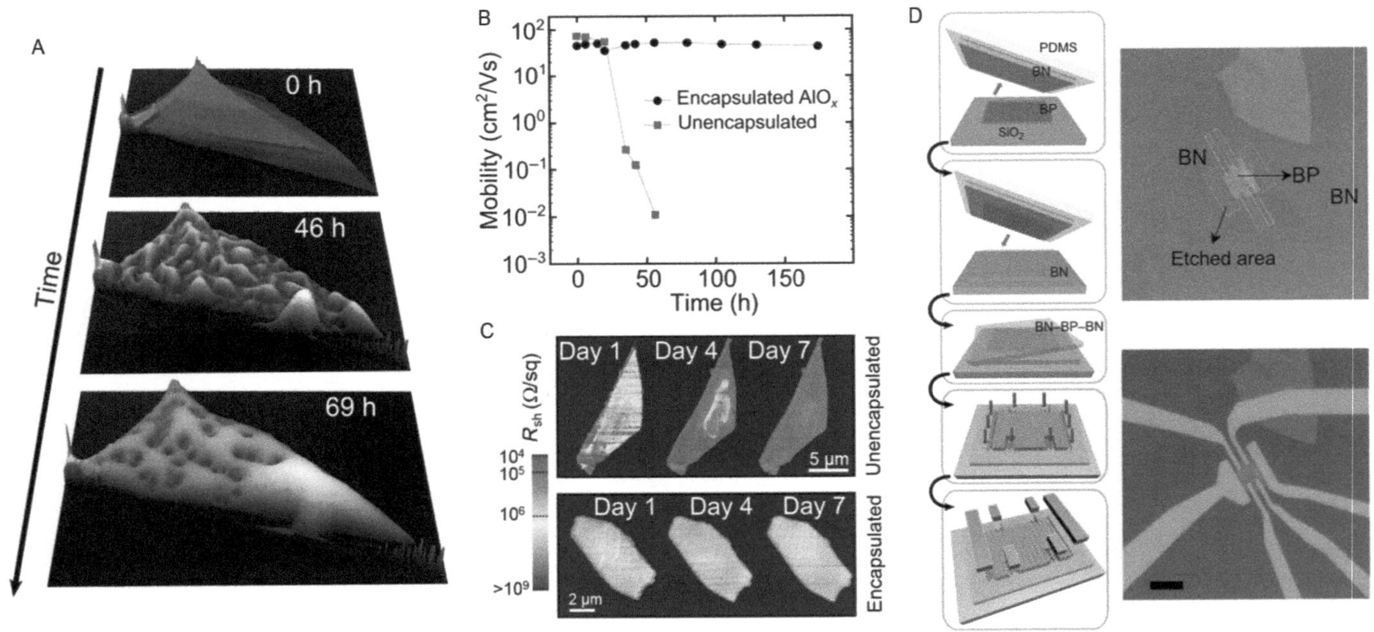

Fig. 9 See legend on opposite page.

stages of fabrication increasing the longevity of end devices (Avsar et al., 2015; Kim et al., 2015; Wood et al., 2014). Several encapsulation techniques have been reported to date including polymer dielectrics (Engel et al., 2014; Tayari et al., 2015), atomic layer deposited (ALD) oxides (Kim et al., 2015; Na et al., 2014; Wood et al., 2014; Zhu et al., 2015), and van der Waals heterostructures using boron nitride and graphene flakes (Avsar et al., 2015; Cao et al., 2015; Chen et al., 2015; Doganov et al., 2015; Gillgren et al., 2014; Li et al., 2015b,c). Fig. 9B and C shows the results of encapsulating black phosphorus using an ALD AlOx capping layer (Kim et al., 2015; Wood et al., 2014). While uncapped FETs break down in a few days, FETs with encapsulation show characteristics (mobility) that do not change over several days (Fig. 9B) and using the unique microwave impedance microscopy measurement technique reveals little to no degradation of encapsulated flakes on Si/SiO_2 substrates (Fig. 9C). Complete encapsulation using boron nitride (bottom and top encapsulation) is a more advanced route toward passivation which has been shown to improve device properties for graphene and MoS_2 by providing an atomically flat, defect-free substrate for 2D

Fig. 9 Environment-induced degradation of black phosphorus and its encapsulation as a route to preserve its pristine properties. (A) Atomic force microscopy topography images acquired as a function of time. Due to the high hygroscopic character of black phosphorus it uptakes moisture from air. (B) Time evolution of the charge carrier mobility of black phosphorus field-effect transistors. Unencapsulated devices rapidly degrade upon air exposure, while devices encapsulated with an aluminum oxide layer show almost time-independent performance for several days. (C) Microwave impedance microscopy map of the local sheet resistance of two black phosphorus flakes: one unencapsulated and one encapsulated with a thick film of aluminum oxide. While the encapsulated device remains unaltered for several days, the unencapsulated device resistance strongly changes with time, starting at the edges of the flake and growing toward the center. (D) Sandwiching of black phosphorus between h-BN flakes to fabricate high-quality encapsulated devices with edge electrical contacts. *Panel (A): Adapted from Island, J.O., Steele, G.A., van der Zant, H.S.J., Castellanos-Gomez, A., 2015. Environmental instability of few-layer black phosphorus. 2D Mater. 2 (1), 011002. Panel (B): Adapted from Wood, J.D., Wells, S.A., Jariwala, D., Chen, K.-S., Cho, E., Sangwan, V.K., Liu, X., Lauhon, L.J., Marks, T.J., Hersam, M.C., 2014. Effective passivation of exfoliated black phosphorus transistors against ambient degradation. Nano Lett. 14 (12), 6964–6970. Panel (C): Adapted from Kim, J.-S., Liu, Y., Zhu, W., Kim, S., Wu, D., Tao, L., Dodabalapur, A., Lai, K., Akinwande, D., 2015. Toward air-stable multilayer phosphorene thin-films and transistors. Sci. Rep. 5, 8989. Panel (D): Adapted from Chen, X., Wu, Y., Wu, Z., Han, Y., Xu, S., Wang, L., Ye, W., Han, T., He, Y., Cai, Y., et al., 2015. High-quality sandwiched black phosphorus heterostructure and its quantum oscillations. Nat. Commun. 6, 7315. This figure has been reproduced from Castellanos-Gomez, A., 2015. Black phosphorus: narrow gap wide applications. J. Phys. Chem. Lett. 6 (21), 4280–4291 with permission.*

materials (Cui et al., 2015; Wang et al., 2013) (see Fig. 9D for fabrication details and resulting device) (Chen et al., 2015). The improved performance in these fully encapsulated devices has led to observations of Shubnikov–de Haas oscillations and the integer quantum Hall effect (Cao et al., 2015; Chen et al., 2015; Gillgren et al., 2014; Li et al., 2015b,c). The realization of these milestones is a testament to the advanced stage of experimental efforts toward stability and performance of black phosphorus devices.

7. SUMMARY

In summary, this chapter has discussed the recent advances on black phosphorus providing an organized selection of the relevant literature on this topic. The main characteristics of black phosphorus such as a strong thickness-dependent band structure and marked in-plane anisotropy have been discussed, and the recent works applying black phosphorus have been summarized. We consider this chapter a good reference for researchers starting work in the field of black phosphorus as it will help them catch up with the state of the art on this interesting and fast developing topic.

ACKNOWLEDGMENTS

J.O.I. acknowledges support from the Dutch Organization for Fundamental Research on Matter (NWO/OCW). A.C.-G. acknowledges financial support from the BBVA Foundation through the fellowship "I Convocatoria de Ayudas Fundacion BBVA a Investigadores, Innovadores y Creadores Culturales" (Semiconductores ultradelgados: hacia la optoelectronica flexible), from the MINECO (Ramón y Cajal 2014 Program, RYC-2014-01406), from the MICINN (MAT2014-58399-JIN), and from the Comunidad de Madrid (MAD2D-CM Program (S2013/MIT-3007)).

REFERENCES

Appalakondaiah, S., Vaitheeswaran, G., Lebègue, S., Christensen, N.E., Svane, A., 2012. Effect of van der Waals interactions on the structural and elastic properties of black phosphorus. Phys. Rev. B 86 (3), 035105.

Asahina, H., Morita, A., 1984. Band structure and optical properties of black phosphorus. J. Phys. C Solid State Phys. 17 (11), 1839–1852.

Avsar, A., Vera-Marun, I.J., Tan, J.Y., Watanabe, K., Taniguchi, T., Castro Neto, A.H., Özyilmaz, B., 2015. Air-stable transport in graphene-contacted, fully encapsulated ultrathin black phosphorus-based field-effect transistors. ACS Nano 9 (4), 4138–4145.

Barton, R.A., Ilic, B., van der Zande, A.M., Whitney, W.S., McEuen, P.L., Parpia, J.M., Craighead, H.G., 2011. High, size-dependent quality factor in an array of graphene mechanical resonators. Nano Lett. 11 (3), 1232–1236.

Baugher, B.W.H., Churchill, H.O.H., Yang, Y., Jarillo-Herrero, P., 2014. Optoelectronic devices based on electrically tunable P-N diodes in a monolayer dichalcogenide. Nat. Nanotechnol. 9 (4), 262–267.

Bridgman, P.W., 1914. Two new modifications of phosphorus. J. Am. Chem. Soc. 36 (7), 1344–1363.
Bunch, J.S., van der Zande, A.M., Verbridge, S.S., Frank, I.W., Tanenbaum, D.M., Parpia, J.M., Craighead, H.G., McEuen, P.L., 2007. Electromechanical resonators from graphene sheets. Science 315 (5811), 490–493.
Buscema, M., Groenendijk, D.J., Steele, G.A., van der Zant, H.S.J., Castellanos-Gomez, A., 2014a. Photovoltaic effect in few-layer black phosphorus PN junctions defined by local electrostatic gating. Nat. Commun. 5, 4651.
Buscema, M., Groenendijk, D.J., Blanter, S.I., Steele, G.A., van der Zant, H.S.J., Castellanos-Gomez, A., 2014b. Fast and broadband photoresponse of few-layer black phosphorus field-effect transistors. Nano Lett. 14 (6), 3347–3352.
Buscema, M., Island, J.O., Groenendijk, D.J., Blanter, S.I., Steele, G.A., van der Zant, H.S.J., Castellanos-Gomez, A., 2015. Photocurrent generation with two-dimensional van der Waals semiconductors. Chem. Soc. Rev. 44 (11), 3691–3718.
Cakir, D., Sevik, C., Peeters, F.M., 2015. Remarkable effect of stacking on the electronic and optical properties of few layer black phosphorus. arXiv. 1506.04707.
Cao, Y., Mishchenko, A., Yu, G.L., Khestanova, K., Rooney, A.P., Prestat, E., Kretinin, A.V., Blake, P., Shalom, M.B., Balakrishnan, G., et al., 2015. Quality heterostructures from two dimensional crystals unstable in air by their assembly in inert atmosphere. Nano Lett. 15 (8), 4914–4921.
Castellanos-Gomez, A., Vicarelli, L., Prada, E., Island, J.O., Narasimha-Acharya, K.L., Blanter, S.I., Groenendijk, D.J., Buscema, M., Steele, G.A., Alvarez, J.V., et al., 2014a. Isolation and characterization of few-layer black phosphorus. 2D Mater. 1 (2), 025001.
Castellanos-Gomez, A., Buscema, M., Molenaar, R., Singh, V., Janssen, L., van der Zant, H.S.J., Steele, G.A., 2014b. Deterministic transfer of two-dimensional materials by all-dry viscoelastic stamping. 2D Mater. 1 (1), 011002.
Chaves, A., Low, T., Avouris, P., Çakır, D., Peeters, F.M., 2015. Anisotropic exciton Stark shift in black phosphorus. Phys. Rev. B 91 (15), 155311.
Chen, X., Wu, Y., Wu, Z., Han, Y., Xu, S., Wang, L., Ye, W., Han, T., He, Y., Cai, Y., et al., 2015. High-quality sandwiched black phosphorus heterostructure and its quantum oscillations. Nat. Commun. 6, 7315.
Chuang, S., Battaglia, C., Azcatl, A., McDonnell, S., Kang, J.S., Yin, X., Tosun, M., Kapadia, R., Fang, H., Wallace, R.M., et al., 2014. MoS_2 P-type transistors and diodes enabled by high work function MoOx contacts. Nano Lett. 14 (3), 1337–1342.
Cui, X., Lee, G.-H., Kim, Y.D., Arefe, G., Huang, P.Y., Lee, C.-H., Chenet, D.A., Zhang, X., Wang, L., Ye, F., et al., 2015. Multi-terminal transport measurements of MoS2 using a van der Waals heterostructure device platform. Nat. Nanotechnol. 10 (6), 534–540.
Deng, Y., Luo, Z., Conrad, N.J., Liu, H., Gong, Y., Najmaei, S., Ajayan, P.M., Lou, J., Xu, X., Ye, P.D., 2014. Black phosphorus-monolayer MoS2 van der Waals hetero-junction P-N diode. ACS Nano 8 (8), 8292–8299.
Doganov, R.A., O'Farrell, E.C.T., Koenig, S.P., Yeo, Y., Ziletti, A., Carvalho, A., Campbell, D.K., Coker, D.F., Watanabe, K., Taniguchi, T., et al., 2015. Transport properties of pristine few-layer black phosphorus by van der Waals passivation in an inert atmosphere. Nat. Commun. 6, 6647.
Du, Y., Liu, H., Deng, Y., Ye, P.D., 2014. Device perspective for black phosphorus field-effect transistors: contact resistance, ambipolar behavior, and scaling. ACS Nano 8 (10), 10035–10042.
Engel, M., Steiner, M., Avouris, P., 2014. Black phosphorus photodetector for multispectral high-resolution imaging. Nano Lett. 14 (11), 6414–6417.

Favron, A., Gaufrès, E., Fossard, F., Phaneuf-L'Heureux, A.-L., Tang, N.Y.-W., Lévesque, P.L., Loiseau, A., Leonelli, R., Francoeur, S., Martel, R., 2015. Photooxidation and quantum confinement effects in exfoliated black phosphorus. Nat. Mater. 14 (8), 826–832.

Fei, R., Yang, L., 2014. Strain-engineering the anisotropic electrical conductance of few-layer black phosphorus. Nano Lett. 14 (5), 2884–2889.

Flores, E., Ares, J.R., Castellanos-Gomez, A., Barawi, M., Ferrer, I.J., Sánchez, C., 2015. Thermoelectric power of bulk black-phosphorus. Appl. Phys. Lett. 106 (2), 022102.

Fontana, M., Deppe, T., Boyd, A.K., Rinzan, M., Liu, A.Y., Paranjape, M., Barbara, P., 2013. Electron-hole transport and photovoltaic effect in gated MoS2 Schottky junctions. Sci. Rep. 3, 1634.

Furchi, M.M., Polyushkin, D.K., Pospischil, A., Mueller, T., 2014. Mechanisms of photoconductivity in atomically thin MoS2. Nano Lett. 14 (11), 6165–6170.

Ge, S., Li, C., Zhang, Z., Zhang, C., Zhang, Y., Qiu, J., Wang, Q., Liu, J., Jia, S., Feng, J., et al., 2015. Dynamical evolution of anisotropic response in black phosphorus under ultrafast photoexcitation. Nano Lett. 15 (7), 4650–4656.

Gehring, P., Urcuyo, R., Duong, D.L., Burghard, M., Kern, K., 2015. Thin-layer black phosphorous/GaAs heterojunction P-N diodes. Appl. Phys. Lett. 106 (23), 233110.

Gillgren, N., Wickramaratne, D., Shi, Y., Espiritu, T., Yang, J., Hu, J., Wei, J., Liu, X., Mao, Z., Watanabe, K., et al., 2014. Gate tunable quantum oscillations in air-stable and high mobility few-layer phosphorene heterostructures. 2D Mater. 2 (1), 011001.

Groenendijk, D.J., Buscema, M., Steele, G.A., Michaelis de Vasconcellos, S., Bratschitsch, R., van der Zant, H.S.J., Castellanos-Gomez, A., 2014. Photovoltaic and photothermoelectric effect in a double-gated WSe2 device. Nano Lett. 14 (10), 5846–5852.

Hanlon, D., Backes, C., Doherty, E., Cucinotta, C.S., Berner, N.C., Boland, C., Lee, K., Lynch, P., Gholamvand, Z., Harvey, A., et al., 2015. Liquid exfoliation of solvent-stabilised black phosphorus: applications beyond electronics. arXiv. 1501.01881.

He, J., He, D., Wang, Y., Cui, Q., Bellus, M.Z., Chiu, H.-Y., Zhao, H., 2015. Exceptional and anisotropic transport properties of photocarriers in black phosphorus. ACS Nano 9 (6), 6436–6442.

Hong, T., Chamlagain, B., Lin, W., Chuang, H.-J., Pan, M., Zhou, Z., Xu, Y.-Q., 2014. Polarized photocurrent response in black phosphorus field-effect transistors. Nanoscale 6 (15), 8978–8983.

Island, J.O., Blanter, S.I., Buscema, M., van der Zant, H.S.J., Castellanos-Gomez, A., 2015a. Gate controlled photocurrent generation mechanisms in high-gain In2Se3 phototransistors. Nano Lett. 15 (12), 7853–7858.

Island, J.O., Steele, G.A., van der Zant, H.S.J., Castellanos-Gomez, A., 2015b. Environmental instability of few-layer black phosphorus. 2D Mater. 2 (1), 011002.

Jiang, J.-W., 2015a. The third principle direction besides armchair and zigzag in single-layer black phosphorus. Nanotechnology 26 (36), 365702.

Jiang, J.-W., 2015b. Thermal conduction in single-layer black phosphorus: highly anisotropic? Nanotechnology 26 (5), 055701.

Jiang, J.-W., Park, H.S., 2014. Mechanical properties of single-layer black phosphorus. J. Phys. D Appl. Phys. 47 (38), 385304.

Jiang, T., Yin, K., Zheng, X., Yu, H., Cheng, X.-A., 2015. Black phosphorus as a new broadband saturable absorber for infrared passively Q-switched fiber lasers. arXiv. 1504.07341.

Kang, J., Wood, J.D., Wells, S.A., Lee, J.-H., Liu, X., Chen, K.-S., Hersam, M.C., 2015. Solvent exfoliation of electronic-grade two-dimensional black phosphorus. ACS Nano 9 (4), 3596–3604.

Keyes, R.W., 1953. The electrical properties of black phosphorus. Phys. Rev. 92 (3), 580–584.

Kim, J.-S., Liu, Y., Zhu, W., Kim, S., Wu, D., Tao, L., Dodabalapur, A., Lai, K., Akinwande, D., 2015. Toward air-stable multilayer phosphorene thin-films and transistors. Sci. Rep. 5, 8989.

Koenig, S.P., Doganov, R.A., Schmidt, H., Castro Neto, A.H., Özyilmaz, B., 2014. Electric field effect in ultrathin black phosphorus. Appl. Phys. Lett. 104 (10), 103106.

Kulish, V.V., Malyi, O.I., Persson, C., Wu, P., 2015. Adsorption of metal adatoms on single-layer phosphorene. Phys. Chem. Chem. Phys. 17 (2), 992–1000.

Laskar, M.R., Nath, D.N., Ma, L., Lee, E.W., Lee, C.H., Kent, T., Yang, Z., Mishra, R., Roldan, M.A., Idrobo, J.-C., et al., 2014. P-type doping of MoS2 thin films using Nb. Appl. Phys. Lett. 104 (9), 092104.

Li, L., Yu, Y., Ye, G.J., Chen, X.H., Zhang, Y., 2013. Electronic properties of few-layer black phosphorus. In: *APS March Meeting*. American Physical Society, College Park, MD. abstract ID BAPS.2013.MAR.J6.13. http://meetings.aps.org/Meeting/MAR13/Session/J6.13.

Li, L., Yu, Y., Ye, G.J., Ge, Q., Ou, X., Wu, H., Feng, D., Chen, X.H., Zhang, Y., 2014a. Black phosphorus field-effect transistors. Nat. Nanotechnol. 9 (5), 372–377.

Li, Y., Yang, S., Li, J., 2014b. Modulation of the electronic properties of ultrathin black phosphorus by strain and electrical field. J. Phys. Chem. C 118 (41), 23970–23976.

Li, X., Deng, B., Wang, X., Chen, S., Vaisman, M., Karato, S., Pan, G., Larry Lee, M., Cha, J., Wang, H., et al., 2015a. Synthesis of thin-film black phosphorus on a flexible substrate. 2D Mater. 2 (3), 031002.

Li, L., Ye, G.J., Tran, V., Fei, R., Chen, G., Wang, H., Wang, J., Watanabe, K., Taniguchi, T., Yang, L., et al., 2015b. Quantum oscillations in a two-dimensional electron gas in black phosphorus thin films. Nat. Nanotechnol. 10 (7), 608–613.

Li, L., Yang, F., Ye, G.J., Zhang, Z., Zhu, Z., Lou, W.-K., Li, L., Watanabe, K., Taniguchi, T., Chang, K., et al., 2015c. Quantum hall effect in black phosphorus two-dimensional electron gas. arXiv. 1504.07155.

Liang, L., Wang, J., Lin, W., Sumpter, B.G., Meunier, V., Pan, M., 2014. Electronic bandgap and edge reconstruction in phosphorene materials. Nano Lett. 14 (11), 6400–6406.

Liu, H., Neal, A.T., Zhu, Z., Luo, Z., Xu, X., Tománek, D., Ye, P.D., 2014. Phosphorene: an unexplored 2D semiconductor with a high hole mobility. ACS Nano 8 (4), 4033–4041.

Low, T., Roldán, R., Wang, H., Xia, F., Avouris, P., Moreno, L.M., Guinea, F., 2014a. Plasmons and screening in monolayer and multilayer black phosphorus. Phys. Rev. Lett. 113 (10), 106802.

Low, T., Engel, M., Steiner, M., Avouris, P., 2014b. Origin of photoresponse in black phosphorus phototransistors. Phys. Rev. B 90 (8), 081408.

Lu, W., Ma, X., Fei, Z., Zhou, J., Zhang, Z., Jin, C., Zhang, Z., 2015. Probing the anisotropic behaviors of black phosphorus by transmission electron microscopy, angular-dependent Raman spectra, and electronic transport measurements. Appl. Phys. Lett. 107 (2), 021906.

Luo, Z.-C., Liu, M., Guo, Z.-N., Jiang, X.-F., Luo, A.-P., Zhao, C.-J., Yu, X.-F., Xu, W.-C., Zhang, H., 2015a. Microfiber-based few-layer black phosphorus saturable absorber for ultra-fast fiber laser. arXiv. 1505.03035.

Luo, Z., Maassen, J., Deng, Y., Du, Y., Lundstrom, M.S., Ye, P.D., Xu, X., 2015b. Anisotropic in-plane thermal conductivity observed in few-layer black phosphorus. arXiv. 1503.06167.

Mak, K.F., Lee, C., Hone, J., Shan, J., Heinz, T.F., 2010. Atomically thin MoS_{2}: a new direct-gap semiconductor. Phys. Rev. Lett. 105 (13), 136805.

Morita, A., 1986. Semiconducting black phosphorus. Appl. Phys. A 39 (4), 227–242.

Na, J., Lee, Y.T., Lim, J.A., Hwang, D.K., Kim, G.-T., Choi, W.K., Song, Y.-W., 2014. Few-layer black phosphorus field-effect transistors with reduced current fluctuation. ACS Nano 8 (11), 11753–11762.

Novoselov, K.S., Geim, A.K., Morozov, S.V., Jiang, D., Zhang, Y., Dubonos, S.V., Grigorieva, I.V., Firsov, A.A., 2004. Electric field effect in atomically thin carbon films. Science 306 (5696), 666–669.

Novoselov, K.S., Jiang, D., Schedin, F., Booth, T.J., Khotkevich, V.V., Morozov, S.V., Geim, A.K., 2005. Two-dimensional atomic crystals. Proc. Natl. Acad. Sci. U.S.A. 102 (30), 10451–10453.

Ong, Z.-Y., Cai, Y., Zhang, G., Zhang, Y.-W., 2014. Strong thermal transport anisotropy and strain modulation in single-layer phosphorene. J. Phys. Chem. C 118 (43), 25272–25277.

Ovchinnikov, D., Allain, A., Huang, Y.-S., Dumcenco, D., Kis, A., 2014. Electrical transport properties of single-layer WS2. ACS Nano 8 (8), 8174–8181.

Popović, Z.S., Kurdestany, J.M., Satpathy, S., 2015. Electronic structure and anisotropic Rashba spin-orbit coupling in monolayer black phosphorus. Phys. Rev. B 92 (3), 035135.

Pospischil, A., Furchi, M.M., Mueller, T., 2014. Solar-energy conversion and light emission in an atomic monolayer P-N diode. Nat. Nanotechnol. 9 (4), 257–261.

Qiao, J., Kong, X., Hu, Z.-X., Yang, F., Ji, W., 2014. High-mobility transport anisotropy and linear dichroism in few-layer black phosphorus. Nat. Commun. 5, 4475.

Qin, G., Qin, Z., Yue, S.-Y., Cui, H.-J., Zheng, Q.-R., Yan, Q.-B., Su, G., 2014a. Strain enhanced anisotropic thermoelectric performance of black phosphorus. arXiv. 1406.0261.

Qin, G., Yan, Q.-B., Qin, Z., Yue, S.-Y., Cui, H.-J., Zheng, Q.-R., Su, G., 2014b. Hinge-like structure induced unusual properties of black phosphorus and new strategies to improve the thermoelectric performance. Sci. Rep. 4, 6946.

Qin, G., Yan, Q.-B., Qin, Z., Yue, S.-Y., Hu, M., Su, G., 2015. Anisotropic intrinsic lattice thermal conductivity of phosphorene from first principles. Phys. Chem. Chem. Phys. 17, 4854.

Radisavljevic, B., Radenovic, A., Brivio, J., Giacometti, V., Kis, A., 2011. Single-layer MoS2 transistors. Nat. Nanotechnol. 6 (3), 147–150.

Ribeiro, H.B., Pimenta, M.A., de Matos, C.J.S., Moreira, R.L., Rodin, A.S., Zapata, J.D., de Souza, E.A.T., Castro Neto, A.H., 2015. Unusual angular dependence of the Raman response in black phosphorus. ACS Nano 9 (4), 4270–4276.

Rodin, A.S., Carvalho, A., Castro Neto, A.H., 2014. Excitons in anisotropic two-dimensional semiconducting crystals. Phys. Rev. B 90 (7), 075429.

Ross, J.S., Klement, P., Jones, A.M., Ghimire, N.J., Yan, J., Mandrus, D.G., Taniguchi, T., Watanabe, K., Kitamura, K., Yao, W., et al., 2014. Electrically tunable excitonic light-emitting diodes based on monolayer WSe2 P-N junctions. Nat. Nanotechnol. 9 (4), 268–272.

Rudenko, A.N., Katsnelson, M.I., 2014. Quasiparticle band structure and tight-binding model for single- and bilayer black phosphorus. Phys. Rev. B 89 (20), 201408.

Schuster, R., Trinckauf, J., Knupfer, M., Büchner, B., 2015. Anisotropic particle-hole excitations in black phosphorus. Phys. Rev. Lett. 115, 026404.

Splendiani, A., Sun, L., Zhang, Y., Li, T., Kim, J., Chim, C.-Y., Galli, G., Wang, F., 2010. Emerging photoluminescence in monolayer MoS2. Nano Lett. 10 (4), 1271–1275.

Sresht, V., Pádua, A.A.H., Blankschtein, D., 2015. Liquid-phase exfoliation of phosphorene: design rules from molecular dynamics simulations. ACS Nano 9 (8), 8255–8268.

Tayari, V., Hemsworth, N., Fakih, I., Favron, A., Gaufrès, E., Gervais, G., Martel, R., Szkopek, T., 2015. Two-dimensional magnetotransport in a black phosphorus naked quantum well. Nat. Commun. 6, 7702.

Tran, V., Soklaski, R., Liang, Y., Yang, L., 2014. Layer-controlled band gap and anisotropic excitons in few-layer black phosphorus. Phys. Rev. B 89 (23), 235319.

Wang, Z., Feng, P.X.-L., 2015. Design of black phosphorus 2D nanomechanical resonators by exploiting the intrinsic mechanical anisotropy. 2D Mater. 2 (2), 021001.

Wang, L., Meric, I., Huang, P.Y., Gao, Q., Gao, Y., Tran, H., Taniguchi, T., Watanabe, K., Campos, L.M., Muller, D.A., et al., 2013. One-dimensional electrical contact to a two-dimensional material. Science 342 (6158), 614–617.

Wang, X., Jones, A.M., Seyler, K.L., Tran, V., Jia, Y., Zhao, H., Wang, H., Yang, L., Xu, X., Xia, F., 2015a. Highly anisotropic and robust excitons in monolayer black phosphorus. Nat. Nanotechnol. 10 (6), 517–521.

Wang, Z., Jia, H., Zheng, X., Yang, R., Wang, Z., Ye, G., Chen, X.H., Shan, J., Feng, P., 2015b. Black phosphorus nanoelectromechanical resonators vibrating at very high frequencies. Nanoscale 7, 877–884.

Wei, Q., Peng, X., 2014. Superior mechanical flexibility of phosphorene and few-layer black phosphorus. Appl. Phys. Lett. 104 (25), 251915.

Wood, J.D., Wells, S.A., Jariwala, D., Chen, K.-S., Cho, E., Sangwan, V.K., Liu, X., Lauhon, L.J., Marks, T.J., Hersam, M.C., 2014. Effective passivation of exfoliated black phosphorus transistors against ambient degradation. Nano Lett. 14 (12), 6964–6970.

Wu, J., Mao, N., Xie, L., Xu, H., Zhang, J., 2015. Identifying the crystalline orientation of black phosphorus using angle-resolved polarized Raman spectroscopy. Angew. Chem. 127 (8), 2396–2399.

Xia, F., Wang, H., Jia, Y., 2014a. Rediscovering black phosphorus as an anisotropic layered material for optoelectronics and electronics. Nat. Commun. 5, 4458.

Xia, F., Wang, H., Xiao, D., Dubey, M., Ramasubramaniam, A., 2014b. Two-dimensional material nanophotonics. Nat. Photonics 8 (12), 899–907.

Yang, Z., Hao, J., Yuan, S., Lin, S., Yau, H.M., Dai, J., Lau, S.P., 2015a. Field-effect transistors based on amorphous black phosphorus ultrathin films by pulsed laser deposition. Adv. Mater. 27 (25), 3748–3754.

Yang, J., Xu, R., Pei, J., Myint, Y.W., Wang, F., Wang, Z., Zhang, S., Yu, Z., Lu, Y., 2015b. Optical tuning of exciton and trion emissions in monolayer phosphorene. Light Sci. Appl. 4 (7), e312.

Yasaei, P., Kumar, B., Foroozan, T., Wang, C., Asadi, M., Tuschel, D., Indacochea, J.E., Klie, R.F., Salehi-Khojin, A., 2015. High-quality black phosphorus atomic layers by liquid-phase exfoliation. Adv. Mater. 27 (11), 1887–1892.

Youngblood, N., Chen, C., Koester, S.J., Li, M., 2015. Waveguide-integrated black phosphorus photodetector with high responsivity and low dark current. Nat. Photonics 9 (4), 247–252.

Yuan, H., Liu, X., Afshinmanesh, F., Li, W., Xu, G., Sun, J., Lian, B., Curto, A.G., Ye, G., Hikita, Y., et al., 2015. Polarization-sensitive broadband photodetector using a black phosphorus vertical P-N junction. Nat. Nanotechnol. 10 (8), 707–713.

Zhang, C.D., Lian, J.C., Yi, W., Jiang, Y.H., Liu, L.W., Hu, H., Xiao, W.D., Du, S.X., Sun, L.L., Gao, H.J., 2009. Surface structures of black phosphorus investigated with scanning tunneling microscopy. J. Phys. Chem. C 113 (43), 18823–18826.

Zhang, Y., Ye, J., Matsuhashi, Y., Iwasa, Y., 2012. Ambipolar MoS2 thin flake transistors. Nano Lett. 12 (3), 1136–1140.

Zhang, S., Yang, J., Xu, R., Wang, F., Li, W., Ghufran, M., Zhang, Y.-W., Yu, Z., Zhang, G., Qin, Q., et al., 2014. Extraordinary photoluminescence and strong temperature/angle-dependent Raman responses in few-layer phosphorene. ACS Nano 8 (9), 9590–9596.

Zhu, W., Yogeesh, M.N., Yang, S., Aldave, S.H., Kim, J., Sonde, S.S., Tao, L., Lu, N., Akinwande, D., 2015. Flexible black phosphorus ambipolar transistors circuits and AM demodulator. Nano Lett. 15 (3), 1883–1890.

Ziletti, A., Carvalho, A., Campbell, D.K., Coker, D.F., Castro Neto, A.H., 2015. Oxygen defects in phosphorene. Phys. Rev. Lett. 114 (4), 046801.

INDEX

Note: Page numbers followed by "*f*" indicate figures, and "*t*" indicate tables.

D

E

F

G

H

W

CONTENTS OF VOLUMES IN THIS SERIES

Volume 4 **Physics of III–V Compounds**

Volume 5 **Infrared Detectors**

Volume 6 **Injection Phenomena**

Volume 11 **Solar Cells**

H. J. Hovel, Introduction; Carrier Collection, Spectral Response, and Photocurrent; Solar Cell Electrical Characteristics; Efficiency; Thickness; Other Solar Cell Devices; Radiation Effects; Temperature and Intensity; Solar Cell Technology

Volume 12 **Infrared Detectors (II)**

W. L. Eiseman, J. D. Merriam, and R. F. Potter, Operational Characteristics of Infrared Photodetectors
P. R. Bratt, Impurity Germanium and Silicon Infrared Detectors
E. H. Putley, InSb Submillimeter Photoconductive Detectors
G. E. Stillman, C. M. Wolfe, and J. O. Dimmock, Far-Infrared Photoconductivity in High Purity GaAs
G. E. Stillman and C. M. Wolfe, Avalanche Photodiodes
P. L. Richards, The Josephson Junction as a Detector of Microwave and Far-Infrared Radiation
E. H. Putley, The Pyroelectric Detector – An Update

Volume 13 **Cadmium Telluride**

K. Zanio, Materials Preparations; Physics; Defects; Applications

Volume 14 **Lasers, Junctions, Transport**

N. Holonyak, Jr., and M. H. Lee, Photopumped III–V Semiconductor Lasers
H. Kressel and J. K. Butler, Heterojunction Laser Diodes
A. Van der Ziel, Space-Charge-Limited Solid-State Diodes
P. J. Price, Monte Carlo Calculation of Electron Transport in Solids

Volume 15 **Contacts, Junctions, Emitters**

B. L. Sharma, Ohmic Contacts to III–V Compounds Semiconductors
A. Nussbaum, The Theory of Semiconducting Junctions
J. S. Escher, NEA Semiconductor Photoemitters

Volume 16 **Defects, (HgCd)Se, (HgCd)Te**

H. Kressel, The Effect of Crystal Defects on Optoelectronic Devices
C. R. Whitsett, J. G. Broerman, and C. J. Summers, Crystal Growth and Properties of $Hg_{1-x}Cd_xSe$ Alloys
M. H. Weiler, Magnetooptical Properties of $Hg_{1-x}Cd_xTe$ Alloys
P. W. Kruse and J. G. Ready, Nonlinear Optical Effects in $Hg_{1-x}Cd_xTe$

Volume 17 **CW Processing of Silicon and Other Semiconductors**

J. F. Gibbons, Beam Processing of Silicon
A. Lietoila, R. B. Gold, J. F. Gibbons, and L. A. Christel, Temperature Distributions and Solid Phase Reaction Rates Produced by Scanning CW Beams

Volume 18 **Mercury Cadmium Telluride**

Volume 19 **Deep Levels, GaAs, Alloys, Photochemistry**

Volume 20 **Semi-Insulating GaAs**

Volume 21 **Hydrogenated Amorphous Silicon**

Part A

Volume 22 **Lightwave Communications Technology**

Part A

Part B

Part C

Part D

Volume 26 **III–V Compound Semiconductors and Semiconductor Properties of Superionic Materials**

Volume 27 **High Conducting Quasi-One-Dimensional Organic Crystals**

Volume 28 **Measurement of High-Speed Signals in Solid State Devices**

Volume 34 **Hydrogen in Semiconductors**

Volume 35 **Nanostructured Systems**

Volume 36 **The Spectroscopy of Semiconductors**

A. K. Ramdas and S. Rodriguez, Piezospectroscopy of Semiconductors
O. J. Glembocki and B. V. Shanabrook, Photoreflectance Spectroscopy of Microstructures
D. G. Seiler, C. L. Littler, and M. H. Wiler, One- and Two-Photon Magneto-Optical Spectroscopy of InSb and $Hg_{1-x}Cd_x$ Te

Volume 37 **The Mechanical Properties of Semiconductors**

A.-B. Chen, A. Sher, and W. T. Yost, Elastic Constants and Related Properties of Semiconductor Compounds and Their Alloys
D. R. Clarke, Fracture of Silicon and Other Semiconductors
H. Siethoff, The Plasticity of Elemental and Compound Semiconductors
S. Guruswamy, K. T. Faber, and J. P. Hirth, Mechanical Behavior of Compound Semiconductors
S. Mahajan, Deformation Behavior of Compound Semiconductors
J. P. Hirth, Injection of Dislocations into Strained Multilayer Structures
D. Kendall, C. B. Fleddermann, and K. J. Malloy, Critical Technologies for the Micromatching of Silicon
J. Matsuba and K. Mokuya, Processing and Semiconductor Thermoelastic Behavior

Volume 38 **Imperfections in III/V Materials**

U. Scherz and M. Scheffler, Density-Functional Theory of sp-Bonded Defects in III/V Semiconductors
M. Kaminska and E. R. Weber, E12 Defect in GaAs
D. C. Look, Defects Relevant for Compensation in Semi-Insulating GaAs
R. C. Newman, Local Vibrational Mode Spectroscopy of Defects in III/V Compounds
A. M. Hennel, Transition Metals in III/V Compounds
K. J. Malloy and K. Khachaturyan, DX and Related Defects in Semiconductors
V. Swaminathan and A. S. Jordan, Dislocations in III/V Compounds
K. W. Nauka, Deep Level Defects in the Epitaxial III/V Materials

Volume 39 **Minority Carriers in III–V Semiconductors: Physics and Applications**

N. K. Dutta, Radiative Transition in GaAs and Other III–V Compounds
R. K. Ahrenkiel, Minority-Carrier Lifetime in III–V Semiconductors
T. Furuta, High Field Minority Electron Transport in p-GaAs
M. S. Lundstrom, Minority-Carrier Transport in III–V Semiconductors
R. A. Abram, Effects of Heavy Doping and High Excitation on the Band Structure of GaAs
D. Yevick and W. Bardyszewski, An Introduction to Non-Equilibrium Many-Body Analyses of Optical Processes in III–V Semiconductors

Volume 40 **Epitaxial Microstructures**

E. F. Schubert, Delta-Doping of Semiconductors: Electronic, Optical and Structural Properties of Materials and Devices
A. Gossard, M. Sundaram, and P. Hopkins, Wide Graded Potential Wells

Volume 41 **High Speed Heterostructure Devices**

Volume 42 **Oxygen in Silicon**

Volume 43 **Semiconductors for Room Temperature Nuclear Detector Applications**

Volume 44 **II–IV Blue/Green Light Emitters: Device Physics and Epitaxial Growth**

Volume 45 **Effect of Disorder and Defects in Ion-Implanted Semiconductors: Electrical and Physiochemical Characterization**

Volume 46 **Effect of Disorder and Defects in Ion-Implanted Semiconductors: Optical and Photothermal Characterization**

Volume 47 **Uncooled Infrared Imaging Arrays and Systems**

Volume 48 **High Brightness Light Emitting Diodes**

Volume 49 **Light Emission in Silicon: from Physics to Devices**

Volume 50 **Gallium Nitride (GaN)**

Volume 51A **Identification of Defects in Semiconductors**

Volume 51B **Identification Defects in Semiconductors**

Volume 52 **SiC Materials and Devices**

Volume 53 **Cumulative Subjects and Author Index Including Tables of Contents for Volumes 1–50**

Volume 54 **High Pressure in Semiconductor Physics I**

Volume 55 **High Pressure in Semiconductor Physics II**

Volume 56 **Germanium Silicon: Physics and Materials**

Volume 57 **Gallium Nitride (GaN) II**

Volume 58 **Nonlinear Optics in Semiconductors I**

Volume 68 **Isotope Effects in Solid State Physics**

Volume 69 **Recent Trends in Thermoelectric Materials Research I**

Volume 70 **Recent Trends in Thermoelectric Materials Research II**

Volume 71 **Recent Trends in Thermoelectric Materials Research III**

Volume 72 **Silicon Epitaxy**

Volume 73 **Processing and Properties of Compound Semiconductors**

Volume 79 **Semiconducting Chalcogenide Glass II**

Volume 80 **Semiconducting Chalcogenide Glass III**

Volume 81 **Conducting Organic Materials and Devices**

Volume 82 **Semiconductors and Semimetals**

Volume 83 **Semiconductors and Semimetals**

Volume 84 **Semiconductors and Semimetals**

Volume 85 **Semiconductors and Semimetals**

Volume 86 **Advances in Semiconductor Lasers**

Volume 87 **Advances in Photovoltaics: Volume 1**

Volume 88 **Oxide Semiconductors**

Volume 89 **Advances in Photovoltaics: Part 2**

Volume 90 **Advances in Photovoltaics: Part 3**

Volume 91 **Defects in Semiconductors**

FSC
www.fsc.org
MIX
Papier | Fördert
gute Waldnutzung
FSC® C083411

Kolibri 360 GmbH
Ferdinand-Jühlke-Straße 7
99095 Erfurt, Deutschland
produktsicherheit@kolibri360.de